PUBLICATION No. 8
of the Mathematics Research Center
United States Army
The University of Wisconsin

NONLINEAR PROBLEMS

NONLINEAR PROBLEMS

Proceedings of a Symposium Conducted by the Mathematics Research Center, United States Army, at the University of Wisconsin, Madison, April 30-May 2, 1962

edited by
RUDOLPH E. LANGER

MADISON · THE UNIVERSITY
OF WISCONSIN PRESS · 1963

Published by
THE UNIVERSITY OF WISCONSIN PRESS
430 Sterling Court, Madison 6, Wisconsin

Prepared for the camera by Phyllis J. Kern
Printed in the United States of America by
Cushing-Malloy, Inc., Ann Arbor, Michigan

Library of Congress Catalog Card Number 63-8971

FOREWORD

This volume preserves in print the lectures and the abstracts of the contributed papers of the sixth Symposium conducted by the Mathematics Research Center, United States Army. The symposium was held at the University of Wisconsin, April 30 to May 2, 1962.

The purpose of this, as of all symposiums, was to assemble a substantial number of effective researchers in a frontier field of live current interest, and to afford them and other attendants a forum for the presentation and discussion of the results of some recent investigations. Modern technology, with its great refinement of instrumentation and its increasing venturesomeness, has made it abundantly evident that the formulations of natural laws in which troublesome non-linear terms are suppressed or neglected to achieve workable linearizations, often lead to faulty results or to sacrifices of precision which are no longer tolerable. Results for the equations in their unmodified non-linear forms are needed. The mathematician, on his part, is at the same time impelled to round out a theory which still invokes restrictive hypotheses. Non-linear theory is still in a juvenile state; is still largely fragmentary. One hopefully believes that its rounding-out may be hastened by surveys of advances made such as this symposium sought to achieve.

The symposium was held in the University's "Wisconsin Center Building." The attendants were greeted on behalf of the University by Professor John E. Willard, Dean of the Graduate School. It was closed by some words on behalf of the United States Army, spoken by Dr. John A. Tierney of the Army Research Office, Washington, D.C. There were fifteen invited lectures of forty minutes duration, and contributed papers, to each of which a time allowance of twenty minutes was allotted. This volume presents the lectures in full and the contributed papers through authors' abstracts.

We extend thanks to all participants, and in especial measure to those who responded by travel from great distances overland and oversea.

The symposium program was arranged by the committee

P. M. Anselone	Calvin Wilcox
H. F. Bueckner	L. C. Young.

PREFACE

The great successes of the applied mathematicians of the eighteenth and nineteenth centuries in constructing effective theories for physical phenomena, were, in the main, attributable to the availability of the principle of superposition. In accordance with this, elementary solutions of the pertinent mathematical equations could be combined to yield more flexible ones, namely ones which could be brought into adjustment with the auxiliary conditions that characterize the particular phenomenological manifestation. The equations to which that principle is applicable are the linear ones. With the neglect of elements that could then be judged to be of subordinate importance, the laws of nature could be formulated linearly, and the theories that were based upon such idealizations will forever be useful, imposing and admirable. Nor has the last word in linear theory by any means been said.

Mathematicians, however, have an unfailing urge to stray from trodden paths into the unpaved borderlands, and to step from the linear is to plunge into the domain of the non-linear. This is a difficult terrain, in which the going has been found rough and slow. But it is a domain in which the demand for a clearing of paths is rising. As in all virgin mathematical territory, so here, the challenge to master difficulties for the glory of the human spirit manifests itself. In this case however, it is firmly backed up also by more mundane considerations. For technology in its trends toward both refinement and magnifications of scope and complexity has been pointing with increasing insistence to the fact that in the formulation of natural laws modern requirements of precision forbid the suppression of non-linear elements—that suitable formulations are, almost without exception, non-linear.

Present day research in mathematical analysis is, therefore, heavily engaged with non-linear problems. Important results—in the main special ones—have been found, and are being found; known general principles are still few. Certainly, in a history of non-linear methods many chapters remain to be written. For such a history the papers that are combined in this volume, setting forth as they do, the findings of dedicated researchers in the field, may well shape up as valuable reference material.

In Euclid's elements many theorems of the early chapters on rectilinear geometry are gems, yet they are fully matched by ones in the later chapters on the circle and the conic sections.

Rudolph E. Langer

CONTENTS

NONLINEAR PROBLEMS

PETER D. LAX

Nonlinear hyperbolic systems of conservation laws

Introduction: This talk gives a brief survey of the subject of the title. Although much good work has been done in this field in the last ten years, the main problems—existence and uniqueness—still await solution and the really difficult case of several space variables is still untouched by theorems of any sort, although much important and suggestive work has been done by aerodynamicists.

As we shall see below, the main problems of the theory arise naturally and are easily formulated. From what we know there is every reason to expect that there is a harmonious theory awaiting development, one of the relatively few nonlinear theories in the large in partial differential equations.

The articles [4], [5], and [2] by Oleinik, Rozhdestvenskii, and the author contain fairly comprehensive reviews and detailed mathematical treatment of the literature which has appeared since the pioneering paper of E. Hopf [1]. The present paper will not repeat the details which have already appeared in these foregoing papers, and in the bibliography we will list only those papers which have appeared in the last few years.

1. A system of conservation laws is a system of equations of the form

$$u_t + f(u)_x = 0 \qquad (1)$$

where u is a vector function of x, t with n components, and f is a nonlinear vector function with n components of the vector variable u. A similar concept can be defined in the case of several space variables x.

By a solution of (1), we mean a bounded, measurable vector function $u(x, t)$ which satisfies (1) in the sense of distributions. This definition can be put in two slightly different forms:

I) For every suitable interval $I = (a, b)$

$$-\frac{d}{dt}\int_I u\,dx = f(u(x))\Big|_a^b .$$

II) For all suitable curves C in the x, t plane, the value of the integrals

$$\int_C u\,dx + f\,dt$$

depends only on the endpoints of C. By a suitable interval we mean one at whose endpoints u is well determined, and by a suitable curve one along which u is well determined almost everywhere.

The forward initial value problem is to find a solution of (1) whose value at $t = 0$ is prescribed[1] in some reasonably broad class of permissible initial values. Our aim is to construct solutions which exist for all time $t \geq 0$.

If the differentiation in the second term of (1) is carried out, we obtain a quasilinear system of differential equations:

$$u_t + Au_x = 0 \tag{2}$$

where $A = A(u) = \text{grad}\, f$ is an $n \times n$ matrix. We assume that (2) is strictly hyperbolic, i.e. that for each u, $A(u)$ has n real and distinct eigenvalues. For hyperbolic systems there is a theory of the initial value problem, but it is a theory in the small, i.e. the solutions are constructed only for a finite time range and only for reasonably smooth initial data. Moreover, this is all that can be expected within the framework of differentiable solutions. To illustrate the kind of breakdown that can take place, consider a single quasilinear equation

$$v_t + a(v)v_x = 0 , \tag{3}$$

v a scalar variable, with prescribed initial values $v(x, 0) = \phi(x)$. The equation (3) asserts that the directional derivative of v in the direction

$$\frac{dx}{dt} = a(v) \tag{4}$$

is zero, i.e. that v is a constant along curves which are trajectories of (4). These curves are called characteristics. Since v is a constant along each characteristic, the slopes (4) are constant and so the characteristics are straight lines.[2] Through each point of the initial line we can construct characteristic lines, which brings us to the difficulty at the root of the matter:

Let y_1 and y_2 be two points on the initial line $t = 0$, say $y_1 < y_2$; denote by ϕ_1 and ϕ_2 the value of ϕ at these points respectively. Suppose that $a(\phi_1) > a(\phi_2)$; then the characteristics issuing from $(y_1, 0)$ and $(y_2, 0)$ intersect at the positive time

$$t_0 = \frac{y_2 - y_1}{a(\phi_1) - a(\phi_2)}$$

At the point of intersection there is an incompatibility: v cannot be equal to y_1 and y_2 both. This proves that the given initial value problem has no differentiable solution beyond the time t_0

We shall show now on an example how to continue such solutions beyond the time t_0 as distribution solutions of the given conservation law. The example will be a discontinuous solution, one which is differentiable outside of a finite number of smooth curves C_j across which it has a jump discontinuity. It is easy to show that such a function is a distribution solution of the conservation law (1) if and only if it is a genuine solution of (2) outside the curves and if across every line of discontinuity the jump conditions (Rankine-Hugoniot condition)

$$s[u] = [f] \tag{5}$$

are satisfied, where s is the speed of propagation of the discontinuity and [ ] denotes the jump of the quantity in the brackets across the line of discontinuity.

For our example we shall take the single conservation law

$$u_t + (\tfrac{1}{2}u^2)_x = 0 \tag{6}$$

and the initial values

$$\phi(x) = \begin{cases} 1 & \text{for } x < 0 \\ 0 & \text{for } 0 < x . \end{cases}$$

Clearly the discontinuous function

$$u(x, t) = \begin{cases} 1 & \text{for } x < t \\ 0 & \text{for } t < x \end{cases} \tag{7}$$

satisfies the jump conditions (5) and so is a distribution solution of (6).

The next example shows that not all discontinuous solutions are acceptable. Consider the same conservation law as before and the discontinuous initial value problem

$$\phi(x) = \begin{cases} 0 & \text{for } x < 0 \\ 1 & \text{for } 0 < x . \end{cases}$$

The discontinuous function

$$u_1(x, t) = \begin{cases} 0 & \text{for } x < 2t \\ 1 & \text{for } 2t < x \end{cases} \tag{8}$$

has the prescribed initial values, clearly satisfies the jump condition, and so is a solution of (6). On the other hand the function

$$u_2(x, t) = \begin{cases} 0 & \text{for } x \le 0 \\ x/t & \text{for } 0 \le x \le t \\ 1 & \text{for } t \le x \end{cases}$$

is continuous, piecewise differentiable, satisfies equation (6) except along the lines $x = 0$ and $x = t$, and takes on the same prescribed initial values. Since u_2 is continuous, it is the preferred solution and u_1 is not wanted. The question is: how to eliminate in a systematic way such undesirable discontinuous solutions.

A clue to the answer lies in the analysis presented before of the development of discontinuities in differentiable solutions by the collision of characteristics. In the acceptable discontinuous solution (7) the role of the line of discontinuity $x = 2t$ is to keep the characteristics issuing from the negative x-axis from colliding with the ones issuing from the positive x-axis; in the unwanted discontinuous solution (8) the characteristics issuing from the initial line never collide. Thus we are led to define an acceptable discontinuity as one where the line of discontinuity intersects the forward characteristics drawn at either side of it. Denoting by u_ℓ and u_r the states on the left and right side of a point on a line of discontinuity, the above statement can be expressed as follows:

$$a(u_\ell) > s > a(u_r) \quad . \tag{9}$$

This condition is easily extended to systems of any order as follows:

There exists an index k such that

$$\begin{aligned} \lambda_{k-1}(u_\ell) &< s < \lambda_k(u_\ell) \\ \lambda_k(u_r) &< s < \lambda_{k+1}(\mu_r) \quad , \end{aligned} \tag{10}$$

where $\lambda_1, \ldots, \lambda_n$ denote the eigenvalues of $A(u)$ arranged in increasing order. Condition (10) is sometimes called the entropy condition since in the case of gas dynamics it is equivalent to requiring that the entropy of gas increase after crossing a discontinuity line.

A discontinuity of a solution satisfying the entropy condition (10) is called a shock, of index k. Observe that (10) distinguishes the positive time direction from the negative one; this is to be expected because our original problem, the <u>forward</u> initial value problem, already did so.

The main problem can now be formulated as follows:

<u>Prove that there exists for all</u> $t \ge 0$ <u>exactly one solution of a</u>

given system of conservation with prescribed initial data all of whose discontinuities are shocks.

We shall describe briefly the state of this problem.

2. The theory of the forward initial value problem is fairly complete for a single conservation law. At first we impose the condition that the corresponding quasilinear equation (3) be genuinely nonlinear in the sense that $a(u)$, the coefficient of u_x, does depend on u:

$$\frac{da}{du} \neq 0 \qquad \text{for all } u .$$

Thus

$$\frac{d^2 f}{du^2} = \frac{da}{du} \neq 0$$

which makes the function f convex or concave. Here we shall take it to be convex.

Since a is a strictly monotonic function of u it can be inverted:

$$b(a(u)) \equiv u \quad . \tag{11}$$

Define[3] g as the integral of b:

$$\frac{dg(a)}{da} = b(a) .$$

Let $\phi(x)$ be an arbitrary bounded, measurable initial function with, say, compact support. The solution of

$$u_t + f(u)_x = 0$$

with initial value ϕ can—somewhat surprisingly—be written down explicitly. The explicit formula involves Φ, the integral of ϕ:

$$\Phi(y) = \int_0^y \phi(x)\,dx .$$

Consider the function

$$\Phi(y) + tg\left(\frac{x-y}{t}\right) ; \tag{12}$$

for fixed t and x this is a continuous function of y which tends to infinity with increasing $|y|$, on account of the convexity of g. It is not hard to show that for given t for all but a denumerable set of values of x the function (12) has a unique minimum. Denote by $y_0 = y_0(x,t)$ that value of y where the minimum occurs:

Theorem: The function

$$u(x,t) = b\left(\frac{x-y_0(x,t)}{t}\right)$$

is a solution of the conservation law with initial value ϕ and all whose discontinuities are shocks.

For a verification, see [2].

Oleinik has shown, following earlier work of Germain and Bader, that the above solution is the only one with initial value ϕ which satisfies the entropy condition.

Subsequently, Kalashnikov has shown that—somewhat surprisingly—the condition of genuine nonlinearity can be removed. For this class of equations the entropy condition has to be suitably reformulated.

3. We turn now to systems; first of all we need to formulate an analogue of the condition that the associated quasilinear equation should be genuinely nonlinear, i.e. an analogue of the condition $da/du \neq 0$. An analysis carried out in [2] shows that the appropriate condition[4] is

$$r_k \cdot \operatorname{grad} \lambda_k \neq 0 \quad \text{for all } u \text{ and } k = 1, \ldots, n\ ; \tag{13}$$

where $r_k = r_k(u)$ is the right k^{th} eigenvector of the coefficient matrix A.

It is an open question whether condition (13) guarantees that the initial value problem can be solved for all initial states $\phi(x)$ which differ by a sufficiently small amount from a constant state. The answer is yes in the very special case when ϕ consists of two constant states, i.e.

$$\phi(x) = \begin{matrix} u_\ell & \text{for } x < 0 \\ u_r & \text{for } 0 < x\ , \end{matrix} \tag{14}$$

where u_ℓ and u_r are two constant states sufficiently close to each other. One can construct a solution with such initial values which is centered, i.e. it depends only on the ratio x/t and consists of $n+1$ constant states separated by shock waves or centered differentiable solutions.[5] For details see [2].

In order for solutions to exist when the initial data have large variations one very likely must impose in addition to (13) some conditions in the large. One such condition has been proposed by Rozhdestvenskii, see [5], for systems of two equations for two unknowns. It requires that for any pair of points u^1 and u^2 in the u-plane there should exist a point u^3 such that

$$f(u^1) - f(u^2) = A(u^3)(u^1 - u^2)\ . \tag{15}$$

In addition it is required that $\lambda_1(u^1) \neq \lambda_2(u^2)$, and that the corresponding eigenvectors are linearly independent. Under these conditions Kusnetzov has proved that the initial value problem (14) has a unique solution for arbitrary initial states u_ℓ and u_r , not necessarily close to each other. Other than this result no existence theorems have been proved for systems.

We list now three methods which have been proposed for constructing solutions:

A) <u>The viscosity method</u>: Enlarge (1) by adding second order terms.

$$u_t + f_x = \varepsilon u_{xx} . \qquad (16)$$

Denote by u_ε the solution with initial value ϕ of the above parabolic system. If one can prove that such a solution u_ε exists for all time, and if as ε tends to zero u_ε approaches almost everywhere boundedly some limit u , then u will be a solution of the original system (1). It can be shown that all discontinuities of this limit would be shocks.

Suppose the equations (1) describe the motion of some conservative physical system. If we alter the physical description of the system to include dissipative features, such as viscosity, heat conduction, ohmic loss, etc., the equations have to be modified by the inclusion of terms of second order very much as in equation (16), except that the additional terms are more complicated, even nonlinear. This shows that the study of equations like (16), and the manner in which their solution approaches those of (1) as the strength of the dissipative mechanisms tends to zero, is of great interest in itself.

B) <u>The finite difference method</u> consists of approximating the differential conservation law (1) by a difference equation. In particular we consider the following class of difference equations:

$$u(x, t+h) = g(u_{-\ell+1}, \ldots, u_\ell) - g(u_{-\ell}, \ldots, u_{\ell-1}) \qquad (17)$$

where u_j abbreviates $u(x-hj, t)$, h is the mesh size, and g is a function of 2ℓ vector arguments. In order for (17) to be consistent with (1) g must satisfy the consistency relation

$$g(u, \ldots, u) = f(u) .$$

Note that the right side of (17) is a perfect difference; thus our difference equation is also in conservation form. Thanks to this one can prove that if u_h , the solution of (17) with initial value ϕ , tends to a limit as h tends to zero, this limit is a solution of the conservation equations (1).

Approximation schemes of the form (17) are of great practical importance since they provide a means of finding approximate solutions. In [3] a specific choice for g is proposed and we give a heuristic analysis of the stability of the resulting difference scheme.

C) Mixed differential-difference schemes: Such schemes were considered by Douglis, who replaced the space derivative by a left difference quotient but retained the time derivative. The resulting system is a set of ordinary differential equations:

$$u_t = [f(u(x-h)) - f(u(x))]/h .$$

To justify the use of the left difference quotient one has to assume that all propagation is to the right, i.e. that all characteristic speeds λ_k are positive.

We turn now to the question of uniqueness. Oleinik has proved uniqueness for systems of the form

$$u_t + p(v)_x = 0 , \quad v_t - u_x = 0 ,$$

p a convex function, and Rozhdestvenskii and Kalashnikov for systems satisfying condition (15). However, unlike in the work of Oleinik, these authors consider only solutions whose first derivatives are bounded outside of the discontinuities. This is unsatisfactory since solutions will in general contain centered rarefaction waves arising from the collision of shocks, and first derivatives of such centered waves are unbounded.

Another proof of such uniqueness theorems is contained in Douglis [6].

4. We shall discuss briefly the question of how solutions behave for large values of t . We shall take initial values ϕ which are constant outside a finite interval:

$$\phi(x) = u_0 \quad \text{for} \quad |x| > R .$$

We expect the corresponding solution will tend to the constant state u_0 ; the problem is to determine the rate of decay and to give an asymptotic description of the decaying solution. Studies made on the equations of gas dynamics suggest that solutions approach a constant state like $1/\sqrt{t}$, and that the asymptotic description is as follows:

$$u(x,t) = \begin{cases} u_0 + (\frac{x}{t} - \lambda_k) r_k & \text{for } a_k\sqrt{t} < x - \lambda_k < b_k\sqrt{t}, \; k = 1, 2, \dots, n \\ u_0 & \text{otherwise} \end{cases} \tag{18}$$

Here λ_k denote the eigenvalues of $A(u_0)$ and r_k the corresponding eigenvectors, so normalized that

$$r_k \cdot \text{grad } \lambda_k = 1 \ .$$

In light of (13) such a normalization is always possible.

The quantities a_k and b_k are functionals of the initial data ϕ ; at present there are approximate but not rigorous methods for calculating them. From their definition it follows that a_k and b_k are invariants of the motion; from the conservation form of the equations (1) we deduce immediately that the integral of each component of u with respect to x over the whole x-axis is an invariant of the motion. From the analysis above it follows that there are n further nontrivial invariants.

In case of a single equation it is easy to deduce from the explicit formula given in section 2 that formula (18) gives correctly the asymptotic shape of solutions. In this case the nontrivial invariant functional, for f convex, is

$$\operatorname*{Max}_{y} \int_{-\infty}^{y} \phi(x)\, dx \ .$$

NOTES

1. As we shall see, we will be dealing mostly with piecewise continuous solutions, so the concept of initial value is well defined. In general it can be defined in a weak sense.
2. This circumstance is a great help in analyzing single quasilinear equations; it no longer is true for systems.
3. g can be defined directly in terms of f as its Legendre transform.
4. If this condition is violated, i. e. if $r_k \cdot \text{grad } \lambda_k = 0$ for some k , then there will be solutions discontinuous along a curve which is characteristic with respect to each side. These are called contact discontinuities and they play an important role in gas dynamics.
5. Called centered simple waves.

BIBLIOGRAPHY

1. E. Hopf, The partial differential equation $u_t + uu_x = \mu u_{xx}$, Comm. Pure Appl. Math. 3(1950), 201-230.
2. P. D. Lax, Hyperbolic systems of conservation laws II, Comm. Pure Appl. Math. 10(1957), 537-566.

3. P. D. Lax and Burton Wendroff, Systems of conservation laws, Comm. Pure Appl. Math. 13(1960), 217-237.

4. O. A. Oleinik, Discontinuous solutions of non-linear differential equations, Uspekhi Mat. Nauk 12(1957), 3-73. (Russian)

5. B. L. Rozhdestvenskii, Discontinuous solutions of hyperbolic systems of quasilinear equations, Uspekhi Mat. Nauk (1959), 53-111. (English translation)

6. Douglis, A., The continuous dependence of generalized solutions of non-linear partial differential equations upon initial data, Comm. Pure Appl. Math. 14(1961), 267-284.

C. L. DOLPH

The extant nonlinear mathematical theory of plasma oscillations

Introduction

Although much has been written about plasma physics in recent years, the author is unaware of any work that attempts an assessment of the nonlinear aspects of the theory from a mathematical viewpoint. To be sure, this theory exists only in a very rudimentary form and falls far short of being capable of giving definite answers to the problems of real physical interest; nevertheless, the fact that almost all of the matter in the universe (estimated at 99.9 per cent) exists in the plasma or fourth state makes it worthy of attention by mathematicians. In many ways the current situation is reminiscent of that which existed fifteen years ago in both nonlinear oscillation theory and compressible flow theory in that both theories consisted in the main of a few isolated examples which could be discussed to some degree. It is reasonable to suppose that a mathematical concentration on the problems of plasma physics would produce results as scientifically and technologically rewarding as those that have occurred in the above fields.

As a concrete example, the travelling wave amplifier and backward wave oscillator are highly developed and useful microwave devices and as yet no really satisfactory nonlinear mathematical theory for these exist except in a technical or design sense. While it would be too time consuming to describe this problem in detail here, the basic equations have been formulated on several conceptual levels and are readily available. Thus, the transmission line approximation to the slow wave structure is due to Brillouin (9) who treated some nonlinear effects in the cold plasma approximation (Cf. also (10)). The field approach via a perfectly conducting sheath, conducting only in the direction of a helix has been formulated by Chu and Jackson (18) and the linear approximation discussed in the cold plasma approximation. Finally the transport approach has been formulated by Watkins and Rynn (49) and discussed in the linear approximation. What is lacking is a full nonlinear treatment of either the lumped circuit or field approximation with the nonlinear transport equation. The problem involves the simultaneous consideration of a linear wave system driven by the second time derivative of the density and a transport equation for the determination of the density as a function of the voltage potential of the linear wave system. The development of techniques adequate for this problem would eventually lead to the same type of development for periodic solutions of a class of systems of nonlinear partial differential equations as the techniques developed for the triode oscillator associated with the name of Van der Pol led to our understanding of periodic solutions for a large class of systems of ordinary differential equations.

The word "plasma" as applied to an ionized gas seems to have been introduced by Langmuir (37) in 1929. As it is currently used in physics, a plasma, while a highly ionized gas, is also approximately electrically neutral. The presence of the long-range Columb forces which arise whenever charge separation occurs result in a "medium-like" behavior of the gas. To quote from the pioneering paper of Bohm and Gross (8).

"If electric fields are introduced, either by an external disturbance or by incomplete space-charge neutralization, the highly mobile free charges automatically respond to the forces in such a way as to shield out the fields. One can, therefore, regard a plasma as a medium which tends to remain near a field-free and neutral equilibrium state, resisting efforts to produce deviations from this state, just as a liquid tends to remain near an equilibrium state of definite volume, resisting efforts to produce changes in this volume... .

"If, for example, a given region contains an excess of electrons, they repel each other and therefore begin to move out. By the time neutrality has been established the electrons have gained momentum so that they keep on going and create a deficiency of negative charge which attracts the electrons back in. In time the motion is reversed, and a systematic oscillation of the charged region is set up. For the case in which the random thermal motions of the charges are slow enough to be neglected, Langmuir and Tonks have studied these oscillations and have shown that their angular frequency is given by

$$\omega_p = \left(\frac{4\pi n_0 e^2}{m}\right)^{\frac{1}{2}}$$

where n_0 is the density of charged particles and is their mass. For a typical density of 10^{12} electrons per cm , the plasma frequency is about 10^{10}c. p. s. ... .

"These oscillations are irrotional and, therefore, do not radiate. Transverse plasma oscillations are, however, also possible... . Because the ions are so much heavier than the electrons, their motions will be so small that we can neglect them altogether and assume that they remain at rest. The positive ions can also oscillate, but at a much lower frequency... . For wave lengths much greater than interionic spacing, it is a good approximation to regard the charge of positive ions as uniformly smeared over the region."

Many other types of phenomena also can take place in a plasma. For a physical introduction, the book by Spitzer (44) is most widely quoted, while a good mathematical introduction to the subject can be found in Chandrasekhar (16).

Most theoretical work devoted to plasmas in non-equilibrium states is based on the so-called Landau-Vlasov equations which are as follows: [1]

$$\frac{\partial f_\sigma}{\partial t} + \underline{v} \cdot \nabla f_\sigma + \frac{ze}{m} [\underline{E} + \frac{1}{c} \underline{v} \times \underline{H}] \cdot \nabla_v f_\sigma = (\frac{\partial f_\sigma}{\partial \tau})_{coll} \qquad (1.1)$$

$$\nabla \cdot \underline{E} = 4\pi n \qquad (1.2)$$

$$\nabla \cdot \underline{B} = 0 \qquad (1.3)$$

$$c \nabla \times \underline{B} = 4\pi \underline{J} + \frac{\partial \underline{E}}{\partial t} \qquad (1.4)$$

$$c \nabla \times \underline{E} = \frac{\partial \underline{B}}{\partial t} \qquad (1.5)$$

$$n = \sum_\sigma ze \int f_\sigma d^3 v \qquad (1.6)$$

$$\underline{J} = \sum_\sigma ze \int d^3 v f_\sigma \underline{v} \qquad (1.7)$$

$$\underline{D} = \varepsilon \underline{E} \qquad (1.8)$$

$$B = \mu \underline{H} \qquad (1.9)$$

In these equations it is assumed that the plasma can be fully described by a distribution function $f_k(\underline{r}, \underline{v}, t)$ for each species (usually k is taken to be two with $z = 1$ for ions and -1 for electrons) which determines the number of particles of type k in a gas element of $\underline{r}$, $\underline{v}$ space at time t.

$\underline{E}$, $\underline{D}$, $\underline{B}$, $\underline{H}$ have their usual meaning as in electro-magnetic theory while the zero moment (1.6) defines the density of the plasma, the first moment (1.7) the current in the plasma.

These equations appear to have been guessed at by Vlasov; and since they form a closed set of equations in which $\underline{E}$, $\underline{H}$ determine f_k and are themselves determined by f_k, they are often called self-consistent. Physically the first equations state that the distribution function f_k changes by streaming alone and that the acceleration

$$\frac{ze}{m}\left[\underline{E} + \frac{\underline{V} \times \underline{H}}{c}\right]$$

is determined by the smeared out fields of all the other particles. External fields are usually neglected, but sometimes must be included (Cf. §4).

While most physicists believe that conclusions drawn from this set of equations possess physical content, they are nonetheless subject to several criticisms. In the first place, it is certainly not clear that a collection of single particle distribution functions f_k will furnish an adequate description. In fact, correlations between two particles, three particles, etc. may be necessary for a realistic theory. Recently, however, it has been shown independently by Guernsey (27) and Balesceu (3) that these equations do represent the first approximation to the so-called B-B-G-K-Y hierarchy (Bogolubov-Born-Green-Kirkwood-Yvon) in which higher order correlations are considered. In the second place, the Landau-Vlasov equations in the above form do not furnish a complete physical description for another reason; namely, they do not exhibit irreversibility and unlike the Boltzman equation, the f_k's do not approach a Maxwellian distribution as t goes to infinity. In particular, the H-theorem of statistical mechanics does not hold; and consequently, the solutions of these equations do not approach maximum entropy as t goes to infinity. These equations have been modified by Balesceu (3) and Guernsey (27) for the case of a spatially uniform distribution by replacing the collision term in (1.1) by the first term in the expansion of the electronic charge. The resulting equations do exhibit full irreversibility and the H-theorem is valid for them. The Balescu-Guernsey equations contain, however, a logarithmic divergence due to the fact that their derivation ignores short-range forces. This and many other questions concerning the foundations of plasma physics still remain open.

The rest of this article will treat the Landau-Vlasov equations, in spite of their short-comings. These equations are still much too difficult to be treated in general; and consequently, it is frequently necessary to either consider special situations or introduce further idealizations.

Existence and Uniqueness Theory for Cauchy's Problem

It is clear that a possible approach to the existence problem consists in replacing equations (1.1) by system of ordinary differential equations describing the characteristics or, physically, the particle orbits associated with these equations. These take the form

$$m\frac{d\underline{v}}{dt} = \frac{ze}{m}[\underline{E} + \frac{1}{c}(\underline{v} \times \underline{H})]$$

$$\frac{d\underline{x}}{dt} = \underline{v} \; ; \quad \frac{df}{dt} = (\frac{\partial f}{\partial \tau})\text{collisions}$$

and given the acceleration term, the integration of these equations is known to be completely equivalent to the integration of (1. 1) from Lagrange theory of first order partial differential equation. Physicists according to Chandrasekhar (17) know this result as Jeans Theorem. S. V. Iordanskii (30) seems to have been the first to obtain a result by this method for the simplest Cauchy problem with no collisions where the Landau-Vlasov equations reduce to

$$\frac{\partial f}{\partial t} + v\frac{\partial f}{\partial x} - \frac{e}{m}E(x, t)\frac{\partial f}{\partial v} = 0 \qquad (2.1)$$

$$\frac{\partial E}{\partial x} = -4\pi e\left\{\int_{-\infty}^{\infty} f(x, v, t)\, dv - N_0\right\} \qquad (2.2)$$

m , e represent the mass and charge of the electron while N_0 represents the density of the ion uniform background(which is assumed constant) subject to the conditions

$$f(x, v, 0) = f_0(x, v)$$

and

$$\lim_{x \to -\infty} E(x, t) = 0$$

for a given positive function $f_0(x, v)$. In addition, the plasma was assumed as neutral in that:

$$\int_{-\infty}^{\infty} dx\left\{\int_{-\infty}^{\infty} f_0(x, v)\, dv - N_0\right\} = 0 \qquad (2.3)$$

and Iordanskii assumed that functions $N(v)$, $K(v)$ and $\phi(x)$ exist where K was monotonically decreasing in $|v|$, ϕ bounded, such that

$$\lim_{|x| \to \infty} f_0(x, v) = N(v) \quad , \quad \int_{-\infty} N(v)\, dv = N_0$$

$$\int_{-\infty}^{\infty} v^2 K(v)\, dv < \infty \qquad \int_{-\infty}^{\infty} \phi(x)\, dx < \infty$$

$$|f_0(x, v) - N(v)| < K(v)\phi(x) \qquad (2.4)$$

$$0 < N(v) < K(v)$$

Iordanskii's theorem then states:

THEOREM. The solution of the Cauchy problem for equations (2.1) and (2.2) exists for an arbitrary continuous function $f_0(x, v)$ which satisfies (2.3) and (2.4).

Under some circumstances, obscure from the Doklady note, uniqueness is also claimed. In some, as yet, unpublished work, R. K. Ritt (39) has been able to generalize and sharpen the above result. In place of the above hypotheses, Ritt assumes that

$$f_0(x, v) = N(x)V(v)$$

where $N(x)$ is measurable and bounded such that $0 \leq N(x) < N_0$ where

$$\int_{-\infty}^{\infty} |N(x) - N_0| dx < \infty$$

and where $V(v)$ is essentially Maxwellian; e.g. it is sufficient that the first two moments of $V(v)$ exist, that $V(v)$ be positive semi-definite and even, and that it be continuously differentiable with sign $V(v) = -$ sign v. Further, let $\mathscr{E}$ = [space of all functions $E(x, t) = \int_{-\infty}^{x} p(x', t)\,dx'$ for $p(x, t)$ bounded in x for each fixed t, and in $L_1(-\infty, a)$ for every a].

For each $E(x, t)$ in $\mathscr{E}$, consider next the equations

$$\frac{dx}{dt} = v, \quad \frac{dv}{dt} = E(x, t) \qquad (2.5)$$

subject to

$$x(0) = x_0 \qquad v(0) = v_0$$

and let their solutions be denoted by

$$x = x(x_0, v_0, t)$$

$$y = y(x_0, v_0, t)$$

As in Iordanskii, it can be shown that these solutions have inverses

$$\begin{aligned} x_0 &= x_0(x, v, t) \\ y_0 &= y_0(x, v, t) \end{aligned} \qquad (2.6)$$

and, in fact, that there is a 1-1 mapping from the x, v plane to the x_0, v_0 plane with unit Jacobian.

The given problem is reduced now to showing that a certain nonlinear transformation has a fixed point. For the case considered by Ritt, this transformation is given by

$$F(E) = \int_{-\infty}^{x} \left\{ \int_{-\infty}^{\infty} N[x_0(x', v, t)] V[v_0(x', v, t) \, dv - N(x') \right\} dx' \qquad (2.7)$$

and it is shown to have a fixed point in the uniform topology with norm given by

$$\|E\| = \max_x |E(x, t)|$$

Specifically, Ritt has proven the following:

THEOREM: Under the above hypotheses on $N(x), V(v)$, the transformation $F(E)$ defined by (2.7) has a fixed point E^*. When $V(v)$ is continuously differentiable this fixed point is unique and

(a) $E^* = E(x, t)$ is absolutely continuous in x and differentiable in t almost everywhere;

(b) the limit $E(x, t) = 0$ exists, uniformly on each finite t-interval;

(c) the $\max_x E(x, t) = \|E\|$ exists and is bounded by a continuous function;

(d) if $x_0(x, v, t)$ and $v_0(x, v, t)$ are solutions of the characteristic equations (2.5) such that $x = x_0(x, v, 0)$, $v = v_0(x, v, 0)$ then

$$\frac{\partial E}{\partial x} = \int_{-\infty}^{\infty} \left\{ N[x_0(x, v, t)] V[v_0(x, v, t)] - N(x) \right\} dx$$

(e) $E(x, 0) = 0$.

While the statement of the theorem is involved, the proof is even more complicated since the usual procedures of nonlinear functional analysis do not appear to be directly applicable since uniform estimates do not appear possible. Instead, the method of proof uses a modified Picard process which yields estimates depending upon the first approximation. This process is, however, far from straightforward since at every stage of the iterative process it is necessary to use the map from the x, v space to the x_0, v_0 space as well as its inverse. In order to determine the correct domain in the x_0, v_0 space, only double integrals can be used and then it appears necessary to use the method due to Iordanskii who introduced $W_E(x, t)$ as the chain of regions lying between the straight-line $x_0 = x - vt$ and the image of the line $x' = x$ under the map (2.6) corresponding to the assumed value of E. With the aid of these regions the operator (2.7) can be shown to be equivalent to

$$F(E) = \int_{-\infty}^{\infty} \int_{-\infty}^{\frac{x-x_0}{t}} \Big\{ [N(x_0)-N_0]V(v_0)\,dv_0\,dx_0$$

$$+ \int_{-\infty}^{x} [N_0 - N(x')]\,dx' + \iint_{W_E} N(x_0)V(v_0)\,dv_0\,dx_0$$

for every E in $\mathscr{E}$. From this, an expression similar to that used by Iordanskii, Ritt's sharper estimates and results follow.

It should be remarked that an alternate approach exists in principle. Both Iordanskii and Ritt thought of $E(x, t)$ as known in the transport equation and reduced the problem to a nonlinear functional equation derived from Poisson's equation. It is, of course, equally possible to think of the density as known in Poisson's equation and eliminate the electric field in order to obtain a nonlinear transport equation. In fact, this is the usual procedure in perturbation procedures based on a spatially homogeneous distribution where Fourier transforms are often used. It is conceivable, but by no means certain, that simpler proof might be possible were this to be followed; but the author is unaware of any for the nonlinear case.

In contrast to the case of one-dimensional longitudinal oscillations treated above, the corresponding theory of nonlinear transverse oscillations described by the equations

$$\frac{\partial f}{\partial t} - \frac{e}{m} E_y(x, t) \frac{\partial f}{\partial v_y} + v_x \frac{\partial f}{\partial x} = 0$$

$$\nabla \cdot \underline{E} = 0$$

$$\frac{\partial^2 E_y}{\partial x^2} - \frac{1}{c^2} \frac{\partial^2 E_y}{\partial t^2} = \frac{4\pi}{c} \frac{\partial J}{\partial t}$$

$$v_y = -\frac{e}{c} \int v_y f(x, v, t)\,dv_x\,dv_y$$

is much simpler and can be obtained formally by standard techniques. This has been done by J. Enoch (24) who developed a dispersion relation both for stationary waves and for the general initial value problem. His results are similar to those obtained by Van Kampen (46) in his linearized theory for the stationary case and similar but probably not as significant as those obtained by Landau (36) for the linearized initial value problem for longitudinal oscillations.

Comparable theorems for the more general situation described by the Landau-Vlasov equations are not known to the author. It is clear that even the extension of the above longitudinal case with zero magnetic field to two or three spatial dimensions will be a formidable task since the relationship between the electric field and

and the distribution will involve the solution to a nonlinear Poisson equation as well.

While the concern here is limited primarily to the nonlinear aspects, it should probably be noted that since any distribution function of the form $f_0(v)$ satisfies (1.1) in the absence of collisions, the usual approach is to linearize by assuming that each f_k is of the form

$$f = f_0(\underline{v}) + f_1(\underline{x}, \underline{v}, t) \qquad (2.8)$$

where usually $f_0(v)$ is taken to be the Maxwellian; namely,

$$f_0(v) = N_0\left(\frac{m}{2\pi kt}\right)^{\frac{1}{2}} \exp\left[\frac{-mv^2}{\alpha kt}\right] .$$

In the one-dimensional case where $\underline{B} = 0$, Landau (36) treated the resulting problem by the method of Laplace transformation and discovered a phenomenon now known as Landau damping which has been the subject of much controversy. Van Kampfen (46) treated the same problem by the introduction of singular normal modes on which we will comment briefly in (§3) and later Case (12) was able to establish L_1 type completeness relations for this linearized problem. For initial values which are entire functions of velocity, the procedure of Landau used analytic continuation in the transformed plane and gave rise to asymptotic damping of the perturbed electric field (but not of the perturbed distribution f_1 —reversibility would then imply the existence of a physically unstable situation). Kruskal (35) has attempted to reconcile the singular theory with the results of Landau by linking the process to one of non-uniform convergence and noting that linear combination of singular modes with sufficient smooth amplitude coefficients can be made free of singularities. Most physicists believe that the phenomenon of Landau damping exists and is the result of a phase mixing or wave interference process. However, as Case (12) has demonstrated by the use of specific examples, the final asymptotic form is highly sensitive to the relationship which exists between the standard deviations in velocity of the initial $f_0(v)$ and $f_1(x, v, t)$ and that further it does not seem possible to give an estimate of the time scale necessary for the appearance of the asymptotic form. For further discussion and complete bibliography, see Kino-Crawford (34).

In the event that the magnetic field cannot be neglected, it is customary to add to the above perturbation process the expansions

$$B = \underline{B}_0 + \underline{B}\ (\underline{r}, t)$$
$$E = 0 + E_1(\underline{r}, t)$$

where $\underline{B}_0$ is assumed constant. Under these circumstances,

Bernstein (4) has shown it is possible to uncouple the full set of Landau-Vlasov equations by use of Fourier transforms in $\underline{x}$, Laplace transforms in t , and by constructing a suitable Green's function in velocity space. It is even possible to separate out transverse and longitudinal oscillations and carry out quite a general discussion with the aid of the inversion integral for the Laplace transform.

For some applications such as electronic tubes, it is not sufficient to linearize about a spatially uniform distribution. Under these circumstances one normally linearizes about the steady state distribution, the existence of which follows from Ritt's theorem, and for harmonically varying electric fields it is possible to obtain a linear integral equation for the spatial dependence of the electric field in the manner indicated by Rosenbluth (40) and employed, for example, by Kino and Chodorow (33) in their work on the so-called Langmuir Paradox (cf. 34). A linearization leading to the solution of the wave equation has also been carried out by Ritt (39).

More generally, in the collisionless case so far discussed, Chandrasekhar, Kaufmann, and Watson (17) have developed a perturbation procedure for large magnitude fields while in the opposite extreme where collisions dominate, the usual procedure is to use the hydrodynamic approximation. This consists of the use of the moment equations which can be derived from the Landau-Vlasov equations [Cf. Spitzer (44), Chandrasekhar (16)] and which, while not closed, can be closed at some order by an additional assumption. For example, only the first three are necessary if the existence of a local Maxwellian distribution can be assumed and only the first two are necessary if the plasma can be assumed to be cold, i.e. the limit of a local Maxwellian distribution as the temperature goes to zero. Often in this case one assumes the initial distribution $f(x, v, 0) = \delta(v-v_0)$ where δ is the Dirac delta function. The existing nonlinear theory is largely devoted to this case and it is the cold plasma alone that will be treated in any detail in the hydrodynamic approximation in this paper.

An excellent review article on wave motion in a plasma in the magnetohydrodynamic approximation is that of R. Lust (38). While much of this is devoted to various linear problems, Alfven waves with finite amplitude and hydromagnetic compression waves are treated. Finally, a critique of this theory is given based upon the usual assumptions of infinite plasma conductivity and isotropic pressure.

Uhlenbeck[1] notes that the situation in plasma theory is in sharp contrast to that which exists in kinetic theory where one can derive both the general conservation laws and the hydrodynamical equations from the Boltzman equation as well as the hydrodynamical equations from the general conservation laws. In plasmas on the other hand, only the general conservation laws have so far been derived from the

Landau-Vlasov equations. In his opinion, our current ignorance of characteristic relaxation times for long range forces makes the study of the approach to equilibrium difficult; and in reality, it may involve the onset of turbulence. (For short range forces and binary collisions, it should be recalled that the Maxwellian distribution is closely approached in a time of the order of the mean free path length divided by the mean particle velocity and that this ratio is of the order of the time between collisions (the relaxation time). Since complete equilibrium is achieved in a time which is large compared to that necessary to achieve local equilibrium, the linearization of the Boltzman equation as carried out in the Chapman-Enskog method can be justified. Neither Uhlenbeck nor the author knows of any similar justification or convergence proof for any of the linearized methods described above.)

Nonlinear Steady and Travelling Wave Solutions

The most general analytically tractable plasma problem treated to date seems to be due to Bernstein, Green and Kruskal (5) who found solutions to the Landau-Vlasov equations involving nonlinear electrostatic waves due to both ions and electrons when consideration was limited to a coordinate system in which the wave was at rest. For this case the Landau-Vlasov equations reduce to

$$v\frac{\partial f\pm}{\partial x} \mp \frac{e}{m}\frac{d\phi}{dx}\frac{\partial f\pm}{\partial v}(x, v) = 0 \tag{3.0}$$

where the electric field is given by $E = -\dfrac{d\phi}{dx}$

$$\frac{d^2\phi}{dx^2} = 4\pi e\int_{-\infty}^{\infty} dvf_-(x, v) - 4\pi e\int_{-\infty}^{\infty} dvf_+(x, v)$$

and where the corresponding characteristic equations are:

$$\frac{dx}{v} = \frac{dv}{\mp\frac{e}{m}\frac{d\phi}{dx}}$$

These admit energy integrals

$$E_\pm = \frac{1}{2}m_\pm v^2 \pm e\phi$$

so that any solution of (3.0) is of the form

$$f_\pm = f_\pm(E_\pm)$$

where $f_\pm$ is arbitrary. In terms of energy as the independent variable, Poisson's equation takes the form

$$\frac{d^2\phi}{dx^2} = 4\pi e \left\{ \int_{-e\phi}^{\infty} \frac{dE\, f_-(E)}{[2m\,(E+e\phi)]^{\frac{1}{2}}} - \int_{e\phi}^{\infty} \frac{dE\, f_+(E)}{[2m_+(E-e\phi)]^{\frac{1}{2}}} \right\}$$

and admits the first integral

$$\left(\frac{d\phi}{dx}\right)^2 + 8\pi \int_{e\phi}^{\infty} dE\, f_+(E) \left[2\, \frac{(E-e\phi)}{m_+}\right]^{\frac{1}{2}} + \int_{-e\phi}^{\infty} dE\, f_-(E) \left[2\, \frac{(E+e\phi)}{m_-}\right]^{\frac{1}{2}} = \text{const.} \tag{3.1}$$

which can in principal be solved by quadrature and shown to admit periodic solutions for certain distributions. According to Kruskal (35) the stability of these periodic solutions has not been established.[2] Rather than investigate various periodic solutions, the above authors viewed (3.1) as an integral equation for the trapped electrons by rewriting it in the form

$$\int_{-e\phi}^{-e\phi_{min}} \frac{dE\, f_-(E)}{[2m_-(E+e\phi)]^{\frac{1}{2}}} = g(e\phi) \tag{3.2}$$

$$g(e\phi) = \frac{d^2\phi}{dx^2/4\pi e} + \int_{e\phi}^{\infty} \frac{dE\, f_+(E)}{[2m_+(E-e\phi)]^{\frac{1}{2}}} - \int_{e\phi_{min}}^{\infty} \frac{dE\, f_-(E)}{[2m_-(E+e\phi)]^{\frac{1}{2}}}$$

and assuming that ϕ , $f_+(E)$ were given and $f_-(E)$ known for the untrapped electrons for which $E > -e\phi_{min}$. Since (3.2) is of the convolution type, it can be solved by Laplace transformation subject only to the weak conditions that $g(e\phi_{min}) = 0$ and that the resulting $f(E)$ be non-negative in order to represent a distribution function. Its solution is specifically given by

$$f_-(E) = \frac{(2m_-)^{\frac{1}{2}}}{\pi} \int_{e\phi_{min}}^{-E} dv \frac{dg(v)}{dv} \left[-E - v\right]^{-\frac{1}{2}} \qquad E < -e\phi_{min}$$

Next they investigated this solution in the small amplitude limit for which $e\phi_{max} - e\phi_{min}$ was very much smaller than the mean particle energy obtaining expansions in half-integral powers of $(-e\phi_{min}-E)$, in contrast to the result of linear theory which predicts an expansion linear in the amplitude in the first order. Linear theory involves a singular distribution, on the other hand, for utilizing the expansion (2.8), one has for $f_1 = f$, for example, the linearized equations

$$v \frac{\partial f}{\partial x} + \frac{e}{m} \frac{d\phi}{dx} \frac{\partial f_0}{\partial v} = 0$$

$$\frac{d^2\phi}{dx^2} = 4\pi e \int f_1 \, dv$$

Taking Fourier transforms in x and eliminating the transform of the potential yields

$$\hat{f}_1 = \frac{4\pi e^2}{mk^2} \frac{\partial f_0}{\partial v}$$

where, because of homogenity, the normalization

$$\int f_1 \, dv = 1$$

has been used.

As in the more general case proved by Van Kampfen (46) and Case (12), admitting distributions implies that

$$f_1 = \frac{e^2}{k^2} \frac{\frac{1}{n_0} \frac{\partial f_0}{\partial v}}{v} + \lambda\delta(v)$$

where the unknown constant is to be determined from the normalization condition.

Perhaps the major achievement of these authors consisted of the interpretation of this singularity as being the linear theory's way accounting for the existence of trapped particles. This they accomplished by examining the moments of their nonlinear solution and showing that by suitable expansions only the zeroth order moment—that of the density—could be affected to the first order in $(\phi - \phi_{min})$ by the presence of trapped particles and that the introduction of the above Dirac delta function resulted in a first order distribution of the form

$$f_1(x, v) = f_0(v) - \frac{e(\phi - \phi_{min})}{mv} \frac{\partial f_0(v)}{\partial v} + \lambda\delta(v)$$

which would reproduce all the moments of the exact nonlinear distribution correctly to the first order in the amplitude of the potential. While similar singularities arise in quantum field theory and can be explained or eliminated by the use of non-local theories, the situation here is much more clean-cut and gives rise to many interesting mathematical possibilities for generalization.

A closely related problem and historically the first nonlinear problem treated (Cf. Achiezher and Liubarskii (1)) is that of travelling waves in a cold plasma. This problem may be treated by either the appropriate Landau-Vlasov equation or by the first two moment equations. In order to relate this problem to that treated by Sen (42) who was trying to explain the observation of the second harmonic in the sun's corona, we shall demonstrate the equivalence directly. Let $x' = x-qt$ where q is a constant streaming velocity and assume (a) the ions form a uniform constant background, and (b) that all quantities depend only on x, t in the combination of x', and (c) collisions can be neglected. Under these circumstances the transport equation takes the form $f = f_-$

$$(v-q)\frac{\partial f}{\partial x} + \frac{e}{m}\frac{d\phi}{dx}\frac{df}{dv} = 0$$

and has for its general solution

$$f = f\left[\tfrac{1}{2}(v-q)^2 - \frac{e}{m}\phi\right]$$

where f is arbitrary. Poisson equation becomes

$$\frac{d^2\phi}{dx^2} = -4\pi e n_0 + 4\pi e n_0 \int f\left[\frac{(v-q)^2}{2} - \frac{e}{m}\phi\right] dv$$

By requiring that f reduce to $f_0(v)$ at $\phi = 0$ and making the change of variables

$$\tfrac{1}{2}(v'-q)^2 = \tfrac{1}{2}(v-q)^2 - \frac{e}{m}\phi$$

This can be written as

$$\frac{d^2\phi}{dx^2} = -4\pi e n_0 + 4\pi e n_0 \int \frac{f_0(v')(v'-q)\,dv'}{\left\{(v'-q)^2 - \frac{2e}{m}\phi\right\}^{\frac{1}{2}}}$$

a result first obtained by Bohm and Gross (8). If one now takes $f_0(v) = \delta(v-v_0)$, the above simplifies to

$$\frac{d^2\phi}{dx^2} = -4\pi e n_0 + 4\pi e n_0 \frac{1}{\left[1 - \frac{2e\phi}{m}\Big/(v_0-q)^2\right]^{\frac{1}{2}}} \qquad (3.3)$$

The moment equations appropriate to this problem are

$$m\frac{\partial v}{\partial t} + v\frac{\partial v}{\partial x} = -\frac{e}{m}E \qquad (3.4)$$

$$\frac{\partial n}{\partial t} + \frac{\partial}{\partial x}(nv) = 0 \qquad (3.5)$$

$$\frac{\partial E}{\partial x} = -4\pi e(n_0-n) .$$

Introducing the assumption that all quantities depend only on x' one finds that (3.4) has the integral

$$(v-q)^2 = (q-v_0)^2 - \frac{2e}{m}\phi$$

while (3.5) implies that

$$n(q-v) = n_0(q-v_0)$$

so that once again Poisson equation takes the form (3.3). This equation was integrated once by Sen (42) and twice by Kahlman (32). Achiezher and Liubarskii (1) chose to eliminate the potential from the moment equation and arrive at the same result by integrating the resulting velocity equation. Following Kahlman (32), it is convenient to introduce the dimensionless quantities

$$\psi = \frac{\frac{e}{m}\phi}{(v_0-q)^2}, \qquad k^2 = \frac{\omega_0{}^2}{(v_0-q)^2}$$

so that (3.3) becomes

$$\frac{d^2\psi}{dx^2} = -k^2\{(1-2\psi)^{-\frac{1}{2}}-1\}$$

and possesses the following integrals

$$\psi' = \sqrt{2}\,k\{\sqrt{1-2\psi} + \psi - \sqrt{(1-2\psi)}_{max} + \psi_{max}\}^{\frac{1}{2}}$$

$$\pm(x-qt) = \arccos\frac{\sqrt{1-2\psi}-1}{\sqrt{1-2\psi}_{max}-1} + \sqrt{2}\,\{\sqrt{1-2\psi}+\psi - (\sqrt{1-2\psi}_{max}+\psi_{max})\}^{\frac{1}{2}} \qquad (3.7)$$

[Note the dependence on $(1-2\psi)^{-\frac{1}{2}}$ as in the Bernstein, Greene, Kruskal case.] Now (3.6) is of the type $\psi'' = f(\psi)$ and clearly has for its first approximation the equation $\psi'' + k^2\psi = 0$ if ψ is sufficiently small. Unlike the situation for general equations of this type, the form of the solution (3.7) shows that the only effect of nonlinearity is to change the wave-form and not the period, and that the nonlinear solution is in fact characterized by the same wave number as the solution of the linear equation approximating to it. Achiezher and Liubarskii (1) failed to observe that this solution had a limited validity and in fact breaks down for $\psi_{max} = \frac{1}{2}$. The reason for this is that at this point a shock begins to form due to the electrons overtaking each other. This will be clearer when we reexamine this same problem for a more general point of view in the next section. Kahlman (32) has given a very thorough discussion of this travelling wave solution; and in particular he has demonstrated that it is not a true physical travelling wave in that only those particles which are initially involved are ever involved and thus are of little interest except for, perhaps, the case of streaming gas shot from the sun. However, unlike the case considered by Bernstein, Greene and Kruskal, the stability of these travelling waves for spatially bounded perturbations has been proved by E. A. Jackson (31).

Travelling wave solution for the full set of the Landau-Vlasov equations have been considered by Akhiezer and Polovin (2), in the relativistic hydromagnetic approximation. The assumption that all quantities depend only on $\underline{i} \cdot \underline{r} - qt$ permit this system to be reduced to a system of ordinary differential equations which, under certain conditions admits solutions in terms of elliptic functions. While they are able to discuss the longitudinal and transverse oscillations separately, the general problem which involves the coupling between these modes is very difficult even with the travelling wave assumption.

Nonlinear Oscillations in a Cold Plasma

Many authors beginning with Dawson (20) have treated time-dependent oscillations in a cold plasma. Almost all of the problems successfully discussed depend in the end on the introduction of Lagrangian coordinates. The present author has made an attempt to unify a large class of problems in this area by utilizing the theory of systems of partial differential equations of the first order with equal principal parts as it is, for example, developed in (19). Since this work will be reported on in detail in (22), consideration here will be limited to the discussion of the simplest case and a statement of the generalizations that are now apparent. Consider then a cold plasma described by the equations, appropriate for a uniform ion background and zero magnetic field, in which x and t are the only independent variables.

$$v_t + vv_x = -\frac{e}{m}E \tag{4.1}$$

$$n_t + \frac{\partial(nv)}{\partial x} = 0 \tag{4.2}$$

$$E_x = 4\pi e(n_0 - n) \tag{4.3}$$

$$\nabla\times\underline{H} = 0 \;;\quad \nabla\times\underline{E} = 0 \tag{4.4}$$

$$\frac{\partial E}{\partial t} + 4\pi e(n_0 v_0 - nv) = 0 \tag{4.5}$$

One may make this system into a system of first order ordinary differential equations by any of three slightly different procedures. The first which suggests a natural perturbation process for the more general three dimensional situation consists merely in forming

$$E_t + vE_x = 4\pi e n_0 v - J_+ \qquad (4.6)$$

with the aid of (4.3), (4.5) and (1.7), and noting that the equation of continuity is a consequence of Maxwell's equations. (4.1) and (4.6) now consist of an appropriate system which can be rigorously reduced to the solution of the following system of first order ordinary differential equations of which the first represents the single common characteristic of the system.

$$\frac{dx}{dt} = v \qquad (4.7)$$

$$\frac{dv}{dt} = -\frac{e}{m}E \qquad (4.8)$$

$$\frac{dE}{dt} = 4\pi e n_0 v - J_+ \qquad (4.9)$$

Alternately, one can solve Poisson's equation (4.3) for n and substitute it into the equation of continuity which then takes the form

$$\frac{\partial}{\partial x}\left(\frac{\partial E}{\partial t} + v\frac{\partial E}{\partial x} - 4\pi e n_0 v\right) = 0$$

and admits a first integral

$$\frac{dE}{dt} = \frac{\partial E}{\partial t} + v\frac{\partial E}{\partial x} = 4\pi e n_0 v + F(t)$$

where $F(t)$ is an arbitrary function of the time whose physical significance will be seen to represent the effect of an external field. Finally, one may replace the equation of continuity in its Eulerian form (4.2) by its Lagrangian form which for this simple problem is:

$$n\frac{\partial x}{\partial x_0} = n_0$$

Utilizing this in Poisson's equation (4.3) one obtains

$$\frac{\partial E}{\partial x} = 4\pi e n_0\left(1 - \frac{\partial x_0}{\partial x}\right)$$

or after integration

$$E = 4\pi e n_0 (x - x_0) + G(t)$$

where again $G(t)$ is an arbitrary function of the time. Now it is clear that F , G must represent the effect of an external electric field since one may argue that Poisson's equation does not determine the D.C. consideration component of the electric field, a fact obvious from consideration of Fourier transforms; or, equivalently, no spatial distribution of electric charges is capable of producing a field which depends only on the time and not on position. Regardless of the route taken to considerations of the external field, one obtains the equation of simple harmonic motion for the velocity and the solution for the displacement is simply

$$x = x_0 + A(x_0)\sin[\omega_p t + \phi(x_0)] \tag{4.10}$$

This appears to be a linear result; but, of course, it is not. That is, as in the existence theory discussed in §2, the system of equations (4.7), (4.8), (4.9) determine $x = x(x_0, v_0, t)$, $v = v(x_0, v_0, t)$, $E = E(x_0, v_0, t)$ and in order to obtain a solution to the given problem, it is necessary to invert the first two of these and substitute into the last to obtain $E = g(x, v, t)$. However, in this simple situation, this can be done explicitly by the use of Lagrange's theorem applied to (4.10) (Cf. for example (50) and Sturrock (45).) For this case, the result obtained here is completely equivalent to that obtained by Dawson (20). We refer the reader to (11) and (21) for examples illustrating the breaking of the pseudo-harmonic wave as a function of the amplitude of the initial displacement.

The above procedures can be extended to more general situations. A constant magnetic field can be admitted as well as velocity components in the y and z directions as long as the restriction to one independent spatial variable is retained. The oscillations of the ions can also be considered simultaneously, perhaps with less ambiguity by using the last procedure and employing the Lagrangian form of the equations of continuity for both the ions and electrons. In rectangular coordinates one can explicitly solve the most general problem of this type, but the method applies to any orthogonal coordinate system subject to the restriction above; namely, constant magnetic field and one independent spatial coordinate. In the simplest of these cases, cylindrical and spherical coordinate systems with only two velocity components and zero magnetic field, one obtains equations for an-harmonic oscillations first derived and discussed by Dawson (20). The method used by Kahlman (32) to force the solution given above into a travelling wave will also apply to the most general problem formulated above in rectangular coordinates; and as in his work, one sees: (a) the travelling wave solution breaks down when the Jacobian of the transformation to Lagrangian coordinates vanishes so that the electrons no longer move as a sheet but are permitted to overtake and pass one another, and (b) the travelling wave solution requires the presence of an external field.

Finally, for the general three dimensional situation described by the equations(where again the ions are assumed infinitely heavy for simplicity):

$$\frac{d\underline{v}}{dt} = -\frac{e}{m}[\underline{E} + \frac{\underline{v}}{c} \times H] \qquad (4.11)$$

$$\frac{\partial n}{\partial t} + \nabla \cdot [n\underline{v}] = 0 \qquad (4.12)$$

$$\nabla \times \underline{H} = \nabla \times \underline{E} = 0 \qquad (4.13)$$

$$\nabla \cdot \underline{E} = 4\pi e(n_0 - n) \qquad (4.14)$$

$$\frac{\partial \underline{E}}{\partial t} + 4\pi e(n_0 \underline{v}_0 - n\underline{v}) = 0 , \qquad (4.15)$$

one can use the first procedure to set up a perturbation process by combining (4.14) and (4.15) into the form

$$\frac{d\underline{E}}{dt} = (\underline{v} \cdot \nabla) E - \underline{v}(\nabla \cdot \underline{E}) + 4\pi e n_0 (\underline{v} - \underline{v}_0)$$

For the case of zero magnetic field, and with the substitutions

$$\theta = \omega_0 t \ , \quad \varepsilon = 1/\omega_0$$

one readily obtains the velocity equation

$$\frac{d^2 \underline{v}}{d\theta^2} + \underline{v} = \varepsilon[(\underline{v} \cdot \nabla')\frac{d\underline{v}}{d\theta} - \underline{v}(\nabla' \cdot \frac{d\underline{v}}{d\theta})] \qquad (4.16)$$

In this equation the operator ∇' must be viewed as computed with respect to the Eulerian variable and then expressed in terms of Lagrangian variables so that at each stage in effect a new metric is introduced. Now the vector equation (4.16) can be put into a form where the Cesari-Hall-Gambill method (14) for establishing the existence of periodic solutions can be formally applied. This has been carried out in (21) for the two dimensional example first treated by Dawson (20) by a complicated process which did not eliminate the occurrence of secular terms. For this example, the initial values are assumed to be

$$u_0 = A \sin k_1 x_0 \cos \alpha$$

$$v_0 = B \sin k_2 y_0 \cos \beta$$

and the Cesari-Hall-Gambill method yields the existence of a one parameter family of periodic solutions subject only to the condition

$$k_2^2 B^2 \cos(2k_2 y_0) = k_1^2 A^2 \cos(2k_1 x_0) \tag{4.17}$$

in the second approximation. The shift in the plasma frequency from its normalized value of "1" in (4.16) is given by

$$\tau = 1 + \frac{\varepsilon^2}{12}[k_1^2 A^2 \cos(2k_1 x_0)]$$

and thus it is seen that the frequency is stable to second order in $\varepsilon = \frac{1}{\omega_p}$. The problem is somewhat special in that the complex determining equation of the Cesari-Hall-Gambill method turns out to be purely imaginary so that no unique amplitude is determined as in the Van der Pol equation but merely the compatibility relation (4.17) above, which, it should be noted, is independent of the initial phase angles. While the extension of the Cesari-Hall-Gambill method to systems of first-order partial differential equations with equal principal parts causes no convergence difficulties, the extension to the present situation involving the constant interplay between Eulerian and Lagrangian variables has not yet been rigorously established but it is believed that this can be done.

Physical Interpretation of Nonlinear Effects

In addition to the problems so far discussed, many others which are closely related have recently been reviewed from the physical point of view in (48). Since this paper is available in translation, no attempt will be made here to summarize its contents in detail. However, several remarks in it do provide a complementary picture and are worthy of specific mention. The over-all scope can be gleaned from the abstract (48) which is as follows:

"The paper basically is a review of a number of studies devoted to the theory of nonlinear motions of plasma under conditions where collisions between particles do not play a determining role."

"The problem is formulated in the introduction. It concerns the evolution in time of an initial perturbation of finite amplitude. The resulting physical picture will depend on these competing processes; nonlinear increase of wave steepness, dispersion, absorption and instability. In a number of cases where absorption and instability are insignificant, it is possible to obtain an idea of the character of the nonlinear motions by applying the appropriate linear 'dispersion law'."

"The second section presents certain specific types of nonstationary nonlinear motions permitting each mathematical solution;

namely, nonlinear oscillations of electrons at zero temperature, nonlinear motion of plasma across a strong magnetic field, and ion waves of finite amplitude in non-isothermal plasma where $p_i \ll p_e$. In a number of instances, evolution of the initial perturbation leads to the formation of a multi-component current, some peculiarities of which are discussed in the third section. In the next section, stationary nonlinear waves are described, i.e. waves not changing their form with time. In a particular case, they are the so-called "solitary waves," similar to waves on the surface of heavy liquid in a channel of finite depth. The possibility of the existence of such waves requires linear laws of dispersion of a specific character. The possibility of "solitary" waves of rarefaction is pointed out.

"The question of absorption of waves in rarified plasma is then discussed. An approximate "quasi-linear" method is developed which permits the kinetic considerations in the absorption of waves of finite amplitude to be simplified. The method consists in the representation of the distribution function $f(r, v, t)$ as a sum of rapidly and slowly varying terms. In the equation for the slowly varying term a quadratic average effect of fast oscillations is taken into consideration. This method is applied to two particular problems; the absorption of Langmuir electronic oscillations (in the limit of very small amplitude the equation for wave-damping goes over into the well-known formula for the so-called "Landau damping," and the cyclotronic absorption of transversely polarized waves which propagate along the constant magnetic field.

"In the last section some types of instabilitites of nonlinear motions are shown. In addition to the instabilities associated with multi-component motion, it is shown that waves in a magnetic field (in particular, solitary waves) are unstable if their amplitude exceeds a certain critical value which decreases as the plasma temperature decreases."

In particular, these authors contrast the situation in rarified plasmas with that of conventional gas dynamics where the nonlinear term $(\underline{v} \cdot \nabla)\underline{v}$ in the equations of motion permits that part of the velocity profile corresponding to lower velocities and cause the formation of shocks if heat conduction and viscous effects are neglected. In plasma the nonlinear term is in competition with the dispersion of the plasma which may limit the steepness of the wave-front of the nonlinear motion. In particular, they note that the dissipation effects such viscosity and heat conduction in gas dynamics disturb reversibility in the equations and change the order to the derivatives by an odd number, e.g. the addition of viscosity changes by first-order Eulerian equations to the second-order Navier-Stokes equations. Dispersion effects in contrast do not disturb reversibility and increase the order by an even number of derivatives. However, if the wave amplitude becomes sufficiently large, then the wave may break and region of multi-streaming develop.

Since the relationship between the linear and nonlinear theory involving trapped particles has been discussed in (5) and (7), attention should be called to their "quasi-linear" theory which attempts to take into account such distortion of the "background" to the first order. They liken this process to that used by Van der Pol in the nonlinear theory of oscillations.

Although the Van der Pol or Krylov and Bogolyubov process appears never to have been justified even for ordinary differential equations [Cf. (14), p. 123], it has been widely used in engineerig applications with great success. In view of the possibility that a similar situation might exist here, the author made an attempt to understand the "quasi-linear" theory, a theory quite obscure in the above review paper and apparently not referenced anywhere else. The following then represents a "rational" for the simple one-dimensional case of Langmuir oscillations due to the author and his colleague K. M. Case.

The Vlasov equation distribution function is decomposed into $f = f_0(x, v, t) + f_1(x, v, t)$ where f_0 is thought of as rapidly oscillating and "small" with respect to f_0 . The Vlasov equation in the form

$$\left[\frac{\partial f_1}{\partial t} + v\frac{\partial f_1}{\partial x} + \frac{eE}{m}\frac{\partial f_0}{\partial v}\right] + \left[\frac{\partial f_0}{\partial t} + v\frac{\partial f_0}{\partial x} + \frac{eE}{m}\frac{\partial f_1}{\partial v}\right] = 0 \qquad (5.1)$$

is then "solved" by noting that the term in the second brackets will be small compared to the first. Thus, the total solution should be near the solution of the linear equation

$$\frac{\partial f_1}{\partial t} + v\frac{\partial f_1}{\partial x} + \frac{eE}{m}\frac{\partial f_0}{\partial v} = 0$$

Now imagine, temporarily, that the entire system is enclosed in a box. Then this equation can be solved by Fourier series, each oscillating component of which will be of the form

$$f_k e^{i(kx - \omega t)}$$

and will satisfy

$$(kv - i\omega)f_k + \frac{e}{m}E_k\frac{\partial f_0}{\partial v} = 0$$

from which it follows that

$$f_k = \frac{\frac{-e}{m}E_k\frac{\partial f_0}{\partial v}}{kv - i\omega}$$

(The constant term in the Fourier series can either be thought of as absorbed in f_0 or taken to be zero since eventually the box will be allowed to become infinite.) The slowly varying part f_0 is now determined by (5.1) which reduces to

$$\frac{\partial f_0}{\partial t} + v\frac{\partial f_0}{\partial x} = -\frac{e}{m}E\frac{\partial f_1}{\partial v} \tag{5.2}$$

Now it has been assumed that the real field E can be expanded so that, for real k, we have

$$E = \sum E_k e^{i(kx-\omega t)} = E^* = \sum E_k^* e^{-i(kx-\omega t)}$$

so that (5.2) becomes

$$\frac{\partial f_0}{\partial t} + v\frac{\partial f_0}{\partial x} = \left(\frac{-e}{m}\right)\left[\sum_k E_k^* e^{-i(kx-\omega t)}\right]\sum_\ell \frac{\partial f_\ell}{\partial v} e^{i(\ell x-\omega t)}\Big]$$

$$= \frac{-e}{m}\sum_{k,\ell} E_k^* \frac{\partial f_\ell}{\partial v} e^{i(k-\ell)x}$$

If a space overage over a period in x is now made, remembering that the quantities on the left side are assumed so slowly varying as to be presumed constant during this average process, one obtains finally

$$\frac{df_0}{dt} \triangleq \frac{\partial f_0}{\partial t} + v\frac{\partial f_0}{\partial x} = \frac{-e}{m}\sum_k E_k^* \frac{\partial f_k}{\partial v}$$

$$= \left(\frac{e}{m}\right)^2 \frac{\partial^2 f_0}{\partial v^2}\sum \frac{|E_k|^2}{kv-i\omega} \rightarrow \left(\frac{e}{m}\right)^2 \frac{\partial^2 f_0}{\partial v^2}\frac{1}{2\pi}\int \frac{dk|E(k)|^2}{kv-i\omega}$$

when the box is allowed to become infinite, and this equation is the one-dimensional analogue of equation 43 of (48). This equation is not solved directly but is used to obtain expressions for the interchange between the plasma and the wave and the stationary distribution for f_0 is obtained from it after it has been suitably modified to take account of collisions, but these details cannot be entered into here. In spite of the many conclusions which apparently can be drawn from this form of the quasi-linear approximation, the use of the single arbitrary time frequency spoils in our opinion the analogy between this method and that of Van der Pol. In addition, the complicated nature of the equation which results for f_0, depending upon a principal value integral and as yet unspecified $E(k)$ raises doubts about the self-consistency of the method. It is also not clear that the "rational" presented here is that actually used (48) since the authors of this paper refer to this equation as "a zeroth order term in the expansion of the exact kinetic equation in $1/N_0$, (i.e. according to the ratio of the energy of thermal noise to the thermal energy of the plasma." Nonetheless, we were able to obtain their results through the use of time averages.

Boundary value problems

The theory reported here has been almost exclusively concerned with the initial value problem or, in some cases, with stationary solutions. Even in the linear theory the formulation and solution of boundary value problems involving the Vlasov equations offer formidable difficulties involving not only constructive methods for solution but even in existence and uniqueness. In fact, at this moment, the only pertinent paper seems to be that of Bernstein and Rabinowitz (7) which deals with a nonlinear theory for probes in a plasma.[3] This is not surprising since not very many boundary problems have been solved for neutral gases. A very excellent idea of the difficulties associated with this type of problem can be gleaned from the article by Gross, Jackson and Ziering (26) even though they discuss only neutral gases. However, at least in the linear case, it is possible to report that F. Shur (43), a student of K. M. Case, has succeeded in extending the latter's normal mode theory to include perfect reflection at plane boundaries of a plasma even in the presence of magnetic fields.[4] Case, however, has recently observed that the use of perfect reflectors introduces a non-physical lack of uniqueness. Said another way, imagine one has a plasma confined between two infinite half-planes which are perfectly reflecting and suppose that an external electric field of a fixed frequency is applied. No matter what the frequency is, a resonance phenomena appears which is believed to be spurious physically and due only to the nature of the idealized boundary conditions. Apart from work in this direction, it is most likely that the treatment of most boundary value problems will continue for some time to be based largely on intuitive physical grounds rather than precise mathematical analysis. As an example which is fairly typical of the "state of the art," Dolph and Weil (23) estimated the radar cross-section of the wake of spherical satellite by using the exact neutral distribution derived by Chang (15) for both specular and diffuse reflection by arguing that because of the strong Coloumb forces the electron density would agree with the ion density to the first order and that the ion density would not represent a serious departure from that of neutral particles. The radar cross-section was then computed by treating the individual scatters by taking into account the various phase differences. The distribution problem was also attacked by A. V. Gurevich (29) by completely different methods, but which nevertheless yielded good agreement with the above; and in particular, gave estimates which substantiated the near identity of the ion and electron densities for satellite conditions. Subsequent numerical calculations by Sawchuck (41) of Poisson's equation have furnished additional empirical justification for the semi-analytical method of Dolph and Weil.

One of the basic difficulties associated with boundary value problems involving more than one species of particles is that even in the linear magnetohydrodynamic but not necessarily cold approximation, the reduced "Oseen like" equations do not factor nor separate so that the usual method of building up solutions is not possible. In contrast, this is possible for a sphere not only in hydrodynamics but even for a single fluid magnetohydrodynamic medium. [Cf. Gotch (25) and Van Blerkson [47)]. The fact that it is usually possible, however, to derive a dispersion relation does suggest a possible approximate method for overcoming these difficulties. Just as one can derive both relativistic and non-relativistic quantum mechanics from the quantum mechanical dispersion relation depending upon whether one includes or neglects terms in $1/c$, one can hope to obtain more tractable partial differential systems by suitable approxmations to the exact dispersion relation. In the hope that this method might be of use to others, a sketch of it will be given here for the case of a sphere moving in an electrostatic field, supersonically with respect to the ions, and subsonically with respect to the electrons. In a coordinate system in which the sphere (assumed of velocity v_0) is at rest, the nine equations for the linearized two component fluid theory are;

$$v_0 \frac{\partial v_k^{\pm}}{\partial x_1} + \frac{kT^{\pm}}{m^{\pm}\rho_0} \frac{\partial \rho^{\pm}}{\partial x_k} + \frac{e}{m^{\pm}} \frac{\partial \phi}{\partial x^k} = 0$$

$$\sum \frac{\partial v_k^{\pm}}{\partial x_k} + \frac{1}{\rho_0} \frac{\partial \rho^{\pm}}{\partial x_1} = 0$$

$$\nabla^2 \phi = -4\pi e(\rho^+ - \rho^-)$$

In these equations, the plus signs refer to ions, the minus signs to electrons, x_1 is the streaming direction, ρ_0 the common ambient density, ρ the perturbed density, v_k the velocity components, and T the temperature. The introduction of a Lagrangian type of potential familiar from acoustics, namely, [3]

$$\underline{v}^{\pm} = \frac{\partial}{\partial x_1} \nabla \psi^{\pm}$$

leads to

$$\rho^{\pm} = \frac{-\rho_0}{v_0} \nabla^2 \psi^{\pm}$$

and

$$\phi = \frac{4\pi e \rho_0}{v_0} [\psi^+ - \psi^-] + H(x, y, z)$$

where H is an arbitrary harmonic function. Letting, for simplicity,

$$q_{\pm}^2 = \frac{m^{+} v_0^2}{kT^{\pm}} \quad \text{and} \quad \lambda_{\pm}^2 = \frac{4\pi e^2 \rho_0}{kT^{\pm}} = \frac{1}{R_{\pm}^2}$$

where $R_{\pm}$ is the Debye length of the ions and electrons respectively, one can derive the following system of second-order coupled equation for ψ^+, ψ^- :

$$(\nabla^2 - q_+^2 \frac{\partial^2}{\partial x_1^2} - \lambda_+^2)\psi^+ + \lambda_+^2 \psi^- = 0$$

$$\lambda_-^2 \psi^+ + (\nabla^2 - q_-^2 \frac{\partial^2}{\partial x_1^2} - \lambda_-^2)\psi^- = 0 .$$

The uncoupled fourth-order equation for either quantity is:

$$[\nabla^2 - q_+^2 \frac{\partial^2}{\partial x_1^2}][\nabla^2 - q_-^2 \frac{\partial^2}{\partial x_1^2}]\psi - (\lambda_+^2 + \lambda_-^2)\nabla^2\psi + (\lambda_+^2 q_-^2 + \lambda_-^2 q_2^2) \frac{\partial^2 \psi}{\partial x_1^2} = 0$$

This equation factors exactly only in the special case when

$$T^+ T^- , \quad \lambda_+^2 = \lambda_-^2 \quad \text{and} \quad q_+^2 - 1 = 1 - q_-^2 = a^2$$

Available experimental evidence seems to indicate that $T^+ = T^-$ in the ionosphere so that $\lambda = \lambda_+ = \lambda_-$ is reasonable for any v_0 . Setting $k^2 = k_*^2 + k_1^2$, $a_+^2 = q_+^2 - 1$, and $a_-^2 = 1 - q_-^2$ and taking a three dimensional Fourier transform

$$\hat{\psi} = \frac{1}{(2\pi)^{3/2}} \int e^{-ik_- \cdot x_-} \psi d^3 x$$

one readily obtains the dispersion relation

$$a_+^2 a_-^2 k_1^4 + k_1^2 (k_*^2 + \lambda^2)(a_+^2 - a_-^2) - k_*^2 (k_*^2 + 2\lambda^2) = 0 . \qquad (6.1)$$

This admits the approximate factorization

$$2a_+^2 a_-^2 k_1^2 \cong (k_*^2 + \lambda^2)(a_+^2 - a_-^2) \left\{ - 1 \pm \left[1 + \frac{2a_+^2 a_-^2 k_*^2 (k_*^2 + 2\lambda^2)}{(a_+^2 - a_-^2)^2 (k_*^2 + \lambda^2)^2} \right] \right.$$

In this it should now be noticed that the expression

$$(k_*^2 + 2\lambda^2)/(k_*^2 + \lambda^2) \tag{6.2}$$

has the value 2 if $\frac{k_*^2}{\lambda^2} << 1$, the value 1 if $\frac{k_*^2}{\lambda^2} > 1$ and the value 3/2 if $k_*^2 = \lambda^2$. Thus, if the exponentially damped roots corresponding to $k_*^2 + \lambda^2 = 0$ are neglected, an approximation that will affect the solution only in the neighborhood of the sphere where the Oseen linearization is poorest anyway, it seems reasonable to call (6.2) a constant, say μ. Thus the dispersion relation (6.1) is replaced by the approximation

$$k_*^4 + k_*^2[(\alpha-\gamma\mu)k_1^2 - \beta\lambda^2] - \gamma\mu k_1^2(\alpha k_1^2 - \beta\lambda^2) = 0 \tag{6.3}$$

where

$$\alpha = \frac{a_+^2 a_-^2}{(a_+^2 - a_-^2)\left(1 - \frac{a_+^2 a_-^2 \mu}{(a_+^2 - a_-^2)^2}\right)} \cong \frac{a_+^2 a_-^2}{a_+^2 - a_-^2}$$

$$\beta = \frac{a_+^2 + a_-^2}{a_+^2 a_-^2} \Big/ \alpha = 1 - \frac{a_+^2 a_-^2 \mu}{(a_+^2 - a_-^2)^2} \cong 1$$

$$\gamma = a_+^2 - a_-^2$$

The above steps are now retraced. A coupled system of the form

$$(k_*^2 + Ak_1^2 - B\lambda^2)\hat{\psi}_+ + B\lambda^2\hat{\psi}_- = 0$$

$$D\lambda^2\hat{\psi}_+ + (k_*^2 + Ck_1^2 - D\lambda^2)\hat{\psi}_-^2 = 0$$

has a dispersion relation of the form

$$k_*^4 + k_*^2[(A + C)k_1^2 - (B + D)\lambda^2 + ACk_1^4 - k_1^2(AD + BC)\lambda^2] = 0$$

and this will agree with (6.3) if and only if the following equations hold:

$$A + C = \alpha - \gamma\mu$$

$$AC = -\frac{\alpha\gamma}{\mu}$$

$$B + D = \beta$$

$$Ad + BC = \beta\gamma\mu$$

These equations have, however, the unique solution

$$A = \alpha\ , \quad C = -\gamma\mu\ , \quad B = \beta\frac{(\alpha-\gamma\mu)}{\alpha + \gamma\mu}\ , \quad D = \frac{2\mu^{\gamma}}{a + \gamma\mu}$$

so that, taking inverse transforms, the approximate dispersion relation leads to the coupled system

$$\left[\frac{\partial^2}{\partial x^2} + \frac{\partial^2}{\partial y^2} + \alpha\frac{\partial^2}{\partial z^2} - \beta\frac{(\alpha-\gamma\mu)}{\alpha+\gamma\mu}\lambda^2\right]\psi_+ + \beta\frac{(\alpha-\gamma\mu)}{\alpha+\mu}\lambda^2\psi_- = 0$$

$$\frac{2\gamma\mu}{\alpha + \gamma\mu}\psi_+ + \left(\frac{\partial^2}{\partial x^2} + \frac{\partial^2}{\partial y^2} - \gamma\mu\frac{\partial^2}{\partial z^2} - \frac{2\gamma\mu}{\alpha + \gamma\mu}\lambda^2\right)\psi_- = 0$$

and the factorable uncoupled fourth-order equation

$$\left(\frac{\partial^2}{\partial x^2} + \frac{\partial^2}{\partial y^2} - \gamma\mu\frac{\partial^2}{\partial z^2}\right)\left(\frac{\partial^2}{\partial x^2} + \frac{\partial^2}{\partial y^2} + \alpha\frac{\partial^2}{\partial z^2} - \beta\lambda^2\right)\frac{\partial^2\psi_\pm}{\partial z^2} = 0$$

whose solution may be sought in the usual way by setting

$$\frac{\partial\psi_\pm}{\partial z} = \chi_1^\pm + \chi_0^\pm$$

and requiring that

$$\left(\frac{\partial^2}{\partial x^2} + \frac{\partial^2}{\partial y^2} - \gamma\mu\frac{\partial^2}{\partial z^2}\right)\chi_1^\pm = 0$$

and that

$$\left[\left(\frac{\partial^2}{\partial x^2} + \frac{\partial^2}{\partial y^2} + \alpha\frac{\partial^2}{\partial z^2}\right) - \beta\lambda^2\right]\chi_0^\pm = 0$$

For the sphere these equations must be solved subject to the boundary conditions on the surface $r = a$

$$\underline{r}\cdot\nabla\chi_1^\pm = 0$$

and

$$\underline{r}\cdot\nabla\chi_0^\pm = -v_0\cos\theta$$

At this writing this problem is not yet completed, but the author and

Dr. R. F. Goodrich have made significant progress toward it and hope to report on the solution and its physical consequence in the near future.

As a final remark, I would like to suggest that the use of Fourier transforms with respect to velocity, rather than space offers many intriguing possibilities for obtaining results in both the linear and nonlinear Vlasov equations, in spite of the fact that they seem to have been used in only one paper (6). These transforms, unlike those with respect to space, do not introduce a convolution integral in the acceleration terms in the nonlinear case and in many ways should play a role similar to that of characteristic functions in statistics. In particular, for the measurable physical quantity such as density, current, etc., inverse transforms are not necessary, but since the physical quantities are always moments, it is only necessary to differentiate with respect to the transform parameter and evaluate the resulting expression at the origin. In particular, most physical problems do not require a full knowledge of the distribution function, and I suspect that it is possible to develop methods in the velocity transform space which would yield important information even though a full solution might not be possible.

NOTES

1. In the remaining part of this section, the author owes much to the illuminating lectures of his friend and former colleagues. G. E. Uhlenbeck ["Statistical Mechanics of Plasmas" notes by E. G. Fontheim, Univ. of Mich. Rad. Lab., 1961].
2. Stability has been established, however, when trapping can be neglected. See D. Montgomery, "Stability of Large Amplitude Waves in a One-dimensional Plasma," Phys. Fluids 3, 1960, p. 274.
3. The problem of the plasma with so-called slab symmetry has been discussed by I. B. Bernstein in "Static Equilibria, Stability, and Wave Propagation in Symmetrical Collisionless Plasma" in the Fifth Lockheed Symposium on Magnetohydrodynamics, Stanford University Press, 1961. Bernstein's article also discusses the contents of references (5) and (7).
4. Similar results have been obtained by D. Montgomery and D. Gorman, "The Boltzman-Vlasov Equation in a Bounded Region" in a paper to be published. They replace the boundary conditions by an infinite, spatially-periodic plasma which has the requisite symmetry in velocity space.
5. This method of reduction is adequate also if the first set of the above equations include (1) a scalar viscosity term, (2) an interaction term proportional to its relative velocity of electrons and ions and (c) a gravitational potential.

REFERENCES

1. Akhiezer, L., and Lubarskii, G. Ya., "On the nonlinear theory of electron plasma oscillations," Doklady Akad. Nauk. USSR 80, 1951, p. 193.
2. Akhiezer, A. I., and Polovin, R. V., "Theory of Wave motion of

an electron plasma," Soviet Phys. JETP 3, 1956, p. 696.

3. Balesceu, R., "Irreversible processes in ionized gases," Phys. Fluids, 3, 1960, p. 52.

4. Bernstein, I. B., "Waves in a plasma in a magnetic field," Phys. Rev. 109, 1958, p. 10.

5. Bernstein, I. B., Greene, J. M., Kruskal, M. D., "Exact nonlinear plasma oscillation," Phys. Rev. 103, 1957, p. 546.

6. Bernstein, I. B., Lenard, A., "Plasma oscillations with diffusion in velocity space," Phys. Rev. 112, 1958, p. 1456.

7. Bernstein, I. B., Rabinowitz, I. N., "Theory of electrostatic probes in a low density plasma," Phys. Fluids 2, 112, 1959.

8. Bohm, D., and Gross, E. P., "Theory of plasma oscillations," Phys. Rev. 75, 1949, p. 1851.

9. Brillioun, L., "The travelling-wave tube (discussion of waves of large amplitude)," Journ. Appl. Phys. 20, 1949, p. 1196.

10. __________, "Interaction between electrons and waves," Proc. Nat. Acad. Sci., 41, 1955, p. 401.

11. Buneman, O., "Dissipation of currents in ionized media," Phys. Rev. 115, 1959, p. 503.

12. Case, K. M., "Plasma oscillations," Ann. Phys. 7, 1959, p. 349.

13. __________, "Lectures on Landau Damping," Notes by D. M. Raybin, Univ. Mich. Rad. Lab. Memo 2500-203-M, March, 1960.

14. Cesari, L., "Asymptotic behavior and stability problems in ordinary differential equations," Springer, Berlin, 1959.

15. Chang, C. S. Wang, "Transport phenomena in very dilute gases, II," Univ. of Mich. Res. Inst. CM-654, Dec., 1950.

16. Chandrasekhar, S., "Plasma oscillations," Univ. of Chicago Press, Chicago, 1960.

17. Chandrasekhar, S., Kaufman, A. N., and Watson, K. M., "Properties of ionized gas of low density in a magnetic field," Ann. Phys. 2, 1957, p. 435.

18. Chu, L. J., and Jackson, J. D., "Field theory of travelling-wave tubes," Proc. Inst. Radio Eng. 36, 1948, p. 853.

19. Courant, R., and Hilbert, D., "Methods of mathematical physics-II," Interscience, New York, 1962.

20. Dawson, J. M., "Nonlinear oscillations of a cold plasma," Phys. Rev. 113, 1959, p. 383.

21. __________, "Breaking of finite amplitude waves," Proc. Conf. Plasma Oscillations, June, 1959, Indianapolis, Indiana

22. Dolph, C. L., "A unified theory of the non-linear oscillations of a cold plasma," To appear in Journal of Math. Analysis and Appl. 4, 1962.

23. Dolph, C. L., Weil, H., "On the change in radar cross-section of a spherical satellite caused by a plasma sheath," Plasma and Space Sci. Vol. 6, 1961, p. 123.

24. Enoch, J., "Non-linearized theory of transverse plasma waves," Physics of Fluids, vol. 5, 1962, p. 467.

25. Gotoh, K., "Magnetohydrodynamic flow past a sphere," Journ. Phys. Soc. Japan 15, 1960, p. 189.

26. Gross, E. P., Jackson, E. Q., Ziering, S., "Boundary value problems in kinetic theory of gases," Ann. Phys. 1, 1957, p. 141.

27. Guernsey, R. L., "The kinetic theory of fully ionized gases," Univ. of Mich. Res. Inst. 03114, 1960.

28. __________, "Kinetic equation for a completely ionized gas," Phys. Fluids 5, 1952, p. 322.

29. Gurevich, A. V., "On disturbances in the ionsphere caused by a moving body," (Russian) Iskustrennge Sputniki Zemli, Vol. 171, 1961, p. 101.

30. Iordanskii, S. V., "A solution of the Cauchy problem for the kinetic equation of an electron plasma," Doklady Akad. Nauk, USSR 127, 1959, p. 509.

√31. Jackson, E. A., "Stability of non-linear travelling waves in a cold plasma," Project Matterhorn Rept. Matt - 53, Oct., 1960.

32. Kalman, G., "Non-linear plasma oscillations and non-stationary flow in a zero temperature plasma," Ann. Pnys. 10, 1960, p. 29.

33. Kino, G. S., and Chodorow, M., Prog. Rept. No. 1, Af-33-616-8121, Proj. No. 61-8-7073, W. W. Hansen Lab. of Phys., Stanford, Oct., 1961.

34. Kino, G. S., and Crawford, F. W., "Oscillations and noise in low pressure D. C. discharges," Proc. Inst. Radio Eng., 49, 1961, p. 1768.

35. Kruskal, M., "Landau Damping," The theory of neutral and ionized gases, John Wiley, N. Y., 1960, p. 285.

36. Landau, L., "On the vibration of the electron plasma," Journ. Phys. 10, 1946, p. 25.

37. Langmuir, I., and Tonks, L., "Oscillations in ionized gases," Phys. Rev. 33, 1929, p. 195.

38. Lust, R., "Uber die Ausbreitung von Wellen in einem Plasma," Fortschritte der Physik, 7, 1959, p. 503.

39. Ritt, R. K., "Evolution of an inhomogeneous plasma and associated high frequency radiation." To appear in UM 2764-10-T, Univ. of Mich. Rad. Lab., 1962.

40. Rosenbluth, M. N., "Hydromagnetic basis for treatment of plasma, Lockheed Symposium in Magneto-hydrodynamics, Stanford Unvi. Press, 1959.

41. Sawchuck, W., "Wake of a charged prolate spheriod at angle of attack in a rarefied medium," Univ. of Mich. Rad. Lab. 2764-9T, 1962.

42. Sen, H. K., "Non-linear theory of a space charge wave in moving interacting electron beams with application to solar radio noise," Phys. Rev. 97, 1955, p. 849.

43. Shur, F., Ph. D. Dissertation, Univ. of Mich. in preparation.

44. Spitzer, L., "Physics of ionized gases," Interscience, N.Y., 1956.

45. Sturrock, P., "Generalizations of the Lagrange expansion with application to physical problems," J. Math. Phys. 1, 1960, p. 405.

46. Van Kampen, N. G., "On the theory of stationary waves in a plasma," Physik 21, 1959, p. 949.

47. Van Blerkson, R., "Magnetohydrodynamic flow of a viscous fluid past a sphere," Journ. Fluid Mech., 8, 1960, p. 432.

48. Vedenov, A. A., Velikhov, E. P., Sagdeev, R. Z., "Non-linear oscillations of rarefied plasma," Nuclear Fusion 1, 1962, p. 82. (Russian, English abstract p. 145). Available in translation without figures from Inter. Atomic Energy Agency, Vienna).

49. Watkins, D., and Rynn, D., "Effect of velocity distribution on travelling-wave tube gain," Journ. Appl. Phys. 25, 1954, p. 1375.

50. Whittaker and Watson, "Modern Analysis," p. 132, MacMillan, New York, 1945.

SUPPLEMENTARY BIBLIOGRAPHY

1. Adlam, J. H., Allan, J. E., "The structure of strong collision free hydromagnetic waves," Phil. Mag 3, 1958, p. 448.

2. Davis, L., Lust, R., and Schluter, A., "The structure of hydromagnetic shock waves, I. Non-linear hydromagnetic waves in a cold plasma," Zeit. fur Natur. 13a, 1958, p. 916.

3. Derler, H., "The frequency of non-linear plasma oscillations," Journ. Elect. and Control 11, 1961, p. 3.

4. Fainberg, Ya., "Non-linear theory of slow plasma waves, Journ. Nuclear Energy, Part C, 13, 1960, p. 153.

5. Ferrero, V., "General theory of plasma," Nuovo Cimento, Supp. Vol. 13, 10th ser. 1959, p. 9.

6. Hess, R., "Large signal travelling-wave tube operation concepts and analysis," ASDTR 6-15, Univ. of Calif. July, 1961.

7. Jackson, J. D., "Longitudinal plasma oscillations," Journ. Nuclear Energy, Part C, 1, 1960, p. 171.

8. Kahn, F. D., "Dispersion relations for waves of finite amplitude in two stream plasma," Journ. Fluid Mech. 10, 1961, p. 357.

9. Montgomery, D., "Non-linear time dependent oscillations," Phys. Rev. 123, 1961, p. 1077.

10. __________, "Landau damping to all orders," Phys. Rev. 124, 1961, p. 1309.

11. Nordsieck, A., "Theory of the large signal behavior of travelling-wave amplifiers, Proc. Inst. Radio Engr. 41, 1953, p. 630.

12. Pines, D., "Classical and quantum plasma," Journ. Nuclear Energy, Part C, 2, 1961, p. 5.

13. Polovin, S. P., "Contribution to the theory of simple magnetohydrodynamic waves, Soviet Phys. JETP 12, 1961, p. 326.

14. Rowe, J. E., "A large signal analysis of travelling-wave amplifiers. Theory and general results," Inst. Radio Engr. Prof. Group Trans. Electron Devices, Vol. ED 3, 1956, p. 39.

15. Sturrock, P., "Non-linear effects in electron plasma," Journ. Nuclear Engr., Part C, 2, 1961, p. 158.

16. Tien, P. K., "A large signal theory of the travelling-wave amplifier," Bell System Tech. J. 35, 1956, p. 349.

17. Watkins, D., "Effects of velocity distribution in a modulated beam," Journ. Appl. Phys. 23, 1952, p. 568.

18. Weibel, E., "Oscillations of a non-uniform plasma," Phys. Fluids 3, 1960, p. 399.

19. Yadaralli, S.V., "On some effects of velocity distribution in electron streams," Quart. Appl. Math. 12, 1954, p. 105.

20. Yoshihara, H., "Collisionless plasma shock waves," Phys. of Fluids, 4, 1961, -. 1361.

S. I. PAI

Some considerations on radiation magnetogasdynamics

I. Introduction

In the flow field of a gas at very high temperature, the gas will be ionized and the radiation phenomena are important. A complete analysis of such a high temperature flow of gas should consist of the study of the gasdynamic field, the electromagnetic field and the radiation field simultaneously. Little has been done on such a complete analysis. This paper presents a preliminary attempt for such a complete analysis from macroscopic point of view. The gas will be considered as a continuum in which gasdynamic, thermodynamic and electromagnetic effects are all considered simultaneously.

In the thermodynamic effects there are three basic modes of heat transfer:

(a) The transfer of heat by conduction
(b) The transfer of heat by convection
(c) The transfer of heat by radiation

In "classical gasdynamics" the transfer of heat by radiation is usually neglected. When the temperature of the gas is not too high and the density of the gas is not too low, the transfer of heat by radiation is usually negligibly small in comparison with those by conduction and by convection. However, when the temperature is high, radiation must be considered as an important mode of heat transfer. In this paper we are going to examine the fundamental equations of gas dynamics such that all of these three modes of heat transfer will be included.

Radiation is a far more complicated phenomenon than conduction of heat or convection of heat. Heat convection depends on the velocity field which is the main object of investigation of gasdynamics. Heat conduction depends on the temperature distribution of the medium. For a first approximation the flux of heat by conduction may be represented reasonably well by a single vector which is proportional to the temperature gradient of the medium. However, the state of radiation at a given instant cannot be represented by a single vector. In order to specify completely the state of radiation, the intensity of rays of radiation must be known in all directions, infinite in number, which pass through the point considered. Hence the description of radiation is different greatly from those for conduction and convection, and it is not familiar to the research workers of gasdynamics. We shall review some fundamentals of radiant transfer in sections 2 and 3 before the basic equations of radiation magnetogasdynamics are derived in section 4.

Radiation phenomena may be treated from either microscopic point of view or macroscopic point of view. In this paper, we shall consider only the macroscopic point of view. We shall assume that the microscopic properties of the gas are known. We discuss the resultant effect of the total radiation transfer on the flow field under given conditions. There are a great number of radiation phenomena such as fluorescence, phosphorescence, etc., besides the thermal radiation. We shall, however, consider only the thermal radiation which depends only on the absolute temperature of the medium. Furthermore, the gas considered is dense so that the theory of continuum is applicable to our problems. Our main object is to find the modifications of the classical equations of magnetogasdynamics by the radiation phenomena.

II. Fundamentals of Radiant Energy Transfer

Radiation of heat is a mode of transfer of heat by waves or photons. According to the concept of geometrical optics, we may represent the radiation of heat by heat rays. Each ray travels in a specific direction. In a homogeneous isotropic media, the rays travel in straight lines with the same velocity, velocity of light, in all directions. By dealing with a geometry of lines rather than of waves one can achieve considerable simplifications.

(a) Specific Intensity: Let $d\sigma$ be an arbitrarily oriented small area, P be a point of this area and $\vec{n}$ be the normal of the area at point P. At a given instant of time, there will be heat rays, traversing this element in all the different directions. Let us consider a specific direction along which we draw a line L which makes an angle θ to the normal $\vec{n}$. We take L as the axis of an elementary cone of solid angle $d\omega$ (Fig. 1). Through each point of $d\sigma$ we construct cones having axes parallel to the line L with solid angle at the apex all equal to $d\omega$. These cones define a truncated semi-infinite cone $d\Omega$, whose cross sectional area perpendicular to L at the point P will be $d\sigma \cos\theta$. Let dE_ν be the total amount of energy passing through the area $d\sigma$ inside the cone $d\Omega$ in time dt and in the frequency interval between ν and $\nu + d\nu$. The specific

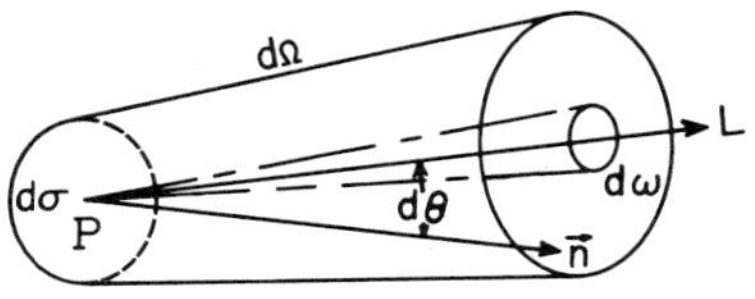

Fig. I Rays in a Radiation Field

intensity of radiation or simply intensity is defined as

$$I_\nu = \lim_{d\sigma,\, d\omega,\, dt,\, d\nu \to 0} \left(\frac{dE_\nu}{d\sigma \cos\theta \, d\omega \, dt \, d\nu} \right) \tag{1}$$

This limit is in general a function of the position of P , the direction L , the time t , and the frequency ν . One of the main objects of radiative transfer is to find the specific intensity I_ν for a given physical problem. If I_ν is independent of the direction L , the radiation field is said to be isotropic, while if I_ν is independent of both the position and the direction L , the radiation field is said to be homogeneous and isotropic.

If we know the intensity I_ν , the amount of energy flowing through the area $d\sigma$ in the frequency interval (ν and $\nu + d\nu$) and in the direction L within an elementary solid angle $d\omega$ and in the time interval dt is

$$dE_\nu = I_\nu \cos\theta \; d\sigma \; d\omega \; d\nu \; dt \; . \tag{2}$$

If we define an integrated intensity I such that

$$I = \int_0^\infty I_\nu d\nu \tag{3}$$

the total amount of energy radiated over the whole spectrum is

$$dE = \int_0^\infty dE_\nu d\nu = I \cos\theta \, d\sigma \; d\omega \; dt \tag{4}$$

(b) <u>The Flux of Radiation</u>: The net flux of radiation across $d\sigma$ per unit area and per unit time in the direction of L is

$$\vec{q}_{R\sigma} = \int \frac{dE}{d\sigma dt} = \int I \cos\theta \, d\omega \tag{5}$$

The direction L can be completely specified by the angular variables θ ($0 \le \theta \le \pi$) and the azimuth angle ϕ ($0 \le \phi \le 2\pi$) (Fig. 2). The elementary solid angle $d\omega$ defined by the range (θ, $\theta + d\theta$) and

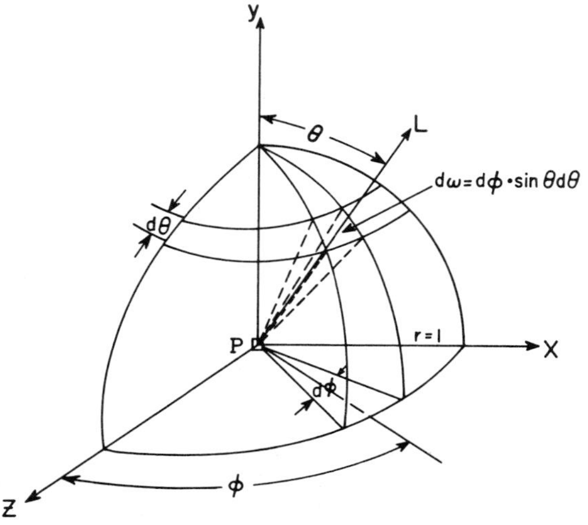

Fig. 2 Spherical Coordinates r, θ, φ

$(\phi\,,\ \phi + d\phi)$ is

$$d\omega = \sin\theta\, d\theta d\phi \tag{6}$$

Hence equation (5) becomes

$$\vec{q}_{R\sigma}(\theta, \phi, r, t) = \int_0^\infty \int_0^{2\pi} \int_0^{\pi} I_\nu(\theta, \phi, r, t) \sin\theta \cos\theta d\theta d\phi d\nu \tag{5a}$$

Let the direction cosines of the line L with respect to the x- , y- and z axis be ℓ , m and n respectively. The net radiation flux across elements of surfaces normal to x- , y- , and z-axis are respectively

$$\begin{aligned} q_{Rx} &= \int I\ell\, d\omega \\ q_{Ry} &= \int I m\, d\omega \\ q_{Rz} &= \int I n\, d\omega \end{aligned} \tag{7}$$

The range of integration is the same as that given in equation (5a). In vector form, we have the resultant radiation flux is

$$\vec{q}_R = i\, q_{Rx} + j\, q_{Ry} + k\, q_{Rz} \tag{7a}$$

where i, j and k are respectively the unit vector along x-, y- and z-axis.

The energy radiated out of a unit volume dx dy dz at P is then

$$Q_R = \nabla \cdot \vec{q}_R = \text{div}\, \vec{q}_R = \frac{\partial q_{Rx}}{\partial x} + \frac{\partial q_{Ry}}{\partial y} + \frac{\partial q_{Rz}}{\partial z} \tag{8}$$

where ∇ is the gradient operator. The term Q_R should be added to the energy equation of classical gasdynamics when radiation effect is considered. This term Q_R is very similar to the heat transfer by conduction $Q_c = \nabla \cdot \vec{q}_c$ in ordinary gasdynamics except that the flux of heat conduction may be expressed as a vector proportional to the temperature gradient and that $\vec{q}_R$ has to be expressed in terms of the triple integral of equation (5a).

(c) The Energy Density of Radiation: The energy density $u_\nu d\nu$ of the radiation in the frequency interval $(\nu, \nu + d\nu)$ at any given point P is the amount of radiant energy per unit volume in the stated interval, which is of transit in the immediate neighborhood of the point considered. The expression of u_ν may be obtained by considering the energy traversing through a small volume around P and is given by

$$u_\nu = \frac{1}{c} \int I_\nu d\omega \tag{9}$$

where c is the velocity of light.

The energy density of the integrated radiation at a given point P is then

$$u_R = \int_0^\infty u_\nu d\nu = \frac{1}{c} \int I\, d\omega \tag{10}$$

This energy density of radiation u_R is similar to the internal energy of gas in the ordinary gasdynamics. We should consider this energy as a part of the internal energy of the gas when a complete relation of the conservation of energy is analyzed.

(d) The Stress Tensor of Radiation: The radiation energy of amount E traversing a medium in a special direction carries with it a momentum E/c, the momentum exerted being in the same direction as the ray of radiation. Hence the rate of transfer of the x-component of the momentum in the x-direction across the element of surface by radiation confined to an element of a solid angle $d\omega$ about a direction with direction cosines ℓ, m and n is

$$d\tau_{Rxx} = \frac{1}{c} I \ell \cdot d\omega \cdot \ell \tag{11}$$

Hence the total rate of transfer of x-component of momentum in the x-direction per unit area of the surface is

$$\tau_{Rxx} = \frac{1}{c} \int I \ell^2 d\omega \qquad (12)$$

It is evident that τ_{Rxx} is the x-x-component of the stress tensor associated with the radiation intensity I . Similarly, we have all the nine components of the stress tensor of radiation as follows:

$$\tau_{Rxx} = \frac{1}{c} \int I\ell^2 d\omega , \quad \tau_{Rxy} = \frac{1}{c} \int I\ell m d\omega , \quad \tau_{Rxz} = \frac{1}{c} \int I\ell n d\omega$$

$$\tau_{Ryx} = \frac{1}{c} \int I\ell m d\omega , \quad \tau_{Ryy} = \frac{1}{c} \int I m^2 d\omega , \quad \tau_{Ryz} = \frac{1}{c} \int I m n d\omega$$

$$\tau_{Rzx} = \frac{1}{c} \int I\ell n d\omega , \quad \tau_{Rzy} = \frac{1}{c} \int I m n d\omega , \quad \tau_{Rzz} = \frac{1}{c} \int I n^2 d\omega \qquad (13)$$

This stress tensor should be considered in the equations of motion of radiation gasdynamics. Equation (13) gives

$$\tau_{Rxy} = \tau_{Ryx} \; ; \quad \tau_{Ryz} = \tau_{Rzy} \; ; \quad \tau_{Rxz} = \tau_{Rzx} \qquad (14)$$

Furthermore, the average radiation pressure at a given point is

$$-\bar{p}_R = \frac{1}{3}(\tau_{Rxx} + \tau_{Ryy} + \tau_{Rzz}) = \frac{1}{3c} \int I d\omega = \frac{1}{3} u_R \qquad (15)$$

If the radiation is isotropic, we have

$$-p_R = \tau_{Rxx} = \tau_{Ryy} = \tau_{Rzz} = \frac{1}{3} u_R$$

and (16)

$$\tau_{Rxy} = \tau_{Ryz} = \tau_{Rzx} = 0$$

Equation (16) will be used in the radiative equilibrium flow.

III. The Transfer Equation of Radiation

Before we know the effects of radiation on the flow field such as those due to radiation energy transfer and radiation stress tensor, we have to know the intensity of radiation I in the flow field. The intensity of radiation is determined by the interaction between the radiation and the matter. Such an interaction is usually expressed in terms of an absorption coefficient and an emission coefficient.

(a) The Absorption Coefficient K_ν: We consider a beam of radiation rays with intensity I_ν traversing through a medium of thickness ds in the direction of the ray. The intensity of radiation is weakened by absorption of an amount dI_ν . The absorption coefficient K_ν is defined by the relations

$$dI_\nu = -K_\nu \rho I_\nu ds \tag{17}$$

where ρ is the density of the absorbing material.

Integration of equation (17) with respect to s gives

$$I_\nu(s) = I_\nu(0) \exp\left\{-\int_0^s K_\nu \rho ds\right\} = I_\nu(0)\exp(-\tau_\nu) \tag{18}$$

where

$$\tau_\nu = \int_0^s K_\nu \rho ds = \text{optical depth of the layer} \tag{19}$$

The optical depth is a measure of the absorption of radiation in a given absorbing medium. The element of optical path is

$$d\tau_\nu = K_\nu \rho ds \tag{20}$$

Sometimes it is convenient to use the optical depth instead of the actual depth s in the analysis of radiative transfer problem.

The absorption coefficient consists of two parts: one is the true absorption in which the radiation energy of frequency ν has been transformed into other forms of energy or into radiation of other frequencies and the other is the scattering in which the energy lost from the incident beam will reappear as scattered radiation in other directions. We have to use microscopic treatment to determine the absorption coefficient. In our treatment of macroscopic analysis, the absorption coefficient of the medium is assumed to be known function of the state of the medium.

(b) The Emission Coefficient j_ν : The energy dE_ν emitted from an element of mass dm in the element solid angle $d\omega$ in time interval dt and in the frequency interval $(\nu, \nu + d\nu)$ may be written as

$$dE_\nu = j_\nu dm\, d\omega\, dt \tag{21}$$

where j_ν is known as the emission coefficient for the frequency ν, which should be calculated from the microscopic analysis such as quantum mechanics. We may assume that it is a known function of the state of the medium in our macroscopic treatment.

The emission coefficient also consists of two parts: one is due to the scattering of the photons and the other is due to the creation of photons from the matter.

(c) The Equation of Transfer: We consider a ray of radiation of intensity I_ν passing through an elementary cylinder of base $d\sigma$ and height ds. We assume that the ray is in the direction of the normal of the surface. The radiant energy passing through the base into the cylinder is

$$dE_i = I_\nu d\omega d\sigma dt d\nu$$

and the corresponding radiant energy coming out of the cylinder is

$$dE_0 = (I_\nu + dI_\nu) d\omega d\sigma dt d\nu$$

where dI_ν is the change of intensity during the path ds. When this ray passes through the cylinder, some of the energy is absorbed by the medium of the cylinder, i. e.,

$$dEa = -K_\nu I_\nu \rho ds d\omega d\sigma dt d\nu$$

and some of the energy will be emitted by the medium of the cylinder, i. e.,

$$dE_m = j_\nu \rho d\sigma ds d\omega dt d\nu$$

In general, the intensity I_ν may change with time t. Hence the radiant energy (u_ν) in the cylinder will change during the time interval dt by the amount

$$dE_t = \frac{1}{c}\frac{\partial I}{\partial t} d\sigma ds d\omega dt d\nu$$

By the conservation of the radiant energy we have

$$dE_0 - dE_i = dE_m + dE_a - dE_t \tag{22}$$

or

$$\frac{\partial I_\nu}{\partial s} = -\frac{1}{c}\frac{\partial I_\nu}{\partial t} + \rho(j_\nu - K_\nu I_\nu) \tag{22a}$$

$$\frac{1}{c}\frac{\partial I_\nu}{\partial t} + \ell\frac{\partial I_\nu}{\partial x} + m\frac{\partial I_\nu}{\partial y} + n\frac{\partial I_\nu}{\partial z} = \rho K_\nu (J_\nu - I_\nu) \tag{23}$$

Equation (23) is the equation of radiative transfer. The function J_ν is known as the source function which is

$$J_\nu = \frac{j_\nu}{K_\nu} = \text{source function} \tag{24}$$

It is interesting to notice that the source function is in general not a scalar but depends on the intensity I_ν. One of the main objects of radiative transfer is to determine the source function J_ν. In general the source function should be determined by microscopic treatment and it is of different expressions for different cases. However, for flow problems where the medium has a definite temperature T, we may apply the assumption of local thermodynamic equilibrium (§5). For these cases, we shall show later that the source function J_ν is a definite function of temperature T and frequency ν.

IV. Fundamental Equations of Radiation Magnetogasdynamics

For radiation magnetogasdynamics we would like to know the following quantities:

(i) The temperature of the plasma T
(ii) The pressure of the plasma p
(iii) The density of the plasma ρ
(iv) The velocity vector of the plasma $\vec{q}$
(v) The elective field strength $\vec{E}$
(vi) The magnetic field strength $\vec{H}$
(vii) The excess electric charge ρ_e
(viii) The electrical current density $\vec{J}$
(ix) The intensity of radiation I_ν

All the vector quantities have three components. Hence there are 17 unknown quantities to be investigated. It is assumed that all properties of the medium such as coefficient of viscosity, specific heat, absorption coefficient of radiation, etc. are known functions of the state variables of the medium. In order to investigate these unknowns, we must find 17 relations connecting them. These relations, known as fundamental equations of radiation magnetogasdynamics, are given below.

(i) Equation of state. The plasma may be considered as a perfect gas whose equation of state is

$$p = R\rho T \tag{25}$$

where R is the gas constant of the plasma.

(ii) Equation of continuity. The conservation of mass of the plasma gives

$$\frac{\partial \rho}{\partial t} + \frac{\partial}{\partial x^i}(\rho u^i) = \beta \cong 0 \tag{26}$$

where u^i is the ith component of the velocity vector $\vec{q}$. The summation convention is used. β is the change of mass due to radiation of nuclear energy which is negligibly small in ordinary flow problems.

We neglect it.

(iii) Equations of motion. The conservation of momentum gives the equation of motion which is

$$\rho \frac{Du^i}{Dt} = -\frac{\partial p}{\partial x^i} + \nabla \cdot \tau + \nabla \cdot \tau_R + F_e^i \tag{27}$$

where τ is the viscous stress tensor whose ij^{th} component may be written as

$$\tau^{ij} = \mu\left(\frac{\partial u^i}{\partial x^j} + \frac{\partial u^j}{\partial x^i}\right) - \frac{2}{3}\mu\left(\frac{\partial u^k}{\partial x^k}\right)\delta^{ij} \tag{28}$$

τ_R is the radiation stress tensor whose nine components are given in equation (13). $\delta^{ij} = 0$, $i \neq j$; $\delta^{ij} = 1$, $i = j$.

F_e^i is the ith component of the electromagnetic force, i.e.,

$$F_e^i = \rho_e E^i + \mu_e(\vec{J} \times \vec{H})^i \tag{29}$$

where μ_e is the magnetic permeability.

We neglect other body forces such as gravitational forces in our analysis.

(iv) Equation of energy. The conservation of energy gives the equation of energy as follows:

$$\frac{\partial}{\partial t}(\rho \bar{e}_m) + \frac{\partial \rho \bar{e}_m u^j}{\partial x^j} = -\frac{\partial u^i p}{\partial x^j} + \frac{\partial u^i \tau^{ij}}{\partial x^j} + \frac{\partial u^i \tau_R^{ij}}{\partial x^j} \tag{30}$$

$$+ Q_c + Q_R + E^j J^j$$

where

$$\bar{e}_m = U_m + \frac{q^2}{2} + \Phi_P + \frac{u_R}{\rho} = \text{total}$$

energy of the plasma per unit mass.

U_m = total internal energy of the plasma per unit mass.

$q^2/2$ = total kinetic energy of the plasma per unit mass.

Φ_P = total potential energy of the plasma per unit mass.

u_R = total radiation energy of the plasma per unit volume which is given in equation (10).

The first term on the left hand side of equation (30) is the rate of change of total energy of the plasma per unit volume and the second term is the energy flow by convection. The first term on the right-hand side of equation (30) is the work done by the pressure of the plasma; the second term is the energy dissipation by the stress tensor τ^{ij} ; the third term is the energy dissipation by the radiation stress tensor of equation (13); the fourth term is the energy change by heat conduction, i.e.,

$$Q_c = \frac{\partial}{\partial x^i}\left(K \frac{\partial T}{\partial x^i}\right) \quad ; \qquad (31)$$

the fifth term is the energy change by radiation flux which is given by equation (8) and the last term is the energy change due to the electromagnetic fields which consists of the work done by the electromagnetic force and the Joule heat, i.e.,

$$E^j J^j = u^j F_e^j + \frac{I_c^2}{\sigma} \qquad (32)$$

where $\vec{I_c} = \vec{J} - \rho_e \vec{q}$ = electrical conduction current and σ is the electric conductivity of the plasma.

(v) Maxwell's equations of electromagnetic field. The electromagnetic fields are governed by Maxwell's equations which are

$$\nabla \times \vec{H} = \vec{J} + \frac{\partial \epsilon \vec{E}}{\partial t} \qquad (33)$$

$$\nabla \times \vec{E} = -\frac{\partial \mu_e \vec{H}}{\partial t} \qquad (34)$$

where ϵ is the inductive capacity. We use the MKS unit system in this paper.

(vi) Equation of conservation of electric charge. The conservation of electric charge in the plasma gives

$$\frac{\partial \rho_e}{\partial t} + \frac{\partial J^i}{\partial x^i} = 0 \qquad (35)$$

(vii) The Equation of electric current density $\vec{J}$. The electric current density $\vec{J}$ is due to the complicated motion of the charged particles in the plasma which should depend on both the electromagnetic fields and all the gasdynamic variables of the plasma. The exact differential equation for $\vec{J}$ is very complicated [8]. However, if the density of the plasma is not too low and the magnetic field strength is not too high we may use the generalized Ohm's law for the equation of electrical current as a first approximation. The generalized Ohm's law is

$$\vec{I_c} = \sigma(\vec{E} + \mu_e \vec{q} \times \vec{H}) \qquad (36)$$

(viii) Equation of radiative transfer. The last relation is the equation of radiative transfer (23). In ordinary flow problem, we do not consider very high frequency phenomena so that the unsteady term $\frac{1}{c}\frac{\partial I_\nu}{\partial t}$ is negligibly small in comparison with the spatial variation terms because the velocity of light is a very large quantity. Hence we shall neglect this term and the equation of radiative transfer becomes

$$\ell \frac{\partial I_\nu}{\partial x} + m \frac{\partial I_\nu}{\partial y} + n \frac{\partial I_\nu}{\partial z} = \rho K_\nu (J_\nu - I_\nu) \tag{37}$$

Our fundamental equations for radiation magnetogasdynamics are equations (25), (26), (27), (30), (33), (34), (35), (36), and (37). The main differences of these equations from those of ordinary magnetogasdynamics are those terms due to radiation transfer which are expressed by complicated integrals, equations (10), (11), (13), etc., which depend on greatly the geometry of the system considered.

V. Radiative Equilibrium and Local Thermodynamic Equilibrium

In order to evaluate the integrals of the radiation terms, some reasonable assumptions should be made. One of the assumptions which we use is the condition of local thermodynamic equilibrium. Since we assume that the plasma has a definite temperature at each point in the space and at every instant, we may also assume that radiative equilibrium exists so that the emission depends only on the local temperature. Under this condition the emission of radiation is given by Planck's radiation function of a black body which is

$$B_\nu = \frac{2h\nu^3}{c^2} \frac{1}{(e^{h\nu/kT} - 1)} \tag{38}$$

where B_ν is the specific intensity of a black body at a constant temperatu[re] T, h is Planck's constant and k is Boltzmann's constant. Since in the flow field of plasma, the temperature T is different at different points, the specific intensity I_ν which comes from a point other than the one c[on]sidered may be different from the intensity of radiation in equilibrium B_ν

Under the assumption of local thermodynamic equilibrium, the radiation transfer equation (37) becomes [2]

$$\frac{dI_\nu}{ds} = \left(\ell \frac{\partial}{\partial x} + m \frac{\partial}{\partial y} + n \frac{\partial}{\partial z}\right) I_\nu = \rho K'_\nu (B_\nu - I_\nu) \tag{39}$$

where

$$K'_\nu = K_\nu (1 - e^{-h\nu/kT}) \tag{40}$$

The reduction of the absorption coefficient in equation (39) is due to the effect of induced emission [2]. We are going to evaluate the various radiation terms in the fundamental equations of radiation magnetogasdynamics.

VI. Case of Small Mean Free Path of Radiation

The radiation terms (5), (10) and (13) can be greatly simplified if the density of the plasma is such that radiation of all wave lengths suffers considerable absorption over a small distance. In other words, the mean free path of radiation ℓ_R is small. The mean free path of radiation is defined as

$$\ell_R = \frac{1}{K_R\rho} \cong \frac{1}{K'_\nu \rho} \tag{41}$$

where K_R is a mean absorption coefficient.

For very large absorption coefficient K'_ν, the solution of equation (39) may be written as

$$I_\nu \cong B_\nu - \frac{1}{\rho K'_\nu}\left(\ell \frac{\partial B_\nu}{\partial x} + m \frac{\partial B_\nu}{\partial y} + n \frac{\partial B_\nu}{\partial z}\right) \tag{42}$$

Substituting equation (42) into the radiation terms we have the following results

(i) Radiation energy density. Equations (9) and (10) give respectively

$$u_\nu = \frac{1}{c} \int I_\nu d\omega = \frac{4\pi}{c} B_\nu \tag{43}$$

and

$$u_R = \int_0^\infty u_\nu d\nu = \frac{4\pi}{c}\int_0^\infty B_\nu d\nu = \frac{8\pi^5 k^4}{15c^3h^3} T^4 = A_R T^4 \tag{44}$$

where A_R is known as Stefan-Boltzmann constant

(ii) Radiation stresses. Equation (13) gives

$$\tau_{Rxx} = \tau_{Ryy} = \tau_{Rzz} = |p_R| = \frac{1}{3} A_R T^4 \tag{45}$$

$$\tau_{Rxy} = \tau_{Ryz} = \tau_{Rzx} = 0$$

Equation (45) gives the magnitude of p_R which is in the opposite direction as the normal stress τ_{Rxx}, etc.

(iii) Radiation transfer. Equation (7) gives

$$q_{Rx} = -\frac{4\pi}{3}\int_0^\infty \frac{1}{K'_\nu \rho}\frac{\partial B_\nu}{\partial x}d\nu = -\frac{4\pi}{3\rho}\int_0^\infty \frac{1}{K'_\nu}\frac{\partial B_\nu}{\partial T}\frac{\partial T}{\partial x}d\nu \tag{46}$$

Now we define a mean absorption coefficient K_R such that

$$\frac{1}{K_R}\int_0^\infty \frac{\partial B_\nu}{\partial T}d\nu = \int_0^\infty \frac{1}{K'_\nu}\frac{\partial B_\nu}{\partial T}d\nu \tag{47}$$

Substituting equation (47) into equation (46), we have

$$q_{Rx} = -\frac{4\pi}{3K_R\rho}\int_0^\infty \frac{\partial B_\nu}{\partial x}d\nu = -\frac{4\pi}{3K_R\rho}\frac{\partial B}{\partial x} = \frac{c}{K_R\rho}\frac{\partial p_R}{\partial x} \tag{48a}$$

Similarly we have

$$q_{Ry} = \frac{c}{K_R\rho}\frac{\partial p_R}{\partial y} \tag{48b}$$

$$q_{Rz} = \frac{c}{K_R\rho}\frac{\partial p_R}{\partial z} \tag{48c}$$

or

$$\vec{q}_R = \frac{c}{K_R\rho}\nabla p_R = -\frac{c\ell_R}{3}\nabla u_R = -D_R\nabla u_R \tag{49}$$

where D_R is known as the diffusion coefficient of radiation.

Substituting equations (44), (45) and (40) into the fundamental equations of section 4, we have the fundamental equation of radiation magnetogasdynamics for opaque medium as follows

$$p = \rho RT \tag{50a}$$

$$\frac{\partial \rho}{\partial t} + \frac{\partial}{\partial x^i}(\rho u^i) = 0 \tag{50b}$$

$$\rho\frac{Du^i}{Dt} = -\frac{\partial p + |p_R|}{\partial x^i} + \nabla\cdot\tau + Fe^i \tag{50c}$$

$$\frac{\partial}{\partial t}(\rho\bar{e}_m) + \frac{\partial\rho\bar{e}_m u^j}{\partial x^j} = -\frac{\partial u^j(p + |p_R|)}{\partial x^j} + \frac{\partial u^i\tau^{ij}}{\partial x^j} + \frac{\partial}{\partial x^j}\left(K\frac{\partial T}{\partial x^j}\right) + \frac{\partial}{\partial x^j}\left(D_R\frac{\partial u_R}{\partial x^j}\right) + E^jJ^j \tag{50d}$$

$$\nabla \times \vec{H} = \vec{J} + \frac{\partial \epsilon \vec{E}}{\partial t} \qquad (50e)$$

$$\nabla \times \vec{E} = -\frac{\partial \mu_e \vec{H}}{\partial t} \qquad (50f)$$

$$\frac{\partial \rho_e}{\partial t} + \frac{\partial J^i}{\partial x^i} = 0 \qquad (50g)$$

$$J^i = \sigma [E^i + \mu_e (\vec{q} \times \vec{H})^i] + \rho_e u^i \qquad (50h)$$

This set of system has been investigated by the author for wave motion [3] and shock wave [4]. The important new parameter of radiation pressure is the radiation pressure number which is defined as

$$\eta = \frac{\text{radiation pressure}}{\text{gas pressure}} = \frac{|p_R|}{p} = \frac{A_R}{3R\rho} T^3 \qquad (51)$$

If η is of the order of unity, the radiation pressure p_R and radiation energy u_R should be considered.

Another important radiation parameter is radiation flux parameter R_F which is defined as

$$R_F = \frac{\text{radiation flux}}{\text{heat conduction flux}} = \frac{4D_R A_R T^4}{KT} = \frac{4c A_R T^3}{3\rho K_R K} = \frac{4c R}{K_R K} \eta \qquad (52)$$

Because c is a large unity, R_F is usually much larger than η. Hence in many engineering problems, even though the radiation pressure and radiation energy density may be neglected, the radiation flux should be considered.

VII. Two-Dimensional Steady Flow of Radiation Magnetogasdynamics when the Mean Free Path of Radiation is Not Small

In the last section, we assume that the mean free path of radiation $\ell_R = \frac{1}{K'_\nu \rho}$ is small, thus we may use equation (42) for the intensity of radiation. If ℓ_R is not small, we should use more terms in the series development of equation (42) or we should solve exactly the radiation transfer equation (30) for given boundary conditions.

In solving the radiation transfer equation (39) the existence of solid bodies in the flow field influences the result because they affect the boundary conditions. In general, we have to solve equation (39) simultaneously with other fundamental equations of radiation magnetogasdynamics. No solution in closed form may be found in the general case. In order to show the exact solution of the radiation transfer equation (39), we consider the two-dimensional steady flow in which there is a straight plane wall (Fig. 3). Such a problem may be considered as an ideal model for the hypersonic flow near the stagnation point of a blunt body.

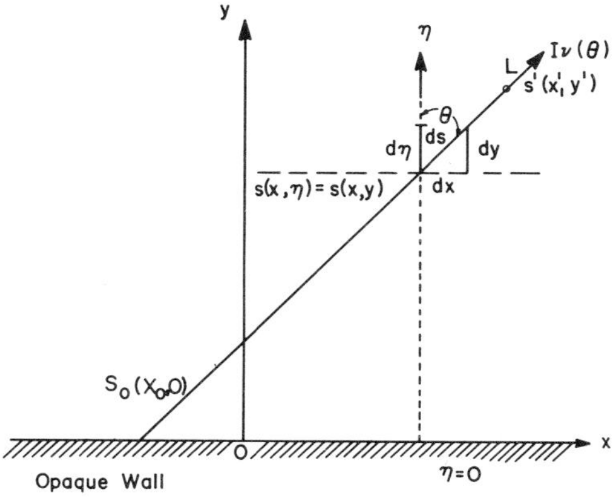

FIG. 3

Two Dimensional flow with an opaque wall

Let $s(x, y)$ represent a point in the flow field, and ds be the elementary length along the ray L which makes an angle θ from the y-axis. Then

$$ds = \sin\theta\, dx + \cos\theta\, dy \tag{53}$$

when $0 \leq \theta \leq \pi/2$, the rays are directed away from the wall while $\pi/2 < \theta < \pi$ the rays are directed toward the wall. The solutions of equation (39) are different for the rays towards the wall from those away from the wall because of the influence of the wall. Hence we have two different solutions for these two regions, i. e.,

(i) For rays directed away from the wall, i. e., $0 \leq \theta \leq \pi/2$, the solution of equation (39) is

$$I_\nu(x, y) = I(x_0, 0)\, e^{-\{\tau(s)-\tau(s_0)\}} + \int_{s_0}^{s} B_\nu(s', \nu)\, e^{-\{\tau(s)-\tau(s')\}} K'_\nu \rho\, ds' . \tag{54}$$

where

$$\tau(s) = \int_0^s K'_\nu \rho\, ds' \rightarrow \tau(s) = \int_b^s K'_\nu \rho\, ds' \tag{55}$$

b is an arbitrary point along ray L. $s_0 = s_0(x, 0)$ is the point on the ray L when $y = 0$, and $s'(x', y')$ is the variable point on the ray L when the integration terms of equation (54) and (55) are carried out. By definition of $\tau(s)$ of equation (55), we have

$$\tau(s) - \tau(s_0) = \tau(s, s_0) = \int_{s_0}^{s} K'_\nu \rho ds'$$
$$\tau(s) - \tau(s') = \tau(s, s') = \int_{s'}^{s} K'_\nu \rho ds' \qquad (55a)$$

$$\tau(s) - \tau(s) = \tau(s, s) = 0 \qquad (55b)$$

We may use $\tau(s, s')$ for the difference of $\tau(s)$ and $\tau(s')$ in the following formula. $I(x_0, 0)$ is the intensity on the opaque wall at the point

(ii) For rays directed toward thw wall, i.e., $\pi/2 < \theta < \pi$ the solution of equation (39) is

$$I_\nu(x, y) = \int_{s}^{\infty} B_\nu(s', \nu) e^{-\tau(s, s')} K'_\nu \rho ds' \qquad (56)$$

where ∞ means $x' = \infty$ and $y' = \infty$.

Now we may calculate the radiation energy density (10), radiation stresses (13) and radiation flux (7) by using the expressions of radiation intensity (54) and (56). In our problem, we assume that the intensity is independent of the azimuth angle ϕ so that

$$\int f d\omega = 2\pi \int_0^{\pi} f \sin\theta \, d\theta \qquad (57)$$

where f may represent any integrand.

Let us consider the case of radiation energy density (10) first. From equations (10), (54) and (56) we have

$$U_R = \frac{2\pi}{c} \int_0^{\infty} d\nu \int_0^{\pi} I_\nu \sin\theta d\theta = \frac{2\pi}{c} \int_0^{\infty} d\nu \int_0^{\pi/2} I(x_0, 0) e^{-\tau(s, s_0)} \sin\theta d\theta$$
$$+ \frac{2\pi}{c} \int_0^{\infty} d\nu \int_0^{\pi/2} \sin\theta d\theta \int_{s_0}^{s} B_\nu(s', \nu) e^{-\tau(s, s')} K'_\nu \rho ds' \qquad (58)$$
$$+ \frac{2\pi}{c} \int_0^{\infty} d\nu \int_{\pi/2}^{\pi} \sin\theta d\theta \int_{s}^{\infty} B_\nu(s', \nu) e^{-\tau(s, s')} K'_\nu \rho ds' .$$

In general, no further simplification of equation (58) can be made because $I(x_0, 0)$, $B_\nu(s', \nu)$ and $(K'_\nu \rho)$ all depend on the angle θ. However, for special problems, equation (58) can be simplified. One of such cases is the flow in the boundary layer where the variations of temperature, density, etc., in the y-direction, are much larger than those in the x-direction. In this case, we may introduce a new variable η (see Fig. 3) such that

$$ds = u\, d\eta \tag{59}$$

where $u = \sec\theta$. We define a new optical thickness

$$t_\nu(y) = \int_0^y K'_\nu \rho d\eta \tag{60}$$

On the surface of the opaque wall, $\tau = 0$, $\eta = 0$. Now we may assume that both $I(x_0, 0) = I(0)$ and B_ν are independent of θ. Then equation (58) reduces to

$$\begin{aligned} u_R = & \frac{1}{c}\int_0^\infty 2\pi I(0)\varepsilon_2(t_\nu)\,d\nu + \frac{1}{c}\int_0^\infty\int_0^{t_\nu} 2\pi B_\nu(t'_\nu, \nu)\varepsilon_1(t_\nu - t'_\nu)\,dt'_\nu d\nu \\ & - \frac{1}{c}\int_0^\infty\int_{t_\nu}^\infty 2\pi B_\nu(t'_\nu, \nu)\varepsilon_1(t'_\nu - t_\nu)\,dt'_\nu d\nu \end{aligned} \tag{61}$$

where

$$t'_\nu = \int_0^\eta K'_\nu \rho d\eta \tag{62}$$

and

$$\varepsilon_n(f) = \int_1^\infty \frac{e^{-(f)u}}{u^n}\, du$$

The approximations of equation (61) have been extensively used in astrophysical problems.

In practical problems, the radiation energy density and the radiation stresses are usually negligible but the radiation flux should be considered if the temperature is not too low. For the approximation of equation (61), only the y-component of radiation flux (7) is important, which is, from equations (7) and approximations of equation (61),

$$q_{Ry} = 2\pi \int_0^\infty d\nu \int_0^\pi I_\nu \sin\theta \cos\theta \, d\theta$$

$$= \int_0^\infty 2\pi I(0) \varepsilon_3(t_\nu) d\nu + \int_0^\infty \int_0^{t_\nu} 2\pi B_\nu(t'_\nu) \varepsilon_2(t_\nu - t'_\nu) dt'_\nu d\nu \qquad (63)$$

$$- \int_0^\infty \int_{t_\nu}^\infty 2\pi B_\nu(t'_\nu) \varepsilon_2(t'_\nu - t_\nu) dt'_\nu d\nu$$

If the wall has a constant temperature T_0, a monochromatic emissivity e_ν and monochromatic reflectivity r_ν, the intensity on the wall will be

$$I(0) = e_\nu B_\nu(T_0) + r_\nu \int_0^\infty 2\pi B_\nu(t'_\nu) \varepsilon_2(t'_\nu) dt'_\nu \qquad (64)$$

Substituting equation (64) into equation (63), we have

$$q_{Ry} = 2\pi \int_0^\infty e_\nu B_\nu(T_0) \varepsilon_3(t_\nu) d\nu + (2\pi)^2 \int_0^\infty \int_0^\infty r_\nu B_\nu(t'_\nu) \varepsilon_2(t'_\nu) dt'_\nu \varepsilon_3(t_\nu) d\nu$$

$$+ \int_0^\infty 2\pi d\nu \int_0^{t_\nu} B_\nu(t'_\nu) \varepsilon_2(t_\nu - t'_\nu) dt'_\nu \qquad (65)$$

$$- \int_0^\infty 2\pi \int_{t_\nu}^\infty B_\nu(t'_\nu) \varepsilon_2(t'_\nu - t_\nu) dt'_\nu d\nu$$

The energy radiated out of a unit volume at $s(x, y)$ is

$$Q_R = \frac{\partial}{\partial y}(q_{Ry}) = \int_0^\infty 2\pi K'_\nu \rho \{ -e_\nu B_\nu(T_0) \varepsilon_2(t_\nu)$$

$$- 2\pi \int_0^\infty r_\nu B_\nu(t'_\nu) \varepsilon_2(t'_\nu) \varepsilon_2(t_\nu) dt'_\nu \qquad (66)$$

$$+ 2B_\nu(t_\nu) - \int_0^\infty B_\nu(t'_\nu) \varepsilon_1(|t_\nu - t'_\nu|) dt'_\nu \} d\nu$$

Even in this simple case, we have double integrals for the heat transfer term of radiation.

VIII. The Gray Gas Approximations

Equations similar to (66) can be greatly simplified for gray-gas in which the absorption coefficient K'_ν is independent of frequency ν. If the variation of K'_ν with ν is small, we may use a "mean" absorption coefficient K_R for all K_ν as a first approximation which is known as gray gas approximation. For a gray gas, the optical thickness t_ν is independent of ν and we may carry out the integration with respect to ν in the radiation integrals such as equation (66), etc. For instance, if we consider the wall is black so that $e_\nu = 1$ and $r_\nu = 0$, equation (66) for gray gas becomes

$$Q_R = K_R \rho \left\{ \int_0^\infty - \frac{A_R c}{2} T_0^4 \varepsilon_1 (| t_\nu - t'_\nu |) dt'_\nu - \frac{A_R c}{2} T_0^4 \varepsilon_2 (t_\nu) + 4 A_R c T^4 \right\} \tag{67}$$

Similar expressions may be found for radiation energy density and radiation stress tensor. Substituting equation (67) and other similar expression for radiation terms into the fundamental equations of radiation magnetogasdynamics, we have a set of integro-differential equations. We plan in the future to solve these sets of equations for simple type of flow and compare them with those simplified equations of section 6. The expression of equation (67) has been discussed by Goulard and Tellep and Edwards for ordinary gasdynamics without electromagnetic fields.

This research was supported in part by the United States Air Force through the Air Force Office of Scientific Research under Contract No. AF 49(638) 401.

REFERENCES

1. Ambartsumyan, V. A. (Ed.) Theoretical Astrophysics, Pergamon Press, 1958.

2. Chandrasekhar, S., Stellar Structure, Dover Publications, Inc., N.Y. 1957.

3. Chandrasekhar, S., Radiative Transfer, Dover Publications, Inc., N.Y. 1960.

4. Goulard, R. and Goulard, M., Energy transfer in the Couette Flow of a Radiant and Chemically Reacting Gas, Heat transfer and Fluid mechanics Institute, 1959.

5. Kourganoff, V., Basic Methods in Transfer Problems, Oxford Press, 1952.

6. Pai, S. I. and Speth, A. I., The Wave Motion of Small Amplitude in Radiation Electromagneto-gasdynamics., Proc. of 6th Midwest Conference on Fluid Mech., Sept. 9-11, 1959, Univ. of Texas, pp. 446-466.

7. Pai, S. I. and Speth, A. I., Shock Waves in Radiation Magneto-gasdynamics, The Physics of Fluids, vol. 4, No. 10, Oct. 1960, pp. 1232-1237.

8. Pai, S. I., Magnetogasdynamics and Plasma Dynamics, Springer-Verlag, Vienna and Prentice Hall, N. J. 1962.

9. Tellep, D. M. and Edwards, D. K., Radiant-energy Transfer in Gaseous Flows, Tech. Report LMSD-288, 139, Missiles and Space Division, Lockheed Aircraft Corp., January 1960.

JAMES SERRIN

The initial value problem for the Navier-Stokes equations

The initial value problem for the Navier-Stokes equations has been the object of a remarkable surge of interest in the last five years. Developments have proceeded with such rapidity, and the results have become so specialized, that it is now no mean task to keep abreast of the field. It seems worthwhile, therefore, to attempt a synthesis of some of these results, to see a pattern of sorts emerging, and to point out some directions for future work. In addition to its expository aspects, the paper also contains several new results, particularly Theorems 3, 6 and 7, as well as some shortened proofs of known results.

For simplicity of presentation I shall confine myself for the most part to a somewhat restricted version of the problem. Consider a fixed (bounded or unbounded) domain Ω in E^n, where $n \geq 2$. This domain may be thought of as a rigid vessel filled with incompressible fluid; the fluid is initially set into motion, and we are to determine its subsequent velocity distribution, subject to the Navier-Stokes equations and the condition of adherence at the boundary of Ω. More precisely, it is required to find a velocity field $u = u(x, t)$ and a pressure field $p = p(x, t)$ which, for $x = (x_1, \ldots, x_n) \in \Omega$ and for $t > 0$, satisfy the differential equations

$$u_t + u \cdot \text{grad } u = -\text{grad } p + \Delta u , \tag{1}$$

$$\text{div } u = 0 , \tag{2}$$

and obey in some sense the initial condition

$$u(x, 0) = u_0(x) , \qquad x \in \Omega \tag{3}$$

and the boundary condition

$$u(x, t) = 0 , \qquad x \in \dot{\Omega} . \tag{4}$$

(More general formulations of the problem include the effect of an external force in equation (1), non-vanishing boundary conditions, and so forth. On the other hand, the fact that we have set the kinematic viscosity equal to 1 is no restriction of generality.) Historically, the interest of mathematicians in the problem (1)-(4) was aroused by Hadamard, Lichtenstein, Villat, and, most particularly, C. W. Oseen. But it is one man, Jean Leray, who really opened the way. In his three fundamental papers of the years 1933-34 are to be found ideas and problems which have remained valuable to this day. In 1951 Eberhard Hopf discovered that the Navier-Stokes equations could be attacked in a simple, elegant fashion using the techniques of functional analysis. His bold introduction of linear methods into an essentially nonlinear problem gave new life to the field, and this trend was further strengthened by the celebrated paper of Kiselev and Ladyshenzkaya in 1957. A veritable flood of papers has followed, working out and extending the methods introduced by these various writers. It is to this situation that we now turn.

In the first section below we consider the basic notion of weak solution, which underlies almost all modern work. This somewhat abstract concept has, rather surprisingly, proved itself to be the key to the problem's defenses. Next we quickly review the fundamental existence theorem of Hopf, and present a somewhat streamlined version of the result of Kiselev and Ladyzhenskaya. An extension of their work to the case n = 4 is also given.

Turning next to the uniqueness and regularity questions which are naturally associated with the existence of weak solutions, we outline in Sections 3 and 4 a new uniqueness theorem, which contains, among other things, the result that in dimension 3 the solutions of Hopf and of Kiselev-Ladyzhenskaya are the same, provided the latter exists (cf. the discussion following Theorem 6). The methods employed here are in the main due to Prodi and Lions, see references [25], [29] and [30]. In Section 6 the interior regularity of a weak solution is proved under assumptions rather similar to those required for the uniqueness theorems. In these sections I have also attempted to relate the work discussed to that of various other writers who have treated similar questions.

Section 5 contains a backward uniqueness theorem, applying to classical solutions of the boundary value problem. The final part of the paper is devoted to a resume of Leray's notion of "epochs of irregularity," and to the discussion of a number of unsolved or open problems.

1. Preliminaries

The program outlined in the introduction is relatively sophisticated requiring a fair amount of mathematical preparation. It is convenient to divide this into three subsections.

1. Let R denote an open region of space-time, which for the purpose of the initial value problem may be considered a cylinder with cross section Ω. Let $u = u(x, t)$ and $v = v(x, t)$ be locally square summable vectors in R. The notation

$$(u, v) = \int u \cdot v \, dx = \int u_i v_i \, dx$$

will be used for the spatial inner product, whenever the integral exists. Since in all applications either u or v can be considered identically zero outside R, there is no need to specify the domain of integration. The product (u, v) as defined is obviously a function of time; when we wish to make this dependence on t more explicit we shall write the product in the form $(u(t), v(t))$.

The vector u will be called weakly divergence free if

$$\int (u, \operatorname{grad} \omega) \, dt = 0 \tag{5}$$

for every C^1 function $\omega = \omega(x, t)$ having compact support in R. Moreover, u will be called strongly differentiable with respect to x if there exists a locally summable tensor, denoted either by u_x or by grad u, such that

$$\int (\varphi_x, u) \, dt = -\int (u_x, \varphi) \, dt \tag{6}$$

for all C^1 vectors φ with compact support in R. Finally, u is called strongly differentiable with respect to time if there exists a locally summable vector, denoted by u_t, such that

$$\int (\varphi_t, u) \, dt = -\int (u_t, \varphi) \, dt$$

for all C^1 vectors φ with compact support in R.

2. Now let D(R) denote the test space of C^∞ vectors which are divergence free and have compact support in R. We have the following basic definition.

DEFINITION 1. A vector $u = u(x, t)$ will be called a weak solution of the Navier-Stokes equations in R if it is locally square summable and weakly divergence free in R, and if

$$\int \{(u, \varphi_t) + (u, \Delta\varphi) + (u, u \cdot \operatorname{grad} \varphi)\} \, dt = 0$$

for all $\varphi \in D(R)$.

For smooth vectors u this formulation is obviously equivalent to the original equations (1) and (2) holding in R with some appropriate pressure $p(x, t)$.

3. In order to deal specifically with the initial value problem (1)-(4)[1], let D denote the test space of C^∞ vector fields $\varphi(x, t)$ which are divergence free and have compact support in $\Omega \times [0, \infty)$, (it should be observed that the vector $\varphi_0 = \varphi(x, 0)$ need not vanish identically). With the notations

$$((u, v)) = \int u_x : v_x dx = \int u_{i,k} v_{i,k} dx$$

and

$$|u|^2 = (u, u) \quad , \quad \|u\|^2 = ((u, u)) \; ,$$

the norm

$$\int_0^T \{|\varphi|^2 + \|\varphi\|^2\} dt \tag{7}$$

is then defined and finite for each $\varphi \in D$ and each T, $0 < T < \infty$.

DEFINITION 2. <u>A vector $u = u(x, t)$ will be said to be in the class V if</u>

i) <u>for each T, $0 < T < \infty$, it is in the closure of D under the norm (7), and</u>

ii) <u>the norm $|u|$ is uniformly bounded in time.</u>

It is evident that any vector $u \in V$ is strongly differentiable with respect to x, is weakly divergence free, and for each fixed T has finite norm (7).[2] Furthermore, u has generalized boundary values zero. (Another way to define the space V would be to consider the vectors $u(x, t)$ as representing a variable point $u(t)$ in an appropriate Hilbert space; cf. [29].)

The initial velocity distributions $u_0 = u_0(x)$ which we shall consider are conveniently defined with reference to the test space D_0 of C^∞ vectors $\psi = \psi(x)$ which are divergence free and have compact support in Ω. In particular, we let W denote the closure of D_0 in the norm $|\psi|$, W^1 the closure in $|\psi| + \|\psi\|$, and so forth.

DEFINITION 3. <u>Let $u_0 \in W$. A vector $u = u(x, t)$ will be called a weak solution of the initial value problem (1)-(4) if $u \in V$ and if</u>

$$\int_0^\infty \{(u, \varphi_t) + (u, \Delta\varphi) + (u, u\cdot \text{grad}\,\varphi)\}dt = -(u_0, \varphi_0)$$

for all $\varphi \in D$. Here φ_0 denotes the vector $\varphi(x, 0)$.

As in Definition 1, it is easy to see that for smooth u and u_0 this formulation is equivalent to (1)-(4). We add that there are procedures other than the above for taking account of the initial data; the present method, however, is convenient and standard, and will be followed henceforth in the paper.

2. Existence Theorems

In this section we shall consider the existence of weak solutions of the initial value problem (1)-(4), in the sense of Definition 3. Excepting the work of Leray, which is briefly reviewed in Sections 7 and 8, the fundamental result for this problem is the following

THEOREM 1 (Eberhard Hopf). For any initial velocity distribution $u_0 \in W$ there exists a weak solution $u \in V$ of the initial value problem. Moreover

$$|u|^2 + 2\int_0^t \|u\|^2 dt \le |u_0|^2 , \quad \lim_{t\to 0} |u-u_0| = 0 . \tag{8}$$

The proof (cf. [6]) is a beautiful application of the technique of Fourier approximation, unfortunately too long to include here. Nevertheless, we may observe that the process succeeds primarily because the original problem (1)-(4) admits a formal energy identity

$$|u|^2 + 2\int_0^t \|u\|^2 dt = |u_0|^2 . \tag{9}$$

[A formal proof of (9) is easily given: multiply both sides of (1) by u and integrate over Ω , to obtain

$$(u, u_t) + (u, u\cdot \text{grad}\, u) = -(u, \text{grad}\, p) + (u, \Delta u) ,$$

Since u is divergence free and vanishes on the boundary of Ω , we have

$$(u, u\cdot \text{grad}\, u) = (u, \text{grad}\tfrac{1}{2}u^2) = 0 , \quad (u, \text{grad}\, p) = 0 ,$$

whence it follows that

$$\frac{1}{2}\frac{d}{dt}|u|^2 = (u, u_t) = (u, \Delta u) = -\|u\|^2 .$$

The identity (9) is an immediate consequence of this last relation.] The fact that the formal procedure supplies the identity (9), while the rigorous proof yields only the inequality (8), should be carefully noted.

The solution which Hopf obtained, in spite of its elegance, is of course not a classically differentiable vector field. In an effort to obtain a solution with stronger regularity properties, Kiselev and Ladyshenskaya in the papers [14] and [15] considered initial data in the space W^2. Their result is as follows.

THEOREM 2 (Kiselev-Ladyzhenskaya). Suppose $n = 2$ or $n = 3$. Then for any initial data $u_0 \in W^2$ there exists a weak solution $u \in V$ of the initial value problem, and a positive number T such that $|u_x|$ and $|u_t|$ are uniformly bounded in the interval $0 \leq t < T$.

The number T depends only on bounds for the initial data; in case $n = 2$, or for sufficiently small data if $n = 3$, we can take $T = \infty$.

It almost goes without saying that this solution possesses in addition the properties (8) noted in Theorem 1. A detailed study of further properties of this solution, and its relation to Hopf's solution will be given in the following sections.

The proof of Theorem 2, as of Theorem 1, depends on Fourier approximation and a type of energy estimate. We shall present here only the formal procedure leading to this estimate. By differentiation of (1) with respect to t, scalar multiplication by u_t, and integration over Ω, there arises

$$\frac{1}{2}\frac{d}{dt}|u_t|^2 = -(u_t, u_t \cdot \operatorname{grad} u) - \|u_t\|^2 , \tag{10}$$

(we omit the straightforward details of the calculation). The first term on the right can be estimated by Hölder's inequality

$$-(u_t, u_t \cdot \operatorname{grad} u) \leq |u_t|_4^2 \|u\| , \tag{11}$$

where $| \ |_4$ denotes the usual norm in $L^4(\Omega)$. Now according to Sobolev's theorem, for any differentiable vector f with compact support we have

$$|f|_4^2 \leq C \begin{cases} |f| \cdot \|f\| & n = 2 \\ |f|^{1/2}\|f\|^{3/2} & n = 3 \\ \|f\|^2 & n = 4 \end{cases} \tag{12}$$

where C depends only on the dimension; for n = 2, 3, and 4 it can be taken to be $2^{-1/2}$, $3^{-3/4}$, and 9/16 , respectively (cf. the appendix at the end of this section). Hence

$$-(u_t, u_t \cdot \operatorname{grad} u) \leq C \begin{cases} |u_t| \cdot \|u_t\| \cdot \|u\| & n = 2 \\ |u_t|^{1/2} \|u_t\|^{3/2} \|u\| & n = 3 . \end{cases} \tag{13}$$

A subsequent application of the inequality $ab \leq a^p/p + b^q/q$ allows us to conclude from (10) and (13) that

$$\frac{1}{2}\frac{d}{dt}|u_t|^2 \leq \begin{cases} 2^{-3}\|u\|^2 |u_t|^2 & n = 2 \\ 2^{-8}\|u\|^4 |u_t|^2 & n = 3 . \end{cases} \tag{14}$$

In case n = 2 , integration of this inequality leads to the estimate

$$|u_t| \leq |u_{to}| \exp\left\{\frac{1}{8}\int_0^t \|u\|^2 dt\right\} .$$

Under the hypothesis $u_0 \in W^2$ it is apparent from (1), (2) and (12) that

$$|u_{to}| \leq |\Delta u_0| + |u_0|_4 |\operatorname{grad} u_0|_4 \leq \text{Const.}$$

But also, by virtue of (9), or (8), the quantity $\int \|u\|^2 dt$ is bounded. Consequently, <u>$|u_t|$ remains uniformly bounded for all $t \geq 0$</u> . In addition, we have formally

$$\|u\|^2 = -\frac{1}{2}\frac{d}{dt}|u|^2 \leq |u| \cdot |u_t| , \tag{15}$$

so that the known bounds on $|u|$ and $|u_t|$ supply us with an obvious estimate for $\|u\|$, that is, for $|u_x|$.

Consider next the case n = 3 . We observe that inequality (15) holds also in this case, whence by (14) we have

$$\frac{d}{dt}|u_t| \leq 2^{-8}|u| \cdot \|u\|^2 |u_t|^2 . \tag{16}$$

This leads by integration to the estimate

$$|u_t| \leq \frac{|u_{to}|}{1 - |u_{to}| \cdot A(t)} , \tag{17}$$

where

$$A(t) = 2^{-8}\int_0^t |u| \cdot \|u\|^2 dt = 2^{-8}3^{-1}\{|u_0|^3 - |u|^3\} .$$

The estimate (17) obviously remains valid only so long as the denominator is positive. Now if the initial data satisfies the condition $|u_0|^3 \cdot |u_{to}| < 768$, then clearly $|u_{to}| \cdot A(t) < 1$ for all time. In this case, the norms $|u_t|$ and $|u_x|$ are uniformly bounded for $t \geq 0$. In general, however, we can only assert that $|u_t|$ remains bounded during a certain finite time interval $0 \leq t \leq T$.

A simple estimate for T can be obtained from the inequality

$$\frac{d}{dt}|u_t| \leq 2^{-8}|u_0|^2|u_t|^3 ,$$

which follows from (15), (16), and the relation $|u| \leq |u_0|$. By integration of this result, we get

$$|u_t|^2 \leq \frac{|u_{to}|^2}{1 - 2^{-7}|u_0|^2|u_{to}|^2 t} ,$$

again valid whenever the denominator is positive. We have therefore proved the following result. In case $n = 3$, and for initial data in W^2, the norms $|u_t|$ and $|u_x|$ are uniformly bounded in the interval

$$0 \leq t \leq 128(|u_0| \cdot |u_{to}|)^{-2} .$$

Moreover, if $|u_0|^3|u_{to}| < 768$ the above norms are bounded for all $t \geq 0$. (If there is a non-unit kinematic viscosity ν, then the numbers 128 and 768 should be replaced by $128\nu^5$ and $768\nu^5$, respectively, according to dimensional considerations.) This completes the formal proof of Theorem 2. A rigorous proof remains a technical matter which can be handled by Hopf's construction. (It is worth pointing out there that the basic difficulty in the case $n = 3$ as compared to the case $n = 2$ is due to the nonlinearity of the differential inequality (16). Unfortunately, there seems to be no direct method for avoiding this difficulty.)

Although the case $n = 4$ was not discussed by Kiselev and Ladyzhenskaya, this too can be treated by similar methods. Indeed, from (10), (11), and (12) we obtain

$$\frac{d}{dt}|u_t|^2 + 2(1-C\|u\|)\|u_t\|^2 \le 0 . \qquad (18)$$

Now suppose the initial data satisfies the inequality

$$|u_0| \cdot |u_{to}| < C^{-2} = 256/81 .$$

Then according to (15) one has $\|u_0\| < C^{-1}$. We assert that actually $\|u\| < C^{-1}$ for all time. Indeed, if this were not so, then there would exist a $\tau > 0$ such that

$$\|u\| < C^{-1} \text{ for } 0 \le t < \tau , \text{ and } \|u(\tau)\| = C^{-1} .$$

But in this case it follows from (18) that $|u_t| \le |u_{to}|$ for $0 \le t \le \tau$. Therefore at the moment $t = \tau$

$$C^{-2} = \|u\|^2 \le |u| \cdot |u_t| \le |u_0| \cdot |u_{to}| < C^{-2} ,$$

which is a contradiction. This completes the (formal) proof of the following result.

THEOREM 3. <u>Suppose $n = 4$. Then for any initial data $u_0 \in W^2$ with sufficiently small norm there exists a solution of the initial value problem which is strongly differentiable with respect to x and t and has uniformly bounded norms $|u_x|$ and $|u_t|$.</u>

For $n \ge 5$ I do not know of any way to establish the existence of a solution with stronger regularity properties than that of Hopf. It seems of the greatest interest to know whether this limitation on dimension is a genuine feature of nonlinear equations of the type (2), or whether it merely indicates an insufficiently developed mathematical attack. In this regard, the possibility of solutions having fractional order time derivatives should not be overlooked; cf. in particular, the interesting paper of J. L. Lions [22].

Further results concerning the regularity of the solutions determined above, and a discussion of the related existence theorems of Leray, Golovkin, Ito, and Sobolevskii, will be found in Sections 6 and 7.

<u>Appendix.</u> The given values for C in inequality (12) are easily derived from the elementary estimate

$$|\omega|_{\frac{n}{n-1}} \le \frac{1}{2\sqrt{n}}|\omega_x|_1 , \quad \left(|\omega_x|_1 = \int |\text{grad } \omega| \, dx\right) .$$

which holds for any differentiable <u>function</u> ω with compact support in E^n (cf. Nirenberg, Annali Scuola Norm. Sup. Pisa (III), 13 (1959), p. 14). If we set $\omega = (f \cdot f)^{2(n-1)/n}$ in this inequality, and apply Schwarz's inequality, there results

$$|f|_4^2 \leq \frac{1}{\sqrt{2}} |f| \cdot \|f\| \qquad n = 2 ,$$

and

$$|f|_4 \leq \frac{3}{4} \|f\| \qquad n = 4 .$$

When $n = 3$ the situation is a little more involved. We have in this case

$$|f|_4^4 = \int |f(x_1)|_4^4 dx_1 ,$$

where $|f(x_1)|_4$ is the L^4 norm of f in the variables x_2 and x_3. Now by what has already been shown

$$|f(x_1)|_4^4 \leq \frac{1}{2} |f(x_1)|^2 \cdot \|f(x_1)\|^2 ,$$

and using the fact that f has compact support,

$$|f(x_1)|^2 \leq |f| \cdot \left| \frac{\partial f}{\partial x_1} \right| .$$

Substituting the last two estimates into the equality for $|f|_4^4$, and performing the x_1 integration, then leads to

$$|f|_4^4 \leq \frac{1}{2} |f| \cdot \left| \frac{\partial f}{\partial x_1} \right| \cdot \left\{ \left| \frac{\partial f}{\partial x_2} \right|^2 + \left| \frac{\partial f}{\partial x_3} \right|^2 \right\} .$$

With obvious meanings for the letters J, A, B, C, this may be written

$$J \leq \frac{1}{2} |f| \cdot A^{1/2} \{B + C\} ,$$

whence by permutation there arises

$$J \leq \frac{1}{2} |f| \cdot (ABC)^{1/6} \{B+C\}^{1/3} \{A+C\}^{1/3} \{A+B\}^{1/3}$$

$$\leq \frac{1}{3\sqrt{3}} |f| \cdot (A+B+C)^{3/2} = \frac{1}{3\sqrt{3}} |f| \cdot \|f\|^3 ,$$

completing the proof of (12).

§3. The energy equality

We shall be concerned in this section with certain rather immediate results concerning the continuity in time of a weak solution of the initial value problem, including the important energy equality. Although some of the results are new (especially when $n \geq 4$), the essential ideas are due to Prodi.

THEOREM 4. Suppose that $u \in V$ is a weak solution of the initial value problem. Then after suitable redefinition of u at a set of values of t of measure zero, we have

$$\int_0^t \{(u, \varphi_t) + (u, \Delta\varphi) + (u, u \cdot \operatorname{grad} \varphi)\} dt = (u, \varphi) - (u_0, \varphi_0) \tag{19}$$

for all $\varphi \in D$ and all $t \geq 0$.

Proof. Let $\theta(t)$ be a smooth function whose graph has the form shown in Figure 1.

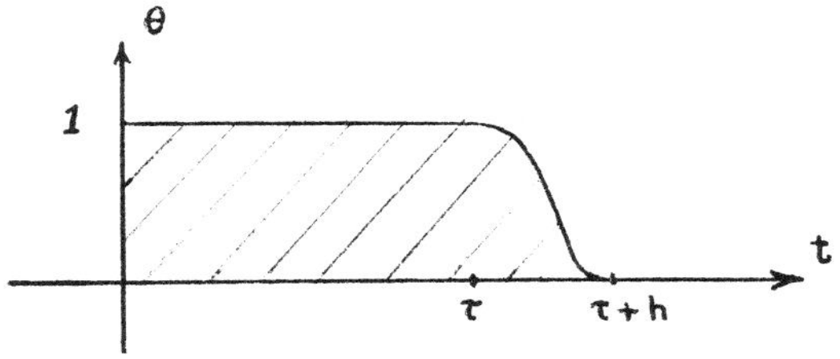

Figure 1

Setting $\Phi(x, t) = \theta(t)\varphi(x, t)$ into the weak form of the Navier-Stokes equations (Definition 3), there results

$$\int_0^\infty \{(u, \varphi_t) + (u, \Delta\varphi) + (u, u \cdot \operatorname{grad} \varphi)\}\theta \, dt = -\int_0^\infty (u, \varphi)\theta_t dt - (u_0, \varphi_0). \tag{20}$$

As $h \to 0$ the integral on the left tends to

$$\int_0^\tau \{(u, \varphi_t) + (u, \Delta\varphi) + (u, u \cdot \operatorname{grad} \varphi)\} dt .$$

To investigate the behavior of the integral on the right, observe that θ_t vanishes outside the interval $(\tau, \tau+h)$, and satisfies $\int \theta_t dt = -1$. Since we may assume further that $\text{Const.}/h \leq \theta_t \leq 0$, it is then a standard procedure to show that this integral tends to $(u(\tau), \varphi(\tau))$

for all τ belonging to the Lebesgue set $\mathscr{L}$ of $u(t)$. We have thus shown that (19) holds for almost all values of t .

Now consider an arbitrary instant of time $\tau \geq 0$. Since $|u(t)|$ is uniformly bounded, it follows from the weak compactness of L^2 that there exists a vector $U(x, \tau)$ and a sequence of values t_j in $\mathscr{L}$, $t_j \to \tau$, such that $u(x, t_j) \to U(x, \tau)$ weakly in L^2 . Since (19) holds for $t \in \mathscr{L}$, it is an easy consequence that

$$\int_0^\tau \{(u, \varphi_t) + (u, \Delta\varphi) + (u, u \cdot \operatorname{grad}\varphi)\}dt = (U(\tau), \varphi(\tau)) - (u_0, \varphi_0) \tag{21}$$

for all φ in D and all $\tau \geq 0$. Now if τ is itself in $\mathscr{L}$ it is not difficult to see that $U(x, \tau)$ must be identical to $u(x, \tau)$, as a function in L^2 , (if $\tau = 0$, then $U(x, 0) = u_2(x)$). Hence the function U is a redefinition of u at most at a set of values of t of measure zero. Replacing U by u and τ by t in (21) completes the proof of Theorem 4.

We remark that the redefined function $u = u(x, t)$ is weakly continuous in L^2 as a function of time. Indeed a necessary and sufficient condition for weak continuity of $u \in V$ is evidently that

$$(u(t), \psi) \to (u(t_0), \psi) \quad \text{as} \quad t \to t_0 ,$$

for all $\psi \in D_0$ and all $t_0 \geq 0$. Since this condition is easily seen to hold by virtue of (19), the assertion is proved.

In what follows we shall regularly assume that all weak solutions under discussion are redefined according to the result of the preceding theorem.

For subsequent work certain spaces $L^{r,r'}$ play a fundamental role. If $g = g(x, t)$ is defined and measurable in a cylindrical domain $S = G \times (T_1, T_2)$ of space time, we set

$$|g(t)|_r = \left(\int |g(x, t)|^r dx\right)^{1/r}$$

and

$$|g|_{r,r'} = \left(\int |g(t)|_r^{r'} dt\right)^{1/4'} ,$$

the integrations being over G and (T_1, T_2) , respectively. Here r and r' are considered to be independent exponents with $1 \leq r, r' \leq \infty$. We note that the norm $|u(t)|_2$ may be alternately denoted simply by $|u|$, as in the earlier sections of the paper.

DEFINITION 4. <u>We say that $g = g(x, t)$ is contained in the class $L^{r,r'}(S)$ if g is defined and measurable in S , and $|g|_{r,r'} < \infty$</u>

The following lemma will play a fundamental, though often implicit, role in the proofs of Theorems 5 and 6 below.

LEMMA 1. <u>Suppose that exponents s and s' satisfy the relation</u>

$$\frac{n}{s} + \frac{2}{s'} = 1 , \qquad (n \le s \le \infty) . \tag{22}$$

<u>Let v and w be contained in the space V , and suppose also that $u \in L^{s,s'}$ on the cylinder $\Omega \times (0, T)$. Then we have</u>

$$\int_0^T (u, v\cdot \operatorname{grad} w)\, dt \le \text{Const.}\, |w_x|_{2,2} |v_x|_{2,2}^{n/s} \{\int |u|_s^{s'} |v|^2 dt\}^{1/s'} ,$$

<u>with the exception of the single case s = n = 2 .</u>

<u>Proof.</u> By Hölder's inequality, if $1/r + 1/s = 1/2$ we have

$$(u, v \cdot \operatorname{grad} w) \le |u|_s |v|_r |w_x|_2 .$$

Also, by Sobolev's theorem

$$|v|_r \le \text{Const.}\, |v|^{2/s'} |v_x|^{n/s} ,$$

whence integrating with respect to t and applying Hölder's inequality once more, we obtain the required conclusion.

THEOREM 5 (Energy equality). <u>Let $u \in V$ be a weak solution of the initial value problem, such that $u \in L^{s,s'}$ on the cylinder $\Omega \times (0, T)$ for a pair of exponents s, s' satisfying (22). Then</u>

$$|u|^2 + 2\int_0^t \|u\|^2 dt = |u_0|^2 , \quad 0 \le t \le T . \tag{23}$$

<u>Remarks</u>. For the cases n = 2, 3 this result is essentially due to Prodi. Our proof is based on a technique introduced by Lions and Prodi in a slightly different context.

We observe that in case n = 2 the assumption $u \in L^{s,s'}$ is redundant. In fact, by Sobolev's inequality any vector $u \in V$ is automatically in $L^{s,s'}$ for <u>every</u> pair of exponents satisfying (22), except $(\infty, 2)$. A similar result fails for $n \ge 3$.

<u>Proof of Theorem 5</u>. The result is obviously true for t = 0 , since we recall that u(x, t) has been redefined to equal $u_0(x)$ when t = 0 . Now let τ be fixed, $0 < \tau \le T$. For an <u>even</u> averaging kernel K(t) with compact support in $|t| \le h$, we set

$$\varphi(x,t) = \int_0^T K(t-t')\, u^k(s,t')\, dt' \ ,$$

where $\{u^k\}$ is a sequence of test functions converging to u in the norm (7). The vector φ obviously may be substituted into relation (19); if we then let $k \to \infty$ there results

$$\int_0^T \{(u, u_{ht}) - ((u, u_h)) + (u, u\cdot \operatorname{grad} u_h)\}\, dt = (u, u_h) - (u_o, u_{ho}) \ , \qquad (24)$$

where

$$u_h = u_h(x,t) = \int_0^T K(t-t')\, u(x,t')\, dt' \ .$$

(The limiting process involved here is a typical one in what follows; it may be justified by standard arguments.) Now

$$\int_0^T (u, u_{ht})\, dt = \int_0^T\!\!\int_0^T K_t(t-t')\,(u(t), u(t'))\, dt'\, dt = 0$$

by the symmetry of K. We may let $h \to 0$ in the remaining terms of (24). By well known properties of the averaging process, one obtains

$$\int ((u, u_h))\, dt \to \int \|u\|^2\, dt, \quad \int (u, u\cdot \operatorname{grad} u_h)\, dt \to \int (u, u\cdot \operatorname{grad} u)\, dt \ .$$

The last integral vanishes since u is weakly divergence free (this is not completely obvious, though we shall omit the demonstration). Finally we have

$$\begin{aligned}
(u(\tau), u_h(\tau)) &= (u(\tau), \int_0^T K(\eta)\, u(\tau-\eta)\, d\eta) \\
&= \int_0^h K(\eta)\,(u(\tau), u(\tau-\eta))\, d\eta \qquad\qquad (h < \tau) \\
&= \int_0^h K(\eta)\,\{|u(\tau)|^2 + \varepsilon(\eta)\}\, d\eta \ ,
\end{aligned}$$

the last step in consequence of the weak continuity of $u(t)$. The limit as $h \to 0$ is therefore $\frac{1}{2}|u(\tau)|^2$, and similarly $\lim(u_o, u_{ho}) = \frac{1}{2}|u_0|^2$. Substituting the appropriate limits into (24), we obtain (23), and Theorem 5 is demonstrated.

<u>Concluding remarks</u>. If $n = 2$ the assumption $u \in L^{s,s'}$ is redundant, as has already been remarked. It follows that <u>all</u> weak solutions in $n = 2$ necessarily obey the energy equality, a fact first discovered by Prodi [30].

If $n \geq 3$, it is not known whether every weak solution $u \in V$ obeys an energy equality. The solution of Kiselev and Ladyzhenskaya, however, satisfies

$$|u|_4 \leq C|u|^{1/4}|u_x|^{3/4} \leq \text{Const.}, \qquad 0 \leq t \leq T ,$$

so that $u \in L^{4,\infty}$ on the interval $(0, T)$. This solution therefore satisfies the hypotheses of Theorem 5, and consequently does have an energy equality. A similar result is valid for the solution constructed in Theorem 3.

It may be added that any solution $u \in L^{s,s'}$, (and in particular the solution of Kiselev and Ladyzhenskaya), is necessarily strongly continuous in time, that is

$$|u(t) - u(t_0)| \to 0$$

as $t \to t_0$. This follows at once from the weak continuity of $u(t)$ together with the continuity of the norm $|u(t)|$. We shall see later that the solutions of Kiselev and Ladyzhenskaya are also Lipschitz continuous in t for each fixed x.

4. Uniqueness and stability

This section contains a general uniqueness theorem for the initial value problem, and some closely related stability theorems. Roughly speaking, the uniqueness theorem states that if a weak solution has a certain mild degree of regularity, then it is unique among all solutions of class V satisfying (8). This result essentially includes uniqueness theorems of Leray [20, Section 32], Prodi [29], and Lions and Prodi [25], though it should be mentioned that the method of proof derives something from all these papers.

THEOREM 6 (Unicity). Suppose $n \leq 4$. Let u and v be weak solutions of the initial value problem, in the sense of Definition 3. Suppose also that

$$|v|^2 + 2\int_0^t \|v\| \, dt \leq |v_0|^2 , \quad 0 \leq t \leq T ,$$

and that $u \in L^{s,s'}$ for a pair of exponents satisfying (22) and also $n < s < \infty$. Then

$$|u-v| \leq |u_0 - v_0| \exp\{\text{Const.} \int_0^t |u|_s^{s'} dt\}$$

in the interval $0 \leq t \leq T$. In particular, if $u_0 = v_0$ then $u \equiv v$.

Proof. (The following demonstration was obtained jointly by myself and Jerome Sather of the University of Minnesota. It may be extended to cover the case of moving boundaries and non-zero boundary values, and, in this generality, an expanded version will be published elsewhere.) By a construction similar to that of Theorem 5, but with some differences in detail, one can show that

$$\int_0^T \{(u, v_{ht}) - ((u, v_h)) + (u, u\cdot \operatorname{grad} v_h)\}dt = (u, v_h) - (u_o, v_{ho}) \tag{25}$$

and

$$\int_0^T \{(v, u_{ht}) - ((v, u_h)) - (u_h, v\cdot \operatorname{grad} v)\}dt = (v, u_h) - (v_o, u_{ho}) \ . \tag{26}$$

(The third term in the integral (26) causes the major difficulty, and will be discussed at some length in the paper referred to above.) Now by virtue of Fubini's theorem and the symmetry of the kernel K it is easy to see that

$$\int_0^T (u, v_{ht})\, dt = -\int_0^T (u_{ht}, v)\, dt \ .$$

Consequently, addition of (25) and (26) yields

$$-\int_0^T \{((u, v_h)) + ((v, u_h)) + (u_h, v\cdot \operatorname{grad} v) - (u, u\cdot \operatorname{grad} v_h)\}dt$$

$$= (u, v_h) + (v, u_h) - (u_o, v_{ho}) - (v_o, u_{ho}) \ .$$

In this identity we may let $h \to 0$. Recalling the calculations of the preceding proof, one sees that the result is

$$-\int_0^T \{2((u, v)) + (u, w\cdot \operatorname{grad} v)\}dt = (u, v) - (u_0, v_0) \ ,$$

where $w = v-u$. Now by hypothesis

$$|v|^2 + 2\int_0^T \|v\|^2 dt \le |v_0|^2 \ ,$$

and by virtue of Theorem 5,

$$|u|^2 + 2\int_0^T \|u\|^2 dt = |u_0|^2 \ .$$

Adding the last three displayed equations (the first being multiplied by 2) then gives

$$|w|^2 + 2\int_0^T \|w\|^2 dt \leq |w_0|^2 + 2\int_0^T (u, w \cdot \operatorname{grad} v)\, dt ,$$
$$= |w_0|^2 + 2\int_0^T (u, w \cdot \operatorname{grad} w)\, dt . \tag{27}$$

So much holds for $n \leq s < \infty$. But by Lemma 1, if $s > n$,

$$\int (u, w \cdot \operatorname{grad} w)\, dt \leq \text{Const.}\ |w_x|_{2,2}^{1+n/s} \left\{ \int |u|_s^{s'} |w|^2 dt \right\}^{1/s'}$$
$$\leq |w_x|_{2,2}^2 + \text{Const.} \int |u|_s^{s'} |w|^2 dt . \tag{28}$$

Consequently

$$|w|^2 \leq |w_0|^2 + \text{Const.} \int_0^t |u|_s^{s'} |w|^2 dt .$$

This differential inequality is easily integrated (set the right hand side equal to a new function z), and one obtains

$$|w| \leq |w_0| \exp \{\text{Const.} \int_0^t |u|_s^{s'} dt\} ,$$

completing the proof.

A similar result is valid when $s = n$, but requires the additional hypothesis that $|u|_{n,\infty}$ be suitably small. This hypothesis is satisfied by the solution of Theorem 3.

Remarks. Let us consider separately the cases $n = 2, 3$, and 4. First, if $n = 2$ the extra hypotheses on u and v are redundant, as follows from Theorem 5 and the accompanying remarks. In other words, for the case of plane flow two solutions u and v in the class V are identifical if and only if $u_0 = v_0$. This result was first proved by Lions and Prodi.

If $n \geq 3$, let us denote by V' the subset of vectors in V which satisfy the energy inequality

$$|u|^2 + 2\int_0^t \|u\|^2 dt \leq |u_0|^2 .$$

Evidently the solutions constructed in Section 2 are all in V' . Theorem 6 may then be rephrased as follows: Suppose that u and v are weak solutions of the initial value problem for the same initial data u_0 , and let both u and v be in the class V' . Suppose also that $u \in L^{s,s'}$ for a pair of exponents satisfying (22) and

$n < s < \infty$. Then $v \equiv u$.

The solution of Kiselev and Ladyzhenskaya ($n = 3$) is in the class $L^{4,\infty}$, as has already been noted. Consequently, for initial data in W^2 it is the only solution in V', the solution constructed by Eberhard Hopf necessarily being the same.

For $n = 4$ a uniqueness theorem holds for the solution constructed in Theorem 3; if that solution exists for given initial data, then it is the only solution in V' .

There is one feature of Theorem 6 which is not quite desirable, namely the restriction to $n \leq 4$. This arises in one of the limiting processes in the proof, and can be removed in any of the following cases: 1) $\Omega \equiv E^n$, 2) Ω is starlike, and 3) the test functions u^k defining u have uniformly bounded norms $|u^k|_{s,s'}$.

There is, finally, a remarkable uniqueness theorem due to Foias for the case $\Omega \equiv E^n$, which assumes no differentiability of the weak solutions in question. On the other hand, a vector $u \in V'$ does not necessarily satisfy the hypotheses of Foias' theorem, so that even in the case $\Omega \equiv E^n$ the result of Foias does not include Theorem 6, nor vice versa.

Stability. The remaining considerations of this section will be directed to the case where Ω is a bounded region, say contained within a sphere of diameter d . It is well known (Poincaré's inequality) that there exists a positive number α, depending only on n, such that

$$(\alpha/d)\,|w| \leq \|w\| \tag{29}$$

for all strongly differentiable vectors w in Ω with generalized boundary values zero.

If we now modify inequality (28) by placing a factor 1/2 in front of the first term on the right, (this can certainly be done simply by increasing the remaining constant), then from (27) and (28) follows

$$\begin{aligned} |w|^2 &\leq |w_0|^2 + \int_0^t \{\text{Const.}\,|u|_s^{s'}\,|w|^2 - \|w\|^2\}dt \\ &\leq |w_0|^2 + \int_0^t \{\text{Const.}\,|u|_s^{s'} - (\alpha/d)^2\}\,|w|^2\,dt\,. \end{aligned}$$

Solving this differential inequality yields

$$|w|^2 \leq |w_0|^2 \exp\{\text{Const.}\int_0^t |u|_s^{s'}\,dt - (\alpha/d)^2 t\}\,.$$

This proof applies, of course, only to the case when both solutions u and v have generalized boundary data zero. However, as we have indicated above, the same methods are valid for the case of non-zero boundary data and moving boundaries. We may assume, therefore, that (30) holds also in this more general case. Perhaps the most important consequence of (30) is the fact that solutions of the initial value problem depend continuously (in the L^2 norm) on the initial data, this condition being one of the requirements of a properly posed problem. Also, from (30) one may derive a number of stability criteria.

I. If for some s, $n < s < \infty$, the basic flow u satisfies

$$|u|_{s,\infty} \leq \kappa < \left(\frac{\alpha}{d\sqrt{\text{Const.}}}\right)^{(s-n)/s}, \qquad 0 < t < T,$$

then the perturbation $w = v - u$ corresponding to any solution $v \in V'$ of the initial value problem, having the same boundary values but possibly different initial data, satisfies

$$|w| \leq |w_0| e^{-\gamma t} \qquad 0 < t < T$$

for some $\gamma > 0$. The basic solution u is thus exponentially stable.

II. Suppose next the special assumption $u \in L^{\infty,\infty}$ that is, the flow speed itself is uniformly bounded, say $\leq U$. In this case inequality (27) certainly holds, and implies the estimates

$$|w|^2 \leq |w_0|^2 + \int_0^t \{U^2 - (\alpha/d)^2\} |w|^2 \, dt$$

and

$$|w|^2 \leq |w_0|^2 \exp\{U^2 - (\alpha/d)^2\}t .$$

This leads to the stability criterion

$$Ud < \alpha, \qquad (31)$$

derived in [32] under more restrictive assumptions on the flows than we have assumed here. It should be observed that the quantity Ud (or, if a non-unit kinematic viscosity ν is present, the quantity Ud/ν) is a Reynolds number for the motion u.

III. The concrete nature of the preceding stability conditions makes it desirable to have a close estimate for the value of the constant α in (29). In [32] it was shown (using the fact that $\operatorname{div} w = 0$) that we can take $\alpha = 5.71$ in case $n = 3$. Weinberger and Payne, making use of eigenvalue theory, have recently proved that the best value for α is 8.98.

For domains which can be enclosed in a cube of edge length d, Velte has found that the best value of α is between 7.70 and 7.91. Finally, for certain particular flows these criteria can be greatly improved by using special techniques (cf. reference [36], and §§4-6 of reference [32]).

5. Other uniqueness theorems

Theorem 6 of the preceding section refers to the case of zero boundary data (or, in the case of moving boundaries, to prescribed velocity at the boundary). If the fluid fills the exterior of a bounded region, however, the condition that the motion have prescribed data at infinity is perhaps not an entirely natural assumption. (The situation may be compared with the initial value problem for the heat equation in an exterior domain, where it is well-known that a sufficient condition for uniqueness is boundedness, or even exponential growth, at infinity). Investigating this situation for the case of classical solutions of the Navier-Stokes equations, Graffi [5] found that the zero condition at infinity could be replaced by the following assumptions:

i) u and u_x are uniformly bounded in x, t;

ii) as $x \to \infty$ the pressure p tends to a limit p_0, with $|p-p_0| \leq \text{Const.}\ |x|^{-1}$.

This remarkable result has been further extended by Graffi to include compressible fluid motions [40], [41]. There is not space to give the details here, though we may remark that a corresponding existence theorem would be highly desirable.

Another problem of interest is the question whether a non-constant solution of the initial value problem (1)-(4) can ever come to rest in a finite time. Although a definitive result is not known, at least for a classical solution in a bounded domain the answer is no. This fact is implicit in the following somewhat more general theorem.

THEOREM 7. Suppose that u and v are (classical) solutions of the Navier-Stokes equations in a bounded region Ω, and let $u = v$ on the boundary of Ω. If $u(x, t_0) = v(x, t_0)$ at some instant t_0, then u and v are identical for all t.

Proof: The method of proof is a simple modification of one established by Lees and Protter [17] for the study of a single parabolic equation of second order. Let us set

$$L\varphi \equiv \Delta\varphi - \varphi_t \quad \text{and} \quad \lambda(t) = t - \tau$$

where τ is some fixed constant, $0 < \tau < 1$. Then, essentially by the argument of [17], pp. 371-372, we have the following result.

LEMMA 2. Let $\varphi = \varphi(x, t)$ be a divergence free vector of class C^2, vanishing on the complete boundary of the space time domain $\Omega \times (-\tau, 0)$. Then for any function $\tilde{p}$ of class C^1 and any integer $m > 0$ we have

$$\int \lambda^{-2m}\{|\varphi|^2 + \|\varphi\|^2\}\, dt \le C \int \lambda^{-2m} |L\varphi - \operatorname{grad} \tilde{p}|^2 dt \ ,$$

where $C = 5(\tau + m^{-1})$. In this inequality it is tacitly understood that the space integration is taken over Ω, and the time integration over the interval $(-\tau, 0)$.

This much being shown, it is now easy to complete the proof. Since uniqueness in the forward time direction certainly holds, we have $u \equiv v$ for all $t \ge t_0$. Therefore, if the theorem were not true there would be an instant of time, say $t = 0$, such that $u \equiv v$ for $t \ge 0$, but $u \not\equiv v$ in any interval $-\varepsilon < t < 0$. Let the magnitudes of u and of grad v be bounded by M in the interval $-1 < t < 0$, so that in this interval

$$|Lw - \operatorname{grad} \tilde{p}| \le 2M^2\{|w|^2 + \|w\|^2\} \ ,$$

where $w = v-u$ and $\tilde{p}$ denotes the difference in pressures for the flows v and u. Now choose τ and m_0 to satisfy the conditions

$$0 < \tau < 1 \quad \text{and} \quad 4CM^2 < 1 \quad \text{for all} \quad m \ge m_0 \ .$$

Finally, let $\theta(t)$ be defined as indicated by the graph in Figure 2.

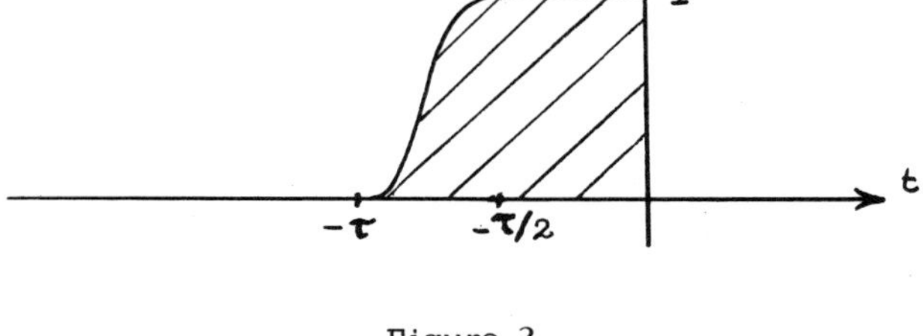

Figure 2

We may set $\varphi = \theta w$ into Lemma 2, with the result

$$\int \lambda^{-2m} \{|\varphi|^2 + \|\varphi\|^2\} dt$$

$$\leq C \int \lambda^{-2m} |L\varphi - \text{grad}\, \tilde{p}|^2 \, dt$$

$$= C \int_{-\tau}^{-\tau/2} \lambda^{-2m} |L\varphi - \text{grad}\, \tilde{p}|^2 \, dt + C \int_{-\tau/2}^{0} \lambda^{-2m} |Lw - \text{grad}\, \tilde{p}|^2 \, dt$$

$$\leq (2/3\tau)^{2m} CD + 2CM^2 \int_{-\tau/2}^{0} \lambda^{-2m} \{|w|^2 + \|w\|^2\} \, dt \ ,$$

where D is some appropriate constant, and $m \geq m_0$. Transferring the final term to the left hand side, and then restricting the integration to the interval $(-\tau/4, 0)$, yields after some simple manipulations

$$\int_{-\tau/4}^{0} |w|^2 dt \leq 2CD(5/6)^{2m} \ , \qquad m \geq m_0 \ .$$

Now if $m \to \infty$ we obtain $w \equiv 0$ for $-\tau/4 < t < 0$, contradicting the fact that w is non-constant on any interval containing the origin. This completes the proof.

6. Regularity of weak solutions

In all treatments of partial differential equations by the methods of functional analysis, it is necessary at some stage to examine the regularity properties of the weak solutions which are obtained. In [34], the author has given a direct and relatively simple proof of the interior regularity of weak solutions of the Navier-Stokes equations, assuming a certain moderate degree of regularity to begin with. We shall review this paper below, but first we must mention the work of various other authors who have also treated the regularity problem, considering it an integral part of their basic existence theorems.

Leray in the important series of papers [18], [19], [20], studied the initial value problem for dimensions $n = 2$ (Ω = bounded convex region, and $\Omega \equiv E^2$) , and $n = 3$ ($\Omega \equiv E^3$) . The existence of smooth solutions of the initial value problem was there established for finite time intervals (if the data was sufficiently small, also for all time) using the fundamental solution of the equations of slow motion, together with the technique of successive approximations. This work of Leray has recently been extended by Golovkin ($n = 2$) and Ito ($n = 3$) , making use of the steady state Green's tensor of F. K. G.

Odquist. Also, P. E. Sobolevskii has announced the construction of a classical solution for a bounded domain in E^3, valid for some finite time interval. Further details will be found at the conclusion of this section, where we discuss the relation of this work to that of reference [34].

Now let R be an open region of space time. We shall say that a vector u is in $\mathscr{L}^{r,r'}$ over R if u is defined and measurable in R and if $u \in L^{r,r'}(S)$ for every space-time cylinder S with compact support in R. We consider weak solutions u of the Navier-Stokes equations in R, in the sense of Definition 1. The condition that u be locally square summable in R is alternately expressed, in the notation just introduced, by the statement $u \in \mathscr{L}^{2,2}(R)$. Now suppose that u is strongly differentiable with respect to x, and $u_x \in \mathscr{L}^{2,2}(R)$. Then in any open cylinder $S = G \times (T_1, T_2)$ with compact support in R it can be shown that

$$\omega(x,t) = \int\int k(x-\xi, t-\tau)\, g(\xi,\tau)\, d\xi d\tau + B(x,t) \quad . \tag{32}$$

Here $\omega = \operatorname{curl} u$ denotes the vorticity of the fluid motion,[3] g is the dyadic tensor $\omega u - u\omega$, B is a solution of the heat equation in S, and $k = \operatorname{grad} K$, where

$$K = K(x,t) = \begin{cases} \text{Const. } t^{-n/2} e^{-|x|^2/4t} & t > 0 \\ 0 & t \le 0 \end{cases}$$

The product kg in the integrand of (32) is to be understood as the tensor sum $k_i g_{ij}$. The proof of (32) is most simply carried out by the process of integral averaging; a <u>formal</u> proof is evident on the basis of the vorticity equation

$$\omega_t - \Delta\omega = \operatorname{div}(\omega u - u\omega) \qquad (n = 3) \;.$$

Making use of the representation formula (32) one can now prove the following interior regularity theorem, (in this result it is worth pointing out the similarity between (33) and the exponent relation (22) which occurred repeatedly in Sections 3 and 4),

THEOREM 8. <u>Let u be a weak solution of the Navier-Stokes equations in an open region R of space time, with $u \in \mathscr{L}^{2,\infty}(R)$, $u_x \in \mathscr{L}^{2,2}(R)$. Suppose further that</u>

$$u \in \mathscr{L}^{s,s'}(R)$$

<u>for a pair of exponents s and s' satisfying the condition</u>

$$\frac{n}{s} + \frac{2}{s'} < 1 \ . \tag{33}$$

Then u is of class C^∞ in the space variables x, and each (spatial) derivative is bounded in compact subregions of R.

Assume in addition that u is strongly differentiable with respect to time. Then u and each of its space derivatives is an absolutely continuous function of time. Moreover, there exists a function $p = p(x, t)$ of class C^∞ in x such that (1) holds almost everywhere in R.

In essence, the proof depends on the following recursion (or bootstrap) argument. Suppose we know that

$$u \in \mathscr{L}^{s, s'}, \qquad \omega \in \mathscr{L}^{\rho, \rho'}$$

where ρ, ρ' are independent exponents ≥ 2. Then $g \in \mathscr{L}^{q, q'}$ for

$$\frac{1}{q} = \frac{1}{s} + \frac{1}{\rho}, \qquad \frac{1}{q'} = \frac{1}{s'} + \frac{1}{\rho'} \ .$$

Hence by virtue of the representation formula (32), and a lemma on the Lebesgue class of convolution integrals, one finds $\omega \in \mathscr{L}^{r, r'}$ where

$$r = \frac{\rho}{1 - \kappa\rho}, \qquad r' = \frac{\rho'}{1 - \kappa\rho'}$$

and

$$\kappa = \left(1 - \frac{n}{s} - \frac{2}{s'}\right)\Big/(n+3) > 0 \ .$$

(put $r = \infty$ if $\kappa\rho \geq 1$, and similarly for r'). But r and r' are larger than ρ, ρ', so that ω is actually in a higher Lebesgue class than originally supposed. This process of improving the class of ω can clearly be repeated as often as one likes, beginning with $\rho = \rho' = 2$. Obviously, after a finite number of steps one will obtain $\omega \in \mathscr{L}^{\infty, \infty}$. This much being shown, a further and more straightforward recursion procedure establishes that u and ω are arbitrarily often differentiable with respect to the space variables. The details of this process, and the verification of the final portion of the theorem, will be omitted (cf. [34]).

For solutions of the initial value problem in $\Omega \equiv E^n$, the conclusion of Theorem 8 can be considerably strengthened without significantly altering the proof. In fact, we have the result,

THEOREM 8'. Let $u \in V'$ be a weak solution of the initial value problem, where $\Omega \equiv E^n$. Suppose further that $u \in L^{s,s'}$ for a pair of exponents satisfying (33). Then u is of class C^∞ in all variables, and is a classical solution of equations (1) and (2).

The improvement of Theorem 8' over Theorem 8 is essentially due to the fact that in the latter case one can give an explicit formula for the function $B(x,t)$. If we have to deal with a bounded region Ω, however, there is apparently no way to obtain an analogous formula for B. Thus, in order to generalize Theorem 8' to this case it seems necessary to use the notion of a Green's function or Green's tensor. This line of attack, if it is successful, should also provide the result that the solution takes on boundary values in the classical sense.

Finally, let us consider Theorems 8 and 8' in connection with the existence theorems of section 2. In both cases $n = 2$ and $n = 3$ the solution of Kiselev and Ladyzhenskaya satisfies the hypotheses of these theorems. It follows that this solution is C^∞ in the space variables, and (by an obvious extension of the final statement of Theorem 8) Lipschitz continuous in time.[4] Moreover, if $\Omega \equiv E^n$ these solutions are C^∞ in all variables.

The weak solutions of Theorem 1 do not possess sufficient regularity properties to apply the above theorems. We are therefore unable to draw any general conclusions concerning the smoothness of these solutions. Nevertheless, it is possible to make certain remarks. First, Hopf's solution obeys the representation formula (32) in a space of any number of dimensions. Second, if $n = 2$ or $n = 3$ and the initial data are smooth enough for the Kiselev-Ladyzhenskaya solution to exist, then according to the remark following Theorem 6, Hopf's solution must be the same and consequently must be of class C^∞ in the space variables.

If the boundary of Ω is smooth, then stronger conclusions are available. In particular, Golovkin and Ito in the papers referred to at the beginning of this section have proved for $n = 2$ and $n = 3$, respectively, the existence of classical solutions continuously taking on the given boundary data (the solution constructed by Ito exists only for a suitable interval of time). By the uniqueness theorem of Section 4 it is apparent that every solution $u \in V'$ with the same initial data will have the same smoothness properties. This fact can, of course, be regarded as a regularity theorem. It would be worthwhile to have a direct proof along the lines indicated in the remarks following Theorem 8'.

7. Open problems

The most important problem remaining unsolved is without doubt that of constructing a solution of the initial value problem in three dimensions which exists and is regular for all $t \geq 0$, no matter what (smooth) initial data may be assigned. As we have seen above, the corresponding problem has been more or less completely solved in the plane: for data in W^2 the solution of Kiselev-Ladyzhenskaya exists for all time, is unique in the class V of weak solutions, and by the results of the preceding section is C^∞ in x and Lipschitz continuous in t . The entire picture is less clear in three dimensions. If the data is not suitably small the solution of Kiselev-Ladyzhenskaya may only exist for a finite time interval $(0, T)$, though certainly a weak solution in the sense of Theorem 1 does persist for all $t \geq T$. There is an important result of Leray which adds a fascinating element of complication to the situation: If $\Omega \equiv E^3$ the weak solution which continues to exist is actually regular except for a set of values of t of measure zero, whose complement is a set of intervals; and the final interval is necessarily of the form $T^* < t < \infty$! In light of this result, one can say in summary that a solution of the initial value problem in E^3 , for smooth initial data, will be regular for a finite time interval, will then possibly pass through a set of epochs of irregularity where branching (non-uniqueness) occurs, and will eventually settle down again to a regular solution which tends uniformly to zero on compact sets.

Whether the behavior indicated above genuinely occurs, or whether it is only apparent and due to an incomplete theory, remains to be decided. The answer will involve either a new existence theorem in case there are no epochs of irregularity, or else an example to show that breakdown actually occurs. Leray proposed the problem of finding such an example (cf. p. 225 of [20]), and in fact gave a possible method for its construction. In spite of this, I am of the opinion that the more likely alternative is the existence of a regular solution for all time.

The reader will have observed in the course of the paper a number of other places where the theory could be improved. We may mention specifically the possibility of results analogous to Theorems 2 and 3 for dimension $n \geq 5$, an extension of Theorem 7 to weak solutions (or at least to solutions with generalized boundary data), and a proof that classical solutions of the Navier-Stokes equations are analytic in the space variables.

A new and wide field opens if one considers regions with moving boundaries, and non-vanishing boundary data. The existence of periodic solutions has been treated by Prodi [30], Yudovich [38] and Serrin [33], though these results do not yet provide a complete theory. Jerome Sather has studied the extension of Hopf's construction to the case of moving boundaries, and he and I together have proved a uniqueness theorem analogous to Theorem 5 for weak solutions of this problem.

Some significant long range contributions are due to Hopf. In an important series of papers [7], [8], [9], [10], he has raised a number of questions concerning the nature of solutions as the viscosity tends to zero. These questions are also intimately related to the asymptotic behaviorof flows as $t \to \infty$ and to the general problem of stability, instability, and turbulence (cf. the opening remarks in [13]). Much work remains before this situation can be clarified.

We may observe finally that the field of classical hydrodynamics blends into more general doctrines, in particular compressible fluid flow and the recently fashionable subject of magnetohydrodynamics. Some results concerning the initial value problem are known for these fields, and following the main references there is a short list of papers of this type. Needless to say, the problems become more complicated as the physical situation does, and, just as for incompressible fluids, there are a multitude of avenues open to the interested mathematician.

NOTES

1. More correctly we should refer to the <u>initial-boundary value problem</u>, but for the sake of brevity we shall use the shorter terminology of the text.
2. We can suppose u to be set identically zero outside Ω, in which case the inner product (u, v) may be taken over either Ω or E^n, the choice being immaterial. Moreover, u is then strongly differentiable and divergence free in the entire half space $E^n \times [0, \infty)$.
3. The definitions of ω and g are here given only for the case $n = 3$. The modifications necessary for general n are explained in [34].
4. A similar result, at least so far as the spatial regularity of u is concerned, was obtained by Ohyama.

REFERENCES

1. C. Foias, Une remarque sur l'unicité des solutions des equations de Navier-Stokes en dimension n, Bull. Soc. Math. France <u>89</u> (1961), pp. 1-8.
2. K. K. Golovkin, The plane motion of a viscous incompressible fluid, Trudy Mat. Inst. Steklov <u>59</u> (1960), pp. 37-86.
3. K. K. Golovkin and O. A. Ladyzhenskaya, On the solution of the non-steady boundary problem for the Navier-Stokes equation, Trudy Mat. Inst. Steklov <u>59</u> (1960), pp. 100-114.

4. K. K. Golovkin and V. A. Solonnikov, On the first boundary value problem for the non-stationary Navier-Stokes equations, Dokl. Akad. Nauk SSSR 140 (1961), pp. 287-290.

5. D. Graffi, Sul teorema di unicità nella dinamica dei fluidi, Annali di Mat. 50 (1960), pp. 379-388.

6. E. Hopf, Über die Anfangswertaufgabe für die hydrodynamischen Grundgleichungen, Math. Nachrichten 4 (1951), pp. 213-231.

7. __________, Ein allgemeiner Endlichkeitssatz der Hydrodynamik, Math. Annalen 117, pp. 764-775 (1941).

8. __________, A mathematical example displaying the features of turbulence, Comm. Pure Applied Math. 1, pp. 303-322, (1948).

9. __________, Statistical hydromechanics and functional calculus, Journ. Rational Mech. Analysis, 1 (1952), pp. 87-124.

10. __________, Repeated branching through loss of stability, an example. Proc. Conference on Differential Equations, Univ. of Maryland (1955).

11. __________, On non-linear partial differential equations, Lecture Series of Symposium on Partial Differential Equations (Berkeley, 1955). Univ. of Kansas 1957.

12. S. Ito, The existence and uniqueness of regular solution of non-stationary Navier-Stokes equation. Journ. Fac. Science, Univ. of Tokyo, 9 (1961), pp. 103-140.

13. J. Kampe de Feriet, Statistical fluid mechanics: Two dimensional linear gravity waves. Conference on Partial Differential Equations and Continuum Mechanics, Univ. of Wisconsin, 1961.

14. A. A. Kiselev and O. A. Ladyzhenskaya, On existence and uniqueness of the solution of the nonstationary problem for a viscous incompressible fluid, Izvestiya Akad, Nauk SSSR 21 (1957), pp. 655-680.

15. O. A. Ladyzhenskaya, Solution "in the large" of non-stationary boundary value problem for the Navier-Stokes system with two space variables, Comm. Pure Applied Math. 12 (1959), pp. 427-433.

16. __________, Mathematical problems in the dynamics of viscous incompressible fluids. Moscow 1961.

17. M. Lees and M. H. Protter, Unique continuation for parabolic differential equations and differential inequalities, Duke Math. Journ. 28 (1961), pp. 369-382.

18. J. Leray, Étude de diverses equations intégrales non-linéares et de quelques problèmes que pose l'hydrodynamique, Journ. Math. Pures Appl. 12 (1933), pp. 1-82.

19. __________, Essai sur les mouvements plans d'un liquide visqueux que limitent des parois, Journ. Math. Pures Appl. 13 (1934), pp. 331-418.

20. __________, Sur le mouvement d'un liquide visqueux emplissant l'espace, Acta Math. 63 (1934), pp. 193-248.

21. L. Lichtenstein, Grundlagen der Hydromechanik. Berlin 1929.

22. J. L. Lions, Quelques résultats d'existence dans des équations aux dérivées partielles non linéaires, Bull. Soc. Math. France 87 (1959), pp. 245-273.

23. __________, Sur la régularité et l'unicité des solutions turbulentes des équations de Navier-Stokes, Rend. Sem. Mat. Padova 30 (1960), pp. 16-23.

24. __________, Equations differentielles - operationnelles, Berlin, 1961. Especially Chapter X.

25. J. L. Lions et G. Prodi, Un théorème d'existence et unicité dans les équations de Navier-Stokes en dimension 2, C.R. Acad. Sci. Paris 248 (1959), pp. 3519-3521.

26. F. K. C. Odquist, Über die Randwertaufgaben der Hydrodynamik zäher Flüssigkeiten, Math. Zeitschrift 32 (1930), pp. 325-375.

27. T. Ohyama, Interior regularity of weak solutions of the time-dependent Navier-Stokes equation, Proc. Japan Acad. 36 (1960), pp. 273-277.

28. C. W. Oseen, Neuere Methoden und Ergebnisse der Hydrodynamik. Leipzig, 1927.

29. G. Prodi, Un teorema di unicità per le equazioni di Navier-Stokes, Annali di Mat. 48 (1959), pp. 173-182.

30. __________, Qualche risultato riguardo alle equazioni di Navier-Stokes nel caso bidimensionale, Rend. Sem. Mat. Padova 30 (1960), pp. 1-15.

31. __________, Teoremi ergodici per le equazioni della idrodinamica. Symposium: Sistemi dinamica e teoremi ergodici, Varenna, 1960. CIME, Instituto Mat. dell' Univ. Roma.

32. J. Serrin, On the stability of viscous fluid motions, Arch. Rational Mech. Analysis 3 (1959), pp. 1-13.

33. __________, A note on the existence of periodic solutions of the

Navier-Stokes equations, Arch. Rational Mech. Analysis 3 (1959), pp. 120-122.

34. __________, On the interior regularity of weak solutions of the Navier-Stokes equations, Arch. Rational Mech. Analysis 9 (1962), pp. 187-195.

35. P. E. Sobolveskii, On the smoothness of generalized solutions of the Navier-Stokes equations, Dokl. Akad. Nauk SSSR 131 (1959), pp. 758-760.

36. T. Y. Thomas, On the uniform convergence of the solutions of the Navier-Stokes equations, Proc. Nat. Acad. Sci. USA 29 (1943), pp. 243-246.

37. W. Velte, Über ein stabilitätskriterium der Hydrodynamik, Arch. Rational Mech. Analysis 9 (1962), pp. 9-20.

SUPPLEMENTARY REFERENCES

38. C. L. Dolph and D. C. Lewis, On the application of infinite systems of ordinary differential equations to perturbations of plane Poiseuille flow, Quart. Applied Math. 16 (1958), pp. 97-110.

39. R. H. Dyer and D. E. Edmunds, A uniqueness theorem in magnetohydrodynamics, Arch. Rational Mech. Analysis 8 (1961), pp. 254-262. Cf. also 9, pp. 403-410 (1962).

40. D. Graffi, Sul teorema di unicità per le equazioni del moto dei fluidi compressibili in un dominio illimitato, Atti Acad. Sci. dell' Istituto Bologna, Sci-Fis. Classe, Series XI, 7 (1960), pp. 1-8.

41. __________, Ancora sul teorema di unicità per le equazioni del moto dei fluidi, Atti Acad Sci. dell' Istituto Bologna, Sci-Fis. Classe, Series XI, 8 (1961), pp. 7-14.

42. R. P. Kanwal, Uniqueness of magnetohydrodynamic flows, Arch. Rational Mech. Analysis 4 (1960), pp. 335-340.

43. O. A. Ladyzhenzkaya and V. A. Solonnikov. On the solution of the non-stationary problem of magneto-hydrodynamics, Dokl. Akad. Nauk SSSR 124 (1959), pp. 26-28. Cf. also Trudy Mat. Inst. Steklov 59 (1960).

44. B. L. Rozhdestvenskii, Discontinuous solutions of hyperbolic systems of quasi-linear equations, Russian Math. Surveys (trans. of Uspekhi Mat. Nauk), 15 (1960), pp. 53-111 in Number 6.

45. J. Serrin, On the uniqueness of compressible fluid motions, Arch. Rational Mech. Analysis 3 (1959), pp. 271-288.

ROBERT FINN

On the Stokes Paradox and related questions

Introduction

We outline in this paper some recent results on time-independent solutions of the Navier-Stokes equations

$$\mu\Delta\vec{w} - \rho\vec{w}\cdot\nabla\vec{w} - \nabla p = 0 \qquad (1)$$
$$\nabla\cdot\vec{w} = 0$$

for which the vector $\vec{w}(x)$ differs by little from a given constant vector $\vec{w}_0$. Interest will center on the case in which the region of definition contains a neighborhood of infinity in two or in three dimensions. In the particular case $\vec{w}_0 = 0$ the perturbation process $\vec{w} \to \vec{w}_0$ is then known to be singular.

In (1) the notation is the usual one of vector analysis. The solutions $[\vec{w}(x), p(x)]$ of these equations have an important interpretation as the (vector) velocity and (scalar) pressure in the steady motion of a viscous incompressible fluid (see, e.g., [1, p. 259]). The constant μ denotes the viscosity coefficient of the fluid, the term $\mu\Delta\vec{w}$ then representing the shearing forces in the motion. The density of the fluid is ρ (assumed constant), and $\rho\vec{w}\cdot\nabla\vec{w}$ corresponds to the inertial reaction of the fluid elements. The first equation (1) thus asserts that these two forces are at each point in equilibrium with the force arising from changes in the pressure p. The second relation, the "continuity equation," asserts the constancy of mass of all fluid elements during the motion. If the fluid is bounded by an inner surface Σ on which $\vec{w}(x) = 0$, and $\vec{w}(x) \to \vec{w}_0$ at infinity, the solution of (1) then represents the flow of a fluid which adheres to the surface Σ and is uniform at infinity. Equivalently, it determines the uniform motion of Σ through a fluid which is at rest at infinity. These physical interpretations are helpful in suggesting properly set problems and in providing motivation, but they are of course unnecessary for the formal mathematical developments.

We study solutions of (1) which are smooth, in the sense that all derivatives occurring in the equations exist and are continuous. Such solutions are known to exist. It will be convenient in the context to assume $\rho = \mu = 1$. This can be achieved by a similarity transformation, and entails no loss of generality. Introducing the new vector $\vec{u} = \lambda^{-1}(\vec{w} - \vec{w}_0)$, with λ a positive parameter, (1) then takes the form

$$\begin{aligned} \Delta \vec{u} - \vec{w}_0 \cdot \nabla \vec{u} - \nabla p &= \lambda \vec{u} \cdot \nabla \vec{u} \\ \nabla \cdot \vec{u} &= 0 , \end{aligned} \qquad (2)$$

where the term $\lambda^{-1} p$ has been replaced by the symbol $p(x; \lambda)$.

The perturbation to the uniform velocity field $\vec{w}(x) \equiv \vec{w}_0$ is then effected by letting $\lambda \to 0$. More precisely, let $[\vec{u}(x; \lambda), p(x; \lambda)]$ be a family of solutions of (2) depending on λ and such that $\vec{u}(x; \lambda)$ satisfies fixed boundary conditions, independent of λ. Then $\vec{u}(x) = \lim_{\lambda \to 0} \vec{u}(x; \lambda)$ and $p(x) = \lim_{\lambda \to 0} p(x; \lambda)$ should exist, be a solution of the linear system (equations of Oseen [1, p. 367])

$$\begin{aligned} \Delta \vec{u} - \vec{w}_0 \cdot \nabla \vec{u} - \nabla p &= 0 \\ \nabla \cdot \vec{u} &= 0 \end{aligned} \qquad (3)$$

and $\vec{u}(x)$ should satisfy the same boundary conditions. We remark the special case $\vec{w}_0 = 0$, in which $\vec{u}(x)$ should appear as solution of the Stokes equations [1, p. 355]

$$\begin{aligned} \Delta \vec{u} - \nabla p &= 0 \\ \nabla \cdot \vec{u} &= 0 . \end{aligned} \qquad (4)$$

It is the purpose of this paper to study this transformation of solutions of (1) into those of (3) in a quantitative way. To do so, we first discuss boundary value problems for the system (4), for which we provide a new and improved interpretation of a singular behavior of the solutions. Our next step is to develop general existence theorems for the nonlinear system (1) in an exterior domain, with particular regard to obtaining solutions which exhibit physically realistic asymptotic behavior at infinity. Such solutions are shown to exist whenever the prescribed data are sufficiently small; this is the principal new contribution we are able to announce. These solutions are further shown to be unique in the class of all solutions which assume the same data and exhibit the same qualitative behavior at infinity. The question remains open, however, whether there exist other solutions which are merely continuous at infinity, even in the class for which the rate of energy dissipation is finite.

Finally we estimate the error incurred by replacing a solution of (2) by a solution of (3) with the same prescribed data. This estimation improves results which were established in [2]. It provides, in particular, new information on the physical validity of the solution of (4) exterior to a closed surface, corresponding to the slow motion of a viscous fluid past an obstacle.

1. The Stokes Paradox

It is not difficult to construct particular solutions of (4) by means of series developments or representation formulas (see, e.g. [3]). A remarkable such solution was discovered by Stokes [4], who determined in closed form a velocity field $\vec{w}(x)$ satisfying (4) exterior to a unit sphere V in three dimensions, such that $\vec{w}(x) = 0$ on the surface S of V and such that $\vec{w}(x)$ tends to a prescribed constant vector $\vec{w}_0$ at infinity. The formula of Stokes is

$$\vec{w}(x) = \vec{w}_0 - \frac{3}{4}\nabla\times\left(r^2\nabla\times\frac{\vec{w}_0}{r}\right) - \frac{1}{4}\nabla\times\nabla\times\frac{\vec{w}_0}{r}, \quad r = |x| . \qquad (5)$$

Assuming that the perturbation process leading to (4) is non-singular, this solution will represent physically the vanishingly slow motion (more precisely, perturbation of the vanishing motion) of a fluid which adheres to the surface of the sphere and tends to the limit $\vec{w}_0$ at infinity. If this motion is viewed from a (Galilean) coordinate frame attached to the velocity vector at infinity, we obtain the physically important (unsteady) slow motion of a sphere through a fluid which is at rest at infinity.

One sees easily that in (5), $\vec{w}(x) = \vec{w}(-x)$, so that the "wake region," which is found behind the sphere in an actual flow, does not appear in this solution. A further difficulty appears if motions in two dimensions are considered. In this case, Stokes found that it is impossible to satisfy the conditions for the development in series of the corresponding solution representing a flow past a circle. More recently, it has been shown (see, e.g. [1, p. 361; 3; 5; 6]), that there are no non-trivial solutions in two independent variables, for which the velocity field vanishes on a closed contour Σ and remains bounded in the exterior $\mathscr{E}$ of Σ. This is the "Stokes Paradox" of hydrodynamics. Because it is natural to expect that problems of physical interest lead to correctly set mathematical questions, the clarification of this paradox is essential to an understanding of the behavior of solutions of (1) corresponding to small prescribed data.

It was pointed out by Oseen [7, p. 165] that in the solution (5) of Stokes, the ratio $\left|\frac{\vec{w}\cdot\nabla\vec{w}}{\Delta\,\vec{w}}\right|$ tends to infinity as $x\to\infty$ in the direction $\vec{w}_0$. Physically, this means the ratio of inertial to viscous forces in the flow becomes unbounded. In other words, the solution of Stokes violates the assumption under which the equations (4) are derived. This non-uniformity in the perturbation process in an infinite region is the apparent source of the singular behavior of the solution. Not every such perturbation is singular, however, and solutions are easily constructed for which no such singularity arises. The discussions in [1, p. 361; 3; 4; 5] make essential use of the condition $\vec{w} = 0$ on Σ , and the relation of this condition to the "paradox" has not been clear. With a view to clarifying this point, the problem is approached from a different point of view in [6], where the Stokes Paradox is obtained as a consequence of a more general result.

THEOREM 1.1: *Let $\vec{w}(x)$ be a solution of (4) in the n-dimensional exterior $\mathscr{E}$ of a closed contour (surface) Σ , and suppose*[1] *$|\vec{w}(x)| = o(r)$ as $x \to \infty$. Then there exists a constant vector*[2] *$\vec{w}_0$ such that*

$$|\vec{w}(x) - \vec{w}_0| = \begin{cases} O(\log r) , & n = 2 \\ O(r^{2-n}) , & n > 2 \ . \end{cases}$$

THEOREM 1.2: *Under the hypotheses of Theorem 1, if*

$$|\vec{w}(x) - \vec{w}_0| = \begin{cases} o(\log r) , & n = 2 \\ o(r^{2-n}) , & n > 2 \end{cases}$$

then no force is exerted on Σ in the motion.

By the force on Σ is meant the integral of the stress tensor over Σ . That is, let $T\vec{w}$ be the tensor defined by the relations

$$(T\vec{w})_{ij} = -p\delta_{ij} + \mu\left(\frac{\partial w_i}{\partial x_j} + \frac{\partial w_j}{\partial x_i}\right) ,$$

$$\delta_{ij} = \begin{cases} 1, & i = j \\ 0 , & i \neq j \end{cases} .$$

Then the force $\mathscr{F}_\Sigma$ is defined by the relation

$$\mathscr{F}_\Sigma = \oint_\Sigma T\vec{w} \cdot \vec{n} dS$$

where $\vec{n}$ is the exterior directed unit normal on Σ .

THEOREM 1.3: Let $\vec{w}(x)$ be a solution of (4) in $\mathscr{E}$, and suppose $|\vec{w}(x)| = o(r)$ as $r \to \infty$. Suppose further that $\vec{w}(x) = 0$ on Σ , and that no force is exerted on Σ . Then $\vec{w}(x) \equiv 0$ in $\mathscr{E}$.

Putting Theorems 1.2 and 1.3 together, we see in particular that if $\vec{w}(x) = 0$ on Σ in a two dimensional motion, and $|\vec{w}(x)| = 0(1)$, then $\vec{w}(x) \equiv 0$ in $\mathscr{E}$. This is the result stated above. The singular behavior of the solution is thus correlated with the forces exerted across fluid interfaces. These forces generate large fluid displacements at infinity, which in turn can be supported only by the presence of infinite velocities. Further, if the velocity vanishes on a closed contour Σ , the frictional forces which are produced by the velocity gradient at the boundary necessarily give rise to a non-vanishing net force on Σ .

Theorems 1.1, 1.2 and 1.3 are sharp, in the sense that "o" cannot be replaced by "0". Their proofs[3] depend on a representation for the solution by means of a fundamental solution tensor $\chi(x, y)$ associated with the system (3). Such a tensor is explicitly known (cf. [7]). In the region $\mathscr{E}_R$ bounded by Σ and by a sphere Σ_R of(large) radius R , we obtain for the solution field $[\vec{w}(x), p(x)]$,

$$\text{a)}\quad \vec{w}(x) = \oint_{\Sigma+\Sigma_R} (\vec{w}\cdot T\chi - \chi\cdot T\vec{w})\cdot\vec{n}\, dS \tag{6}$$

$$\text{b)}\quad p(x) = \oint_{\Sigma+\Sigma_R} (\vec{w}\cdot T\psi - \psi\cdot T\vec{w})\cdot\vec{n}\, dS$$

where $\psi(x, y) = \nabla r_{xy}^{-1}$ is the "fundamental pressure vector" associated with $\chi(x, y)$. The key to the situation lies in the observation[4] that if $|\vec{w}(x)| = o(r)$, then the outer integrals tend in the limit, as $R \to \infty$, to a constant vector and scalar, independent of x . Up to these constants, the asymptotic properties of $\vec{w}(x)$ are thus determined by those of $\chi(x, y)$. Since

$$|\chi(x, y)| = \begin{cases} 0(\log r) , & n = 2 \\ 0(r^{2-n}) , & n > 2 , \end{cases}$$

and $|T\chi| = 0(r^{1-n})$, we find immediately from the mean value theorem that $\mathscr{F}_\Sigma = \oint_\Sigma T\vec{w}\cdot\vec{n}\, dS = 0$ whenever

$$|\vec{w}(x) - \vec{w}_0| = \begin{cases} o(\log r) \quad , & n = 2 \\ o(r^{2-n}) \quad , & n > 2 \end{cases}$$

. To prove Theorem 3, we adjoin the identity, valid for any solution of (4),

$$\oint_{\Sigma+\Sigma_R} (\vec{w} \cdot T\vec{w}) \cdot \vec{n}\, dS = 2 \int_{\mathscr{E}_R} (\operatorname{def} \vec{w})^2\, dV$$

where $(\operatorname{def} \vec{w})$ denotes the tensor $\frac{1}{2}\left(\frac{\partial w_i}{\partial x_j} + \frac{\partial w_j}{\partial x_i}\right)$. If $|\vec{w}(x)| = o(r)$, the condition $\mathscr{F}_\Sigma = 0$ and the above estimates imply that the outer integral vanishes in the limit. But the inner integral vanishes if $\vec{w} = 0$ on Σ , hence $\operatorname{def} \vec{w} \equiv 0$, and we conclude easily that $\vec{w}(x) \equiv 0$ in $\mathscr{E}$.

We remark that the condition $|\vec{w}(x)| = o(r)$ of Theorems 1.1 and 1.3 can be replaced by either of the equivalent conditions,

$$\int_{\mathscr{E}_R} |\vec{w}|^2 r^{-(n+2)}\, dV = o(\log R) \quad , \text{ or } \quad \int_{\mathscr{E}_R} |\nabla \vec{w}|^2\, r^{-n} dV = o(\log R) \quad .$$

These and other extensions of the results of this section can be found in [6].

2. Existence Theorems

Odqvist has considered in [8] the problem of finding a solution $\vec{w}(x)$ of (4) which assumes prescribed data $\vec{w}^*$ on the boundary Σ of the region of definition. He proved the existence and uniqueness of the solution interior to Σ in two and in three dimensions, subject to the (necessary) condition $\oint_\Sigma \vec{w}^* \cdot \vec{n}\, dS = 0$, and in three dimensions he proved also the existence of a solution in the exterior region, which tends to a prescribed vector $\vec{w}_0$ at infinity. Odqvist obtained also general estimates for the Green's tensor[5] $G(x, y)$ for the system (4) in a bounded three dimensional region, analogous to classical estimates of potential theory. These results have been used by Leray [9] and by others [2, 10, 11, 12] to prove the existence in three dimensions of solutions of (1) which assume prescribed data on Σ , both in the interior of Σ and in the exterior region $\mathscr{E}$. (In the latter case the value $\vec{w}_0$ of $\vec{w}$ at infinity is also prescribed).

The three dimensional solutions in $\mathscr{E}$ constructed in the above references are known to have finite Dirichlet Integral[6] , $\int_{\mathscr{E}} |\nabla \vec{w}|^2\, dV < \infty$, and to be continuous at infinity. Further, we have shown in [13] that $\nabla\vec{w} \to 0$ at infinity whenever $\vec{w} \to \vec{w}_0$, and we have proved there the validity of the representation formulas

$$\vec{u}(x) = \oint_{\Sigma} \{\vec{u} \cdot T\chi \cdot \vec{n} - \chi \cdot T\vec{u} \cdot \vec{n} + (\vec{u} \cdot \chi)(\vec{w}_0 \cdot \vec{n})\} dS + \lambda \int_{\mathscr{E}} \chi \cdot \vec{u} \cdot \nabla\vec{u}\, dV \tag{7}$$

$$\vec{u}(x) = \oint_{\Sigma} \{\vec{u} \cdot T\chi \cdot \vec{n} - \chi \cdot T\vec{u} \cdot \vec{n} + (\vec{u} \cdot \chi)(\vec{w}_0 \cdot \vec{n}) + \lambda(\vec{u} \cdot \chi)(\vec{u} \cdot n)\} dS - \lambda \int_{\mathscr{E}} \vec{u} \cdot \vec{u} \cdot \nabla\chi\, dV \tag{8}$$

for the solution $\vec{u}(x) = \lambda^{-1}(\vec{w} - \vec{w}_0)$ of the equivalent system (2). Here $\chi(x, y)$ denotes the fundamental solution tensor for the system (3).

However, no information is available on the rate at which $\vec{w}(x)$ decays to its limiting value, or on whether the solutions exhibit the "wake" region characteristic of observed fluid motions. Experimental evidence indicates that solutions corresponding to small data exist and are unique, but become unstable and perhaps branch as the data increase in magnitude; theoretical confirmation even of the qualitative features of this behavior for the above solutions has been lacking.

By a modification of the approach used in the above references, we are now able to show that if the prescribed data do not differ too much from $\vec{w}_0$, then there exists a solution which is in the indicated respects physically reasonable. We consider a suitably smooth inner bounding surface Σ in three dimensions, on which are prescribed data $\vec{w}^*$. Let δ be a bound for the magnitude of $\vec{w}^* - \vec{w}_0$, together with the tangential derivatives of this vector function on Σ up to third order.

THEOREM 2.1: <u>If δ is sufficiently small, depending only on Σ and on $\vec{w}_0$, there is a solution $[\vec{w}(x), p(x)]$ of (1) in $\mathscr{E}$, such that $\vec{w}(x) = \vec{w}^*$ on Σ, such that $\vec{w}(x) \to \vec{w}_0$ as $r \to \infty$, and such that the flow defined by $\vec{w}(x)$ exhibits a paraboloidal "wake" region in the following sense: Denote by φ the angle made with the direction of $\vec{w}_0$ by an arbitrary ray from the origin. Then interior to the surface $\varphi = r^{1/2}$ there holds $|\vec{w} - \vec{w}_0| = O(r^{-1})$ as $r \to \infty$. On the surface $\varphi = r^{1/2+\sigma}$ there holds $|\vec{w} - \vec{w}_0| = O(r^{-1-2\sigma})$ for $0 \le \sigma \le \frac{1}{2}$, and exterior to the cone $\varphi = r$, we will have</u>

$|\vec{w}-\vec{w}_0| = 0(r^{-2})$. The symbol "0" can be replaced by "o" if and only if no force is exerted on Σ in the motion. In particular, this cannot occur if $w^* \equiv$ const. on Σ . For given Σ , the constant δ can be chosen uniformly for all $\vec{w}_0$ with magnitudes less than some fixed quantity. Thus, in particular, there exists a solution corresponding to data $\vec{w}^* \equiv 0$ if $|\vec{w}_0|$ is sufficiently small.

Estimates on $|\nabla \vec{w}(x)|$ and on $|p(x)|$ near infinity can also be given (cf. [13]).

COROLLARY: Let $\vec{w}^* \equiv 0$. Then for the solution described in Theorem 2.1, there holds

$$\vec{w}_0 \cdot \mathscr{F}_\Sigma = \frac{1}{2} \int_{\mathscr{E}} (\operatorname{def} \vec{w})^2 \, dV ,$$

that is, in the flow past Σ with adherence on the boundary and limiting velocity $\vec{w}_0$, the scalar product of $\vec{w}_0$ with the force on Σ equals the rate at which kinetic energy is dissipated into heat in the motion.

THEOREM 2.2: Let $[\vec{w}(x), p(x)]$ be a solution of (1) in $\mathscr{E}$, which assumes data w^* on Σ , and for which $|\vec{w}-\vec{w}_0| < \frac{\mu}{2\rho} r^{-1}$ uniformly in $\mathscr{E}$. Then $\vec{w}(x)$ is unique among all solutions $[\vec{v}(x), q(x)]$ which assume the same data, and for which there are positive constants C and ϵ such that $|\vec{v}(x) - \vec{w}_0| < Cr^{-\frac{1}{2}-\epsilon}$ in $\mathscr{E}$.

Proof of Theorem 2.2: Let $\vec{\eta} = \vec{w} - \vec{v}$, $\sigma = p - q$. Then $\eta = 0$ on Σ , $\eta \to 0$ at infinity, and

$$\mu \Delta \vec{\eta} - \rho \vec{\eta} \cdot \nabla \vec{\eta} - \nabla \sigma = \rho(\vec{\eta} \cdot \nabla \vec{w} - \vec{w} \cdot \nabla \vec{\eta}) .$$

Thus, multiplying by $\vec{\eta}$ and integrating over the region $\mathscr{E}_R$ bounded by Σ and by a sphere Σ_R of large radius R ,

$$\begin{aligned} -\mu \int_{\mathscr{E}_R} |\nabla \vec{\eta}|^2 dV = \rho \int_R \vec{\eta} \cdot \vec{\eta} \cdot \nabla \vec{w} \, dV - \frac{1}{2} \rho \oint_{\Sigma_R} |\vec{\eta}|^2 (\vec{w} \cdot \vec{n}) \, dS \\ - \mu \oint_{\Sigma_R} \vec{\eta} \cdot \vec{n} \cdot \nabla \vec{\eta} \, dS + \oint_{\Sigma_R} \sigma(\vec{\eta} \cdot \vec{n}) \, dS . \end{aligned} \tag{9}$$

We now apply the general estimates of [13], which show that under the given assumptions, all boundary terms tend in the limit to zero. Further,

$$\int_{\mathscr{E}_R} \vec{\eta} \cdot \vec{\eta} \cdot \nabla \vec{w} \, dV = \int_{\mathscr{E}_R} \vec{\eta} \cdot \vec{\eta} \cdot \nabla(\vec{w} - \vec{w}_0) \, dV$$

$$= -\int_{\mathscr{E}_R} (\vec{w} - \vec{w}_0) \cdot \vec{\eta} \cdot \nabla \vec{\eta} dV$$

$$+ \oint_{\Sigma_R} (\vec{\eta} \cdot \vec{n}) [\vec{\eta} \cdot (\vec{w} - \vec{w}_0)] dS \quad .$$

Again the estimates of [13] show that the surface integral tends to zero. But by hypothesis, $|\vec{w} - \vec{w}_0| < \frac{\mu}{2\rho} r^{-1}$, hence,

$$\left| \int_{\mathscr{E}_R} (\vec{w} - \vec{w}_0) \cdot \vec{\eta} \cdot \nabla \vec{\eta} dV \right|^2 \leq \left| \frac{\mu}{2\rho} \int_{\mathscr{E}_R} r^{-1} \vec{\eta} \cdot \nabla \vec{\eta} dV \right|^2$$

$$\leq \left(\frac{\mu}{2\rho}\right)^2 \int_{\mathscr{E}_R} r^{-2} |\vec{\eta}|^2 dV \int_{\mathscr{E}_R} |\nabla \vec{\eta}|^2 \, dV$$

By the Schwarz inequality. Equality holds if and only if both sides vanish. Since $\vec{\eta} = 0$ on Σ , $\vec{\eta}$ can be extended to a piecewise continuously differentiable vector field over all space, and we may apply Lemma 3.4 of [2] to obtain

$$\int_{\mathscr{E}} r^{-2} |\vec{\eta}|^2 dV \leq 4 \int_{\mathscr{E}} |\nabla \vec{\eta}|^2 dV \quad .$$

Thus, we conclude from (9) that $\int_{\mathscr{E}} |\nabla \vec{\eta}|^2 dV = 0$, and hence $\vec{\eta} \equiv$ const. in $\mathscr{E}$. Since $\vec{\eta} = 0$ on Σ , we have $\vec{\eta} \equiv 0$ in $\mathscr{E}$, the result to be proved.

We outline briefly the proof of Theorem 2.1. Our approach here differs from that used in the known existence theorems for an exterior region, in which the solution is obtained as a limit of solutions defined in a sequence of expanding finite domains. We have found it preferable to seek the solution directly in a class of vector fields defined throughout $\mathscr{E}$, which exhibit some of the qualitative asymptotic structure which is sought for the solution. In this case, however, the integral equation used in the earlier studies, which characterizes the solution as a "small" deviation from a solution of (4), is no longer suitable. We have therefor replaced this equation by another one, which exploits the expected proximity of the solutions of (2) to those of (3) at infinity. The difficulties then consist in the estimation of the kernel at infinity, and the use of these estimates to obtain a priori bounds on the nonlinear operator occurring in the equation.

The integral equation we consider is

$$\vec{u}(x) = \oint_{\Sigma} \vec{u}* \cdot TG\, dS + \lambda \int_{\mathscr{E}} \vec{u} \cdot \vec{u} \cdot \nabla G dV , \qquad (10)$$

for a solution $\vec{u}(x) = \lambda^{-1}(\vec{w}-\vec{w}_0)$ of (2), assuming data $u* = \lambda^{-1}(\vec{w}^*-\vec{w}_0)$ on Σ. Here $G(x, y; \vec{w}_0)$ is the Green's tensor for (3) in the given region. The following lemma leads to the proof of existence and the necessary bounds at infinity for $G(x, y; \vec{w}_0)$:

LEMMA 2.3: Let $\vec{v}(x)$ satisfy $\nabla \cdot \vec{v} = 0$ in $\mathscr{E}$ and also the estimate $|\vec{v}(x)| = 0(r^{-1})$ as $r \to \infty$. Let $\vec{w}^*$ be arbitrary prescribed data on Σ. Then there is a unique solution of the system

$$\begin{aligned} \Delta \vec{w} - \vec{w}_0 \cdot \nabla \vec{w} - \nabla p &= \vec{v} \cdot \nabla \vec{v} \\ \nabla \cdot \vec{w} &= 0 \end{aligned} \qquad (11)$$

in $\mathscr{E}$ such that $\vec{w}(x) = \vec{w}*$ on Σ and $\vec{w}(x) \to 0$ at infinity. If $\vec{v}(x)$ has compact support, the asymptotic properties of $\vec{w}(x)$ at infinity are controlled by those of the fundamental solution tensor $\chi(x, y; \vec{w}_0)$. In any event, there holds always the estimate $|\vec{w}(x)| < Cr^{-1}$ as $r \to \infty$, and C can be chosen uniformly for all $|\vec{w}_0|$ which lie in any finite interval containing the origin.

In a finite region, the demonstration of existence can be reduced to a Fredholm equation of second kind, using as kernel the Green's tensor for (4). The case of an infinite region is obtained by a limiting process, and the validity of representations analogous to (7) and (8) follows by the method of [6]. The properties of $G(x, y; \vec{w}_0)$ are obtained from the case $\vec{v}(x) \equiv 0$, using formal estimates on the fundamental tensor $\chi(x, y; \vec{w}_0)$, which is known explicitly. The estimate for the solution in the general case depends on an auxiliary lemma, in the proof of which lies the chief technical difficulty of our method.

LEMMA 2.4: There is a constant C, depending only on Σ and on $|\vec{w}_0|$, such that

$$\int_{\mathscr{E}} r_{oy}^{-2} |\nabla G(x, y; \vec{w}_0)| dV < C r_{ox}^{-1} \text{ in } \mathscr{E} .$$

The choice of C can be made uniform for all $|\vec{w}_0|$ in any finite interval containing the origin.

Using this lemma, we may construct explicitly a solution of the integral equation (10) in the class of vector fields described in the statement of Theorem 2.1, provided only that the boundary data $\vec{w}^*$ are sufficiently close to $\vec{w}_0$. To do so, let

$$H = \sup_{x \in \mathscr{E}} r_{ox} \int_{\mathscr{E}} r_{oy}^{-2} |\nabla G(x, y; \vec{w}_0)| dV .$$

By Lemma 2.4, $H < \infty$. Let $\vec{W}(x)$ be the solution of (3) such that $\vec{W}(x) = \vec{w}^* - \vec{w}_0$ on Σ and $\vec{W}(x) \to 0$ at infinity. Let

$\lambda = \sup_{x \in \mathscr{E}} r|\vec{W}(x)|$. By Lemma 2.3, $\lambda < \infty$. We may then assert that for any data $\vec{w}^*, \vec{w}_0$ for which $\lambda < (4H)^{-1}$ there will be a solution of (10) in the given class. Further, this solution admits the representation

$$\vec{u}(x) = \vec{u}_0(x) + \sum_{n=1}^{\infty} \vec{u}_n(x) \lambda^n \tag{12}$$

where $\vec{u}(x) = \lambda^{-1}(\vec{w}(x) - \vec{w}_0)$ and $\vec{u}_0(x) = \lambda^{-1}\vec{W}(x)$. The series (12) is uniformly convergent throughout $\mathscr{E}$.

Writing $\vec{v}(x) = r\vec{u}(x)$, $\vec{v}_i(x) = r\vec{u}_i(x)$, (12) takes the equivalent form

$$\vec{v}(x) = \vec{v}_0(x) + \sum_{n=1}^{\infty} \vec{v}_n(x) \lambda^n . \tag{13}$$

Assuming this representation, we are led to the recursion relation

$$\vec{v}_{n+1}(x) = r_{ox} \int_{\mathscr{E}} r_{oy}^{-2} [\vec{v}_0(y) \cdot \vec{v}_n(y) + \vec{v}_1(y) \cdot \vec{v}_{n-1}(y) + \cdots + \vec{v}_n(y) \cdot \vec{v}_0(y)]$$

$$\cdot \nabla G(x, y; \vec{w}_0) dV .$$

The series with constant coefficients

$$V = \sum_{0}^{\infty} V_i \lambda^i , \quad V_0 = 1 , \tag{14}$$

will then be a dominant series for (13), provided

$$V_{n+1} = (V_0 V_n + V_1 V_{n-1} + \ldots + V_n V_0) H .$$

The convergence of (14) thus implies

$$V = 1 + \lambda H V^2 .$$

The solution of this equation has a branch which is analytic in λ in a circle about the origin of radius determined by the vanishing of the discriminant; that is, the series (14), and hence also (13), will converge provided $|\lambda| < (4H)^{-1}$. Thus, in particular, it converges for the chosen value of λ, which is what we set out to prove.

Since the coefficients in (14) are bounded, so are the functions $\vec{v}_i(x)$, and hence $|\vec{u}_i(x)| < Cr^{-1}$, C independent of i. Therefor $|u(x) - \vec{u}_0| = 0(r^{-1})$ at infinity, and also $|\vec{w}(x) - \vec{w}_0| = 0(r^{-1})$. The more precise estimates on asymptotic behavior are obtained by applying results of [13].

Theorem 2.1 can be applied to various cases of physical interest, e.g. the uniform translatory motion of a body in a fluid at rest at infinity, or the uniform rotation of a body about an axis of revolution. The latter case should be readily accessible to calculation, since the Green's tensor to be used is the relatively simple one corresponding to the Stokes equations (4). The method ofproof can in principle be applied also to obtain general qualitative information on behavior of the solution, such as an estimate of the distance to the boundary of th position of maximum speed; however we have not carried out this work

3. Perturbation Theorems

We study two cases, according as the velocity at infinity is varied or not in the perturbation. The latter case, which is the simpler, is described by considering a family of solutions $\vec{u}(x;\lambda)$ of (2) in $\mathscr{E}$ which assume fixed data $\vec{u}^*$ on Σ and vanish at infinity, and letting $\lambda \to 0$. Let $\vec{U}(x)$ be the (unique) solution of (3) subject to the same conditions.

Consider first the solutions $\vec{u}(x;\lambda)$ whose existence is shown in [2, 9, 10], for which information on the rate of decay of $\vec{u}(x;\lambda)$ at infinity is not available. We may nevertheless assert:

THEOREM 3.1: There exists a constant $C > 0$, depending only on Σ, on $\vec{w}_0$, and on $\vec{w}^*$, such that $|\vec{u}(x;\lambda) - \vec{U}(x)| < C\lambda$, uniformly in $\mathscr{E}$. Similar estimates hold for the gradient of the velocity field, and for the pressure.

The proof follows easily from the estimates on the Green's tensor for (3) indicated in §2.

More precise information can be obtained if the perturbed solutions $\vec{u}(x;\lambda)$ lie in the class of solutions with "physically reasonabl behavior at infinity, which are discussed in §2. For any function $f(x) = 0(r^{-1})$ at infinity, let $\|f\| = \sup_{\mathscr{E}} |rf(x)|$. Let $H_0 = \sup_{\mathscr{E},\lambda} r_{ox} \int_{\mathscr{E}} r_{oy}^{-2} |\nabla G| dV$ for λ in some range $0 \leq \lambda \leq \lambda_0$ (see

Lemma 2.4), where $G(x, y; \vec{w}_0)$ is the Green's tensor for (3).

THEOREM 3.2: <u>In the interval</u> $0 \leq \lambda < \min \{\lambda_0, (4H_0 \| U(x) \|)^{-1}\}$, <u>any solution</u> $\vec{u}(x; \lambda)$ <u>of (2) which satisfies</u> $|\vec{u}(x; \lambda)| = 0(r^{-1/2-\epsilon})$ <u>at infinity for some</u> $\epsilon > 0$ <u>can be represented by a convergent expansion</u>

$$\vec{u}(x; \lambda) = \vec{U}(x) + \sum_{i=1}^{\infty} \vec{u}_i(x) \lambda^i$$

<u>throughout</u> $\mathscr{E}$. <u>Further, there is a positive constant</u> C <u>such that for all sufficiently small</u> λ,

$$|\vec{u}(x; \lambda) - \vec{U}(x)| < C \lambda r^{-1} \text{ in } \mathscr{E} .$$

The proof is similar to the proof of Theorem 2.1.

~~~~~~~~~~~

In the expansion we have obtained there is a conceptual similarity to the "outer expansion" of Kaplun, or of Proudman and Pearson[7] (see, e.g. [14, 15]), and Theorems 2.1 and 3.2 assert that this expansion represents the velocity field throughout the flow region. This does not seem to be the case for the "inner expansion" which has been used in [14, 15] and elsewhere to represent the flow near the body. From the point of view of this paper, ths discussion of this part of the flow in [14, 15] is effected by studying a perturbation for which the velocity at infinity is varied. Such a perturbation can be effected by assuming that the solution $\vec{u}(x; \lambda)$ of (2) assumes fixed data $\vec{u}^*$ on Σ and tends at infinity to a fixed limit $\vec{u}_0 \neq 0$. This case includes the classical perturbation which gives rise in two dimensions to the Stokes Paradox.

THEOREM 3.3: <u>The conclusions of Theorem</u> 3.1 <u>are valid also in the case that</u> $\vec{u}(x; \lambda) \to \vec{u}_0 \neq 0$ <u>at infinity.</u>

Our proof of this result is, however, considerably more difficult than the one we can give for Theorem 3.1. It is obtained by a refinement of the method used to prove Theorem 7.4 in [2], and represents a corresponding improvement in that result.

Again we may obtain a better estimate if the perturbation is performed in the class of "physically reasonable" solutions. We are, however, unable to demonstrate in this case the validity of an expansion of the form (12).

THEOREM 3.4: <u>Let</u> $\{\vec{u}(x; \lambda)\}$ <u>be a family of solutions in</u> $\mathscr{E}$ <u>of</u> (2), <u>for</u> $\lambda \to 0$, <u>such that</u> $\vec{u}(x; \lambda) = \vec{u}^*$ <u>on</u> Σ, $\vec{u}(x; \lambda) \to \vec{u}_0$ <u>at infinity, and such that</u> $|\vec{u}(x; \lambda) - \vec{u}_0| = 0(r^{-1/2-\epsilon})$ <u>for some</u> $\epsilon > 0$.
~~~~~~~~~~~

Let $\vec{U}(x)$ be the solution of (3) in $\mathscr{E}$ corresponding to the same data. Then there is a positive constant C such that for all sufficiently small λ, $|\vec{u}(x;\lambda) - \vec{U}(x)| < C \min\{\lambda, r^{-1}\}$ uniformly in $\mathscr{E}$ [8]

We note that Theorems 3.1 to 3.4 do not distinguish the case $\vec{w}_0 = 0$, which is the only case in which the Stokes Paradox can occur. Although there is some evidence that the perturbation process may in some sense be less well behaved when $\vec{w}_0 = 0$, the above results indicate that the chief cause of singular behavior lies in the variation of the velocity at infinity. In fact, Theorem 3.2 shows that for the solutions of principal interest, no singularity can occur even in the case $\vec{w}_0 = 0$, provided this value is not varied with λ.

A particular application of the results of this section is to the "oil-drop" experiment of Millikan for measuring the charge of the electron. In this experiment the solution (5) of the Stokes equation (4) was used to calculate the resistance experienced by an oil drop which falls slowly in the air under the influence of known external forces. Assuming the correctness of the nonlinear system (1) as a description of reality, Theorems 3.3 and 3.4 provide the first formal justification for the use of (4) as an approximation to the actual flow, and yield also an estimate of the possible error incurred by this step in the calculation. In fact, if we denote by $\mathscr{F}_\lambda$ the suitably normalized force on Σ arising from a solution of (2) with data $(\vec{u}^*, \vec{u}_0)$, and by $\mathscr{F}_0$ the force arising from the solution of (3) with the same data, then we obtain, under the hypotheses for either of the Theorems 3.3 or 3.4, the estimate[9] $|\mathscr{F}_\lambda - \mathscr{F}_0| < C\lambda$, as $\lambda \to 0$.

NOTES

1. We write $\alpha = o(\beta)$ when $\alpha\beta^{-1} \to 0$ as β tends to some limit. The notation $\alpha = O(\beta)$ means $\alpha\beta^{-1}$ remains bounded.

2. Even in the case $n = 2$, the vector $\vec{w}_0$ is uniquely determined by a procedure consistent with what occurs in higher dimensional spaces.

3. From a formal point of view these theorems are true for the same reason that a harmonic function $h(x)$ in $\mathcal{E}$ satisfying

$$|h(x)| = \begin{cases} o(\log r), & n = 2 \\ o(r^{2-n}), & n > 2, \end{cases}$$

necessarily satisfies

$|h-h_0| = 0(r^{1-n})$ and $\oint_\Sigma \frac{\partial h}{\partial n} dS = 0$, for some h_0 .

This formal reason is that the fundamental solution tensor $\chi(x, y)$ associated with (4) and the fundamental solution

$$\begin{cases} \log r \ , & n = 2 \\ r^{2-n} \ , & n > 2 \end{cases}$$

of the equation $\Delta u = 0$ have at infinity the same order of growth or decay. The method of proof of Theorems 1.1 to 1.3 yields in particular a new and simple proof of the corresponding properties of harmonic functions. For generalizations to various systems of equations, see [6].

4. This result cannot be obtained by direct inspection of orders of magnitude, but it follows from the fact that successive derivatives of $\chi(x, y)$ will vanish to successively higher orders at infinity. Thus, a derivative of suitably high order of the outer surface integral in (6a) will vanish in the limit as $R \to \infty$. From this one concludes that this integral is a polynomial in x , and hence constant, since $|\vec{w}(x)| = o(r)$. This result determines in turn the asymptotic behavior of $\vec{w}(x)$, which can then be used to estimate the outer integral in (6b) . For details, see [6].

5. The Green's tensor is a fundamental solution tensor, the components of which vanish on Σ .

6. This integral admits a physical interpretation as half the sum of the rate of dissipation of energy into heat, plus the total vorticity in the motion.

7. Our expansion differs from those of [14, 15] in that the coefficients $\vec{u}_i(x)$ contain the boundary conditions—that is, they vanish on the boundary—whereas the coefficients in the expansions in [14, 15] are related to the boundary conditions only indirectly by means of a matching process with the "inner expansion." The representation (12) is not an expansion in powers of the Reynolds' number in the usual sense. The parameter λ has the character of Reynolds' number, but it is proportional to the difference between some characteristic velocity and $\vec{w}_0$, rather than to the characteristic velocity itself.

8. In the case of a sphere, this estimate has been predicted on the basis of formal series development of the solution, by Kaplun and Lagerstrom [15] and, independently, by Proudman and Pearson [14].

9. See also H. Brenner [16].

REFERENCES

1. Kotschin, N. J., Kibel, I. A. and Rose, N. W., "Theoretische Hydromechanik," Band II (translation from the Russian). Akademie-Verlag, Berlin 1955.

2. Finn, R. "On the Steady-State Solutions of the Navier-Stokes Equations," III; Acta Math., 105(1961), 197-244.

3. Finn, R. and Noll, W., "On the uniqueness and non-existence of Stokes flows." Arch. Rat. Mech. Anal., 1(1957), 97-106.

4. Stokes, G. G., "On the effect of the internal friction of fluids on the motion of pendulums," (1851), Math. and Phys. Papers, vol. III, p. 1.

5. Charnes, J., and Krakowski, M., "Stokes' Paradox and Biharmoni Flows," Carnegie Inst. of Tech. Technical Report No. 37 (1953)

6. Chang, I-Dee, and Finn, R., "On the Solutions of a Class of Equations Occurring in Continuum Mechanics, with Application to the Stokes Paradox," Arch. Rat. Mech. Anal., 7(1961), 388-401.

7. Oseen, C. W., "Neuere Methoden und Ergebnisse in der Hydrodynamik." Akademische Verlagsgesellschaft M. B. H., Leipzig 1927.

8. Odqvist, F. K. G., "Die Randwertaufgaben der Hydrodynamik zäher Flüssigkeiten." Stockholm 1928, P. A. Norstedt und Söner. See also Math. Z., 32(1930), 329-375.

9. Leray, J., Étude de diverses équations intégrales non lineaires et de quelques problèmes que pose l'Hydrodynamique. J. Math. Pures Appl., 9(1933), 1-82. See also: Les problèmes non linéaires. Enseignement Math., 35(1936), 139-151.

10. Finn, R., "On steady-state solutions of the Navier-Stokes partial differential equations." Arch. Rat. Mech. Anal., 3(1959), 381-396.

11. Ладыженская, О. А., "Исследование уравнения Навье-Сто в Случае стационарного движения несжимаемой жидкости". Успехи Мат. Наук, 14, 1959, 75-97.

12. Fujita, H., "On the existence and regularity of the steady-state solutions of the Navier-Stokes equations," J. Fac. Sci., Univ. of Tokyo, Sec. 1, IX (1961), 59-102.

13. Finn, R., "Estimates at Infinity for Stationary Solutions of the Navier-Stokes Equations," Bull. Math. de la Soc. Sci. Math.

Phys. de la R. P. R., Tome 3(51), (1959), 387-418. See also: Amer. Math. Soc., Proc. Symposia Pure Math., vol. IV (1961), 143-148.

14. Proudman, I., and Pearson, J. R. A., "Expansion at small Reynolds' numbers for the flow past a sphere and a circular cylinder," J. Fluid Mech., 2 (1957), 237-262.

15. Kaplun, S., and Lagerstrom, P. A., "Asymptotic Expansions of Navier-Stokes Solutions for Small Reynolds Numbers," J. Math. and Mech., 6(1957), 585-593.

16. Brenner, H., "The resistance of an arbitrary particle at small Reynolds numbers," Preliminary report, Dept. of Chemical Engineering, New York University.

This investigation was supported by the Office of Naval Research.

HELMUT H. SCHAEFER

Some nonlinear eigenvalue problems

Since the term "nonlinear" is only the negation of a type of structure known to yield a rich theory, rather than a positive definition, it cannot be expected that there is (or ever will be) any such thing as a well-rounded theory of nonlinear problems in analysis. Thus the investigation of nonlinear problems in analysis (mainly, functional equations) has to be guided either by questions in applications (their vast majority stemming from physics), or by known mathematical structures suggesting new and interesting results; ideally, by both. A very rough classification of results available in nonlinear analysis distinguishes two categories: Theorems of a local nature, giving information on what happens in a neighborhood (suitably to be defined) of a known solution to the problem, and global ones requiring no previous knowledge.

The modest contribution made in this paper belongs to the second category. Our results, divided into three sections, are loosely interconnected: All three sections are concerned with functional equations containing a parameter for certain values of which the existence of non-trivial solutions is established. Another common feature of the present investigations is the systematic use of order concepts, a method apparently growing in importance. While this use of orderings is not so manifest in Section 2 (cf. [12], however), the notion of (partial) order enters sections 1 and 3 quite explicitly. We proceed to give a brief survey of our results.

Section 1 is concerned with a type of functional equation that has been called algebraic by Schmeidler [13]; equation (i) generalizes Schmeidler's homogeneous algebraic integral equations of order n [14]; a somewhat more special approach was made in [9]. The systematic use of the notion of positivity in the algebra $\mathscr{C}(X)$ (X compact Hausdorff) permits to extend to equation (i) those results for Fredholm integral equations that are typical in the presence of a non-negative, continuous kernel (Theorem 1); there is a complete analogue of the theorem of Jentzsch establishing the existence of a positive eigenvalue with a non-negative eigenfunction (Theorem 2). Since powers of the unknown function are essentially involved in (i), $\mathscr{C}(X)$ seems to be the natural domain for this equation; that this domain, X being an arbitrary compact space, is not too narrow can be seen from the relationship of Theorem 2 with the existence of an eigenvalue for a compact, positive Hermitian operator.

Section 2 considers the eigenvalue problem for a linear operator function with polynomial-type dependence of a complex parameter; the literature treating such problems is fairly extensive (see, e.g., [5]). We take up the problem for a class of functions where the coefficients of the operator polynomial are commuting spectral operators with countable (but not necessarily bounded) spectrum; here the theory of spectral measures, recently developed for arbitrary locally convex spaces [12], provides a perfect tool to obtain a full generalization of the corresponding linear case. The author does not know of any interesting application in physics, but the example at the end of the section may give a hint to a reader more familiar with applied problems.

The third section studies a type of problem that does have many applications, and that has been studied as far back as 1922, 1923 by Birkhoff-Kellogg and Uryson respectively, and later by Hammerstein, Krasnosel'skii and others. We consider an eigenvalue problem treated as an application in an earlier paper [10], now in a more systematic way. The result is a global existence Theorem (Theorem 4) for positive solutions, whose essential feature is an application of Schauder's fixed point theorem without the use of a priori estimates. This result is applied to a boundary problem involving the Laplacian and, more concretely, to the longitudinal bending of a compressed rod.

§1. A Functional Equation of Algebraic Type

1. Let X be a compact Hausdorff space with elements s, t, ...; $\mathscr{C}(X)$ the Banach algebra of real-valued, continuous functions on X. Elements of $\mathscr{C}(X)$ will generally be denoted by x, y, ...; e denotes the unit element of $\mathscr{C}(X)$, i.e., the function $s \rightarrow 1$. As usual, $x \geqq y$ means $x(s) \geqq y(s)$ for all $s \in L$, and K denotes the positive cone $\{x : x \geqq 0\}$ of $\mathscr{C}(X)$. Let $n \geqq 1$ be an integer, and P_ν $(0 \leqq \nu \leqq n-1)$ continuous maps of K into K satisfying the two conditions:

(a) P_ν <u>is monotone, i.e.,</u> $P_\nu(x) \geqq P_\nu(y)$ <u>when</u> $x \geqq y$, <u>and positive homogeneous of degree</u> $n - \nu$.

(b) <u>For each bounded subset</u> $B \subset K$, $\{P_\nu(x) : x \in B\}$ <u>is equicontinuous</u>.

We consider the functional equation

$$\kappa^n x^n - \sum_{\nu=0}^{n-1} \kappa^\nu x^\nu P_\nu(x) = y \quad , \tag{i}$$

where κ is a positive real parameter. A solution $x \geqq 0$ of (i) is called maximal if for each $s \in X$, $x(s) = 0$ implies $P_\nu(x)[s] = 0$ $(1 \leqq \nu \leqq n-1)$.

THEOREM 1. *There exists a maximal open interval* $J = (\kappa_0, \infty)$ *such that for each* $\kappa \in J$ *and* $y \in K$, (i) *has a non-negative solution* $x = R(\kappa, y)$. *This solution can be obtained by iteration and is the smallest maximal solution of* (i) *on* $D = J \times K$; *moreover,* R *is a monotone function of* κ^{-1} *and* y, *and lower semi-continuous on* D.

Theorem 2 (below) shows κ_0 to be an eigenvalue of (i) when $\kappa_0 > 0$, and gives conditions for this situation to occur.

2. Proof of Theorem 1. We begin the proof by defining J. Let us note first that the greatest non-negative root ξ_0 of the algebraic equation

$$\xi^n - \sum_{\nu=0}^{n-1} a_\nu \xi^\nu = 0 \tag{1}$$

is well defined and a monotone, continuous function $(a_0, \ldots, a_{n-1}) \to \xi_0$ on the positive orthant of the real coefficient space R^n. Thus if $w \in K$ is fixed, $y \in K$ is fixed and $\kappa \in R^+$ then x, defined by

$$\kappa^n x^n(s) - \sum_{\nu=0}^{n-1} \kappa^\nu x^\nu(s) P_\nu[\ ](s) = y(s)$$

as the greatest real root of the corresponding equation (1) for each $s \in X$, is in K. The mapping $(\kappa, w, y) \to x = \Phi(\kappa, w, y)$ so defined is clearly continuous on $R^+ \times K \times K$ into K. Letting $x_0 = 0$,

$$x_{m+1} = \Phi(\kappa, x_m, y) \qquad (m = 0, 1, 2, \ldots)$$

determines a sequence $\{x_m\} \subset K$ which is obviously monotone.

We define J to be the set of all $\kappa > 0$ such that $\{x_m\}$ is bounded for each $y \in K$ [1]. If $\kappa \in J$, then $\lim x_m = x^*$ is clearly a Baire function on X. By the compactness of P_ν, $\lim P_\nu(x_m)$

(which exists by the monotonicity of P_ν) is an element $z_\nu \in K$, and x^* satisfies the equation

$$\kappa^n (x^*)^n - \sum_{\nu=0}^{n-1} \kappa^\nu (x^*)^\nu z_\nu = y . \tag{2}$$

Now for each $s \in X$, $\{x_m\}$ being monotone, $x^*(s) = 0$ implies $x_m(s) = 0$ for all m, hence $P_\nu[x_m](s) = 0$ and, therefore, $z_\nu(s) = 0$ $(0 \leqq \nu \leqq n-1)$. Hence x^* is the greatest (real) solution of (2) which implies that $x^* \in K$. Thus the convergence $x_m(s) \to x^*(s)$ is uniform on X by Dini's theorem, and we have $z_\nu = P_\nu(x^*)$ since P_ν are continuous on K.

It follows that x^* is a maximal solution for the pair (κ, y); it is the smallest maximal solution for this pair since, if $\tilde{x}$ is another maximal solution for (κ, y), then $x_m \leqq \tilde{x}$ for each m and hence $x^* \leqq \tilde{x}$.

Thus, if $\kappa \in J$, we define $R(\kappa, y) = x^*$. It is not difficult to see that R is a monotone function of κ^{-1} and y on its domain $D = J \times K$; to see that R is lower semi-continuous (in the sense that $R(\kappa, y) \geqq R(\kappa_0, y_0) - \varepsilon e$ for preassigned $\varepsilon > 0$ and (κ, y) in a suitable neighborhood U of (κ_0, y_0), in D) it is sufficient to note that $(\kappa, y) \to x_m$ is continuous on D into K for each m; it follows then from a standard argument that $\sup x_m = R(\kappa, y)$ is semi-continuous from below.

The remainder of the proof consists now in showing that the set J is not empty and indeed an open, semi-infinite interval. Let us remark first that $\kappa \in J$ whenever $R(\kappa, e)$ exists; if $x = R(\kappa, e)$ then, by the form of (i) and the assumed homogeneity of the P_ν, $cx = R(\kappa, c^n e)$ for any $c \in R^+$. Now for each $y \in K$, there exists $c \in R^+$ such that $y \leqq c^n e$ and it follows that $x_m \leqq cR(\kappa, e)$ for all m where $\{x_m\}$ is the sequence formed for (κ, y).

This implies that J is non-empty; for there exists $\kappa_1 > 0$ such that

$$(\kappa_1 x)^n - \sum_{\nu=0}^{n-1} (\kappa_1 x)^\nu P_\nu(e) = e$$

implies $\|x\| \leqq 1$ (equivalently, $x \leqq e$), hence $x_m \leqq e$ for (κ_1, e) $(m = 1, 2, \ldots)$. Moreover, since each mapping $(\kappa, e) \to x_m$ is a monotone function of κ^{-1}, it follows that $\gamma \in J$ whenever $\gamma \geqq \kappa$ and $\kappa \in J$. Therefore, J is a semi-infinite interval of positive real numbers; the proof of Theorem 1 will be complete when we show that J is open.

The remarks of the preceding paragraph imply that, whenever $\kappa \in J$,

$$\| R(\kappa, y) \| \leqq C(\kappa) \| y \|^{1/n} \tag{3}$$

with $C(\kappa) = \| R(\kappa, e) \|$. To prove that J is open on the left, assume that $\lambda^{-1} \in J$ and consider the equation

$$x^n - \sum_{\nu=0}^{n-1} (\lambda+\delta)^{n-\nu} x^\nu P_\nu(x) = (\lambda+\delta)^n e , \tag{4}$$

which is equivalent to (i) (with $y = e$) at $\kappa = (\lambda+\delta)^{-1}$. Expansion with respect to δ yields

$$x^n - \sum_{\nu=0}^{n-1} \lambda^{n-\nu} x^\nu P_\nu(x) = (\lambda+\delta)^n e + \delta(\ldots) + \ldots + \delta^n P_0(u) \tag{5}$$

where u has been substituted for x in the right-hand member. If we assume $u \in K$ to be given then, since $\lambda^{-1} \in J$, (5) has a well-defined solution $x = R(\lambda^{-1}, g(\delta, u))$ where the right-hand member of (5) is denoted by $g(\delta, u)$. We define $\tilde{x}_0 = 0$ and

$$\tilde{x}_{m+1} = R[\lambda^{-1}, g(\delta, \tilde{x}_m)] \qquad (m = 0, 1, \ldots) .$$

We claim that for δ small enough, the sequence $\{\tilde{x}_m\}$ is bounded. It follows from (3) that $\| g(\delta, u) \| \leqq (\lambda+1)^n$ implies

$$\| R[\lambda^{-1}, g(\delta, u)] \| \leqq C(\lambda^{-1})(\lambda+1) \quad ;$$

hence if we choose δ small enough for $\| u \| \leqq C(\lambda^{-1})(\lambda+1)$ to force $\| g(\delta, u) \| \leqq (\lambda+1)^n$, which is clearly possible, then

$$\| \tilde{x}_m \| \leqq C(\lambda^{-1})(\lambda+1)$$

for all m . As above, this implies that $\tilde{x} = \lim \tilde{x}_m$ is a continuous, maximal solution of (4) and, therefore, of (i) for $y = e$ and $\kappa = (\lambda+\delta)^{-1}$. Hence $\lambda^{-1} \in J$ implies $(\lambda+\delta)^{-1} \in J$ for a suitable $\delta > 0$. Consequently, J is open and the proof of Theorem 1 is complete.

REMARK. The properties of R established in Theorem 1 imply that $\kappa \to C(\kappa) = \| R(\kappa, e) \|$ is a decreasing function of κ on J, continuous on the right. Moreover, we must have

$$\lim_{\kappa \to \kappa_0} C(\kappa) = +\infty . \tag{6}$$

This is clear when $\kappa_0 = 0$. Assume that $\kappa_0 > 0$; if $\kappa \to C(\kappa)$ were bounded on J , it would follow from the last part of the preceding proof that for some fixed $\delta_0 > 0$, $\lambda^{-1} \epsilon J$ implies $(\lambda + \delta_0)^{-1} \epsilon J$ for all $\lambda^{-1} > \kappa_0$, which is contradictory.

3. A number $\rho > 0$ is called an eigenvalue of (i) if (i) has a continuous, non-zero solution for $\kappa = \rho$ and $y = 0$. From the homogeneity of the coefficients P_ν it follows that when x_0 (ϵ K) is an eigensolution of (i), then so is cx_0 for any $c \epsilon R^+$.

THEOREM 2. _If_ $\kappa_0 > 0$ (_that is, if_ $J \neq R^+$) _then_ κ_0 _is an eigenvalue of_ (i) _with an eigenfunction_ $x_0 \epsilon K$. _This will occur, in particular, when the following condition is satisfied: There exists a continuous functional_ f _on_ K, _monotone and homogeneous of degree one, with_ $f(e) > 0$ _and such that_

$$f[P_\nu(x)] \geqq \delta f(x^{n-\nu})$$

for some $\delta > 0$, _some_ ν $(0 \leqq \nu \leqq n-1)$ _and all_ $x \epsilon K$. _In this case,_ $\kappa_0 \geqq \delta^{\frac{1}{n-\nu}}$.

Proof. Let $\kappa_0 > 0$. By the remark at the end of the previous section, we have $C(\kappa) \to +\infty$ as $\kappa \downarrow \kappa_0$. Let $\{\kappa_m\}$ denote a decreasing sequence with limit κ_0, let $x_m = R(\kappa_m, e)$ and $z_m = x_m / \| x_m \|$; clearly $\| x_m \| \to \infty$.
Now

$$\kappa_p^n z_p^n - \sum_{\nu=0}^{n-1} \kappa_p^\nu z_p^\nu P_\nu(z_p) = e \| x_p \|^{-n}$$

for $p = 1, 2, \ldots$; on the other hand, since P_ν $(0 \leqq \nu \leqq n-1)$ are compact, $\{z_p\}$ is a bounded, equicontinuous sequence in K. For the set $\{P_\nu(z_p)\}$ is equicontinuous, and the greatest real root of (1) is a uniformly continuous function on each bounded subset of the positive orthant in R^n. Hence if x_0 is an accumulation point of $\{z_p\}$, $x_0 \epsilon K$ is clearly an eigenfunction of (i) for $\kappa = \kappa_0$, of norm one. This completes the proof of the first assertion.

To prove the second part of the theorem, it is sufficient to show that there exists, under the condition indicated, a $\kappa > 0$ such that the sequence $\{x_m\}$ (see No. 2 above) diverges for the pair (κ, e). Since $x_{m+1}(s)$ is the greatest real root of (i) for the coefficients $a_\nu = P_\nu[x_m](s)$ $(1 \leqq \nu \leqq n-1)$, $a_0 = P_0[x_m](s) + 1$ respectively, it follows that

$$(\kappa x_{m+1})^{n-\nu} \geqq P_\nu(x_m) ,$$

hence

$$\kappa^{n-\nu} f(x_{m+1}^{n-\delta}) \geqq f(x_m^{n-})$$

since f is positive homogeneous on K. Since f is monotone, $f(e) > 0$ and $x_m \geqq \kappa^{-1}e$, it follows that $f(x_m^{n-\nu}) > 0$ for all m. (The conditions imposed on f imply, incidentally, that f is non-negative on all of K.) Thus if $\delta\kappa^{\nu-n} > 1$, the sequence $f(x_m^{n-\nu})$ is unbounded which implies, by the monotonicity of f and $\{x_m\}$, that $\{x_m\}$ is unbounded. Hence

$$\kappa_0 \geqq \delta^{\frac{1}{n-\nu}} > 0$$

and the proof of Theorem 2 is complete.

4. We consider some illustrations for Theorems 1 and 2. Let ν and τ be fixed, $0 \leqq \nu \leqq n-1$, $1 \leqq \tau \leqq n-\nu$, and let $\alpha_1, \ldots, \alpha_\tau$ be positive integers satisfying $\alpha_1 + \ldots + \alpha_\tau = n-\nu$. If X^τ is the topological product of τ factors X, $s \to \mu_s$ a norm-continuous map of X into the space of positive Radon measures on X^τ, then

$$P_\nu(x): s \to \mu_s(x^{\alpha_1} \otimes \ldots \otimes x^{\alpha_\tau})$$

determines a mapping $x \to P_\nu(x)$ satisfying conditions (a) and (b) of No. 1. More generally, when $(\alpha_1, \ldots, \alpha_\tau)$ ranges over all complexes of positive integers with sum $n-\nu$, the corresponding sum of mappings P_ν is of the type considered; in particular, so is P_ν when defined by

$$P_\nu[x](s) = \sum_{\alpha_1 + \ldots + \alpha_\tau = n-\nu} \int \cdots \int K_{\nu, \alpha_1 \ldots \alpha_\tau}(s, t_1 \ldots t_\tau)\, x^{\alpha_1}(t_1) \ldots x^{\alpha_\tau}(t_\tau) \times d\mu_1(t_1) \ldots d\mu_\tau(t_\tau) \tag{7}$$

for $s \in X$ where $K_{\nu, \alpha_1 \ldots \alpha_\tau}$ are non-negative, continuous functions on $X^{\tau+1}$, $\mu_1 \ldots \mu_\tau$ positive measures on X. Equations (i) with P_ν as in (7) were called homogeneous algebraic equations of order n in [14] and studied in [9], [14] with varying generality. It is clear that for $n = 1$, (i) contains Fredholm integral equations with non-negative kernel on a compact domain X.

For such equations and under the assumption that the kernel is strictly positive, the theorem of Jentzsch [1] asserts the existence of a positive eigenvalue of maximum modulus with a strictly positive eigenfunction. Generalizations of that theorem and refinements are too numerous[2] to be cited here but for the present, nonlinear cases seem to have been studied, in some generality, only in [8], [10], [14]. It is manifest that Theorem 2 above is a result of this type; let us examine conditions under which $\kappa_0 > 0$ when P_ν $(0 \leqq \nu \leqq n-1)$ are mappings of the form (7). We confine ourselves to giving two criteria guaranteeing that $\kappa_0 > 0$.

(1) If there exists an index $(\nu ; \alpha_1, \ldots, \alpha_T)$ such that

$$\delta = \inf_{s \in X} \int \cdots \int K_{\nu, \alpha_1 \ldots \alpha_T}(s, t_1 \ldots t_T)\, d\mu_1(t_1) \ldots d\mu_T(t_T) > 0$$

then

$$\kappa_0 \geqq \delta^{\frac{1}{n-\nu}} .$$

It suffices to verify that the conditions of Theorem 2 are satisfied by the functional $x \to f(x) = \inf \{x(s) : s \in X\}$; the easy proof will be omitted. Condition (1) is in particular fulfilled when at least one of the kernels K has a positive lower bound.

(2) Assume some P_ν contains a summand of the form $\int K(s, t)\, x^{n-\nu}(t)\, d\mu(t)$ such that the linear transformation $x \to \int K(\cdot, t)\, x(t)\, d\mu(t) = K_\mu[x]$ does not have spectrum $\{0\}$. If λ_0 denotes the greatest positive eigenvalue of K_μ, then $\kappa_0 \geqq \lambda_0^{\frac{1}{n-\nu}}$.

It is well known (see, e.g., [11]) that λ_0 is also an eigenvalue, with (at least one) positive eigenvector ρ, of the adjoint K'_μ of K_μ on the space of Radon measures on X. We take f to be the measure ρ. Then clearly $f(e) > 0$, and

$$f[P_\nu(x)] = \int\int K(s, t)\, x^{n-\nu}(t)\, d\mu(t)\, d\rho(s) = \lambda_0 f(x^{n-\nu})$$

from which the assertion follows.

5. Let A_ν $(\nu = 0, \ldots, n-1)$ be a commuting set of compact positive Hermitian operators, not all 0, on some Hilbert space H; does the eigenvalue problem

$$(\lambda^n - \sum_{\nu=0}^{n-1} \lambda^\nu A_\nu)\, x = 0 \tag{8}$$

have a non-trivial solution $\lambda = \lambda_0$, $x = x_0 \in H$? Suppose first that all but one of the A_ν are of finite rank. Recall that the smallest closed real subalgebra of $\mathscr{L}(H)$ ($\mathscr{L}(H)$ the B-algebra of bounded operators on H, with unit I) containing I and all A_ν, is isomorphic with $\mathscr{C}(X)$ where $X = \Pi_\nu \sigma(A_\nu)$, $\sigma(A_\nu)$ the spectrum of A_ν. If we define $u \to P_\nu(u) = A_\nu u^{n-\nu}$ on $\mathscr{C}(X)$, then P_ν are quickly seen to be compact, hence to satisfy conditions (a) and (b) of No. 1. If $A_\nu \neq 0$ and $0 < \lambda \in \sigma(A_\nu)$, let $\xi \in X$ be such that its ν-th coordinate is equal to λ. Then Theorem 2 is applicable with $f(u) = u(\xi)$, the value of $u \in \mathscr{C}(X)$ at ξ. Hence there exists a non-zero $u_0 \geqq 0$ and $\lambda_0 \geqq \lambda^{\frac{1}{n-\nu}}$ such that $(\lambda_0^n - \sum \lambda_0^\nu A^\nu) u_0^n = 0$; every $x \in H$ in the range of u_0^n (in fact, of u_0) is an eigenvector of (8). The case where A_ν are not of finite rank is easily handled by a limiting process.

It can be justly objected that Theorem 2 is entirely unnecessary to obtain the result of the preceding paragraph; we shall see in the next section that problem (8) is capable of a much simpler and much more complete solution. Our point is, however, to indicate the relationship of the generalized Jentzsch theorem, furnished by Theorem 2 above, with the well-known result that every compact, positive Hermitian operator has at least one positive characteristic value.

§2. An Eigenvalue Problem Nonlinear in the Parameter

1. We recall some known results from the spectral theory of linear operators. Let E denote a (Hausdorff) locally convex vector space; a continuous endomorphism A of E, $A \in \mathscr{L}(E)$, is spectral [12, Def. 4] if there exists a compact space X and a continuous homomorphism ν of the Banach algebra $\mathscr{B}(X)$ of bounded, complex-valued Baire functions on X into the algebra $\mathscr{L}(E)$ (under, say, the topology of bounded convergence) such that $\nu(1) = I$ (the identity operator) and $A = \nu(f) = \int f d\nu$, for some $f \in \mathscr{B}(X)$.[3] This implies that the spectrum $\sigma(A)$ (complement of the domain of local holomorphy of the resolvent) is compact, and that there exists a unique Baire measure μ on $\sigma(A)$ into $\mathscr{L}(E)$ such that $f \to \int f d\mu$ is again a continuous homomorphism on $\mathscr{B}[\sigma(A)]$ into $\mathscr{L}(E)$, and $A = \int t d\mu(t)$. μ is called the complex spectral measure associated with A.

Under certain simple (necessary and sufficient) conditions [12, Th. 2] the tensor product $\mu = \mu_0 \otimes \dots \otimes \mu_{n-1}$ exists for the complex spectral measures associated with each member of a commuting family A_ν ($\nu = 0, 1, \dots, n-1$) of spectral operators on E; that is, there exists a unique spectral measure μ on the topological product $X = \Pi_\nu \sigma(A_\nu)$ such that

$$A_\nu = \int f_\nu d\mu \qquad (\nu = 0, 1, \dots, n-1)$$

where f_ν is the projection of X onto $\sigma(A_\nu)$.

If E is a Hilbert space, an operator $A \in \mathscr{L}(E)$ is spectral if and only if it is similar to a normal operator (a result due to Mackey); normal operators are well known to be spectral. According to the distinction between operators similar to normal or Hermitian ones in Hilbert space, in the general case one distinguishes between complex and real spectral operators, i.e., operators with complex or real spectrum, respectively. Spectral operators with non-negative real spectrum are called positive.

The notion of spectral operator on a locally convex space can be extended to operators with unbounded spectrum [12, Section 5]. Such operators are not necessarily continuous, and not necessarily defined everywhere on E, but they are always closed and have dense domain. The spectrum of such an operator will be understood to be the closure of its finite spectrum on the Riemann sphere (and hence will be compact), with $\mu(\infty) = 0$ for the complex spectral measure associated with A.

We remark finally that the tensor product $\bigotimes_{\nu=0}^{n-1} \mu_\nu$ of the associated complex spectral measures of n (bounded or unbounded) normal operators A_ν on Hilbert space exists if and only if the family $\{A_\nu\}$ commutes [12, p. 141].

2. We shall be concerned with the eigenvalue problem for the operator function

$$\lambda \to A(\lambda) = \lambda^n - \sum_{\nu\ 0}^{n-1} \lambda^\nu A_\nu \ , \qquad \text{(ii)}$$

where λ is a complex parameter and n a positive integer. λ_0 will be called a eigenvalue of (ii) if $A(\lambda_0)x = 0$ for some $0 \neq x \in E$ (E a locally convex space), and λ_0 will be called regular if $\lambda \to A(\lambda)^{-1}$ is holomorphic in a neighborhood of λ_0. The following assumptions are made throughout this section:

(a) A_ν $(\nu = 0, \ldots, n-1)$ are spectral operators on E for whose associated complex spectral measures the tensor product $\mu = \bigotimes_{\nu=0}^{n-1} \mu$ exists;

(b) The spectrum $\sigma(A_\nu)$ $(\nu = 0, \ldots, n-1)$ is countable.

We remark that these conditions imply that the point spectrum of each A_ν is dense in $\sigma(A_\nu)$, since every countable, closed subset of the Riemann sphere is the closure of its subset of isolated points. If some (or all) of the A_ν are not defined everywhere on E (as is necessarily the case if, e.g., E is a Hilbert space and $\sigma(A_\nu)$ unbounded), then the operator function (ii) is, for each (finite) complex λ, defined as the unique spectral extension of

$\lambda^n - \sum_{\nu=0}^{n-1} \lambda^\nu A_\nu$ [12, Propositions 23 and 25].

THEOREM 3. The set $\mathscr{E}$ of eigenvalues of (ii) is non-empty, and at most countably infinite; every complex λ is either regular for (ii) or in the closure of $\mathscr{E}$. There are infinitely many distinct eigenvalues of (ii) if and only if at least one A_ν has this property. Moreover, there exists a representing set $\mathscr{E}_0 = \{\lambda_m\}$ of eigenvalues of (ii) such that every $x \in \mathscr{E}$ is an unconditional (weak) sum, $x = \sum x_m$, where x_m is an element of the eigenspace pertaining to λ_m.[4]

If all A_ν are real spectral operators and if n is odd, or if all A_ν are positive spectral operators and n is arbitrary, then $\mathscr{E}_0$ can be chosen as a set of real numbers, or of real non-negative numbers, respectively.

When all A_ν are compact, then 0 is the only possible accumulation point of $\mathscr{E}$ and $\lambda \to A(\lambda)^{-1}$ is meromorphic on the sphere punctured at 0, with poles of order at most n.

3. The proof of Theorem 3 will follow from several lemmas. We consider the topological product $X = \prod_\nu \sigma(A_\nu)$ and denote by X_0 the support of the product measure $\mu = \bigotimes_\nu \mu_\nu$ (condition (a) above); X_0 is a closed, hence compact subspace of X. An atom of μ is an $s = (\xi_0, \ldots, \xi_{n-1}) \in X_0$ such that $\mu(s) \neq 0$. Since $\mu_\nu(\infty) = 0$, a point with at least one infinite coordinate can never be an atom of μ.

LEMMA 1. The set $\mathscr{E}$ of eigenvalues of (ii) is the totality of roots of all algebraic equations

$$\lambda^n - \sum_{\nu=0}^{n-1} f_\nu(s)\lambda^\nu = 0 \qquad (1)$$

where s is an atom of μ, and f_ν the projection $X \to \sigma(A_\nu)$.

Proof. Let s be an atom of μ, and λ a root of (1). If $\mu(s) = P$, then $f_\nu(s)P = \int f_\nu d\mu \cdot \mu(s) = A_\nu P$, hence

$$\lambda^n P - \sum_{\nu=0}^{n-1} \lambda^\nu A_\nu P = 0$$

which shows λ to be an eigenvalue of (ii).

Conversely, let $\lambda_0 \in \mathscr{E}$. Since $A(\lambda_0)$ is a spectral operator, there exists a projection P, mapping E onto the nullspace of $A(\lambda_0)$; we have $P = \int_\delta d\mu$ where

$$\delta \subset \{s \in X_0 : \lambda_0^n - \sum_0^{n-1} f_\nu(s) \lambda_0^\nu = 0\}$$

consists only of atoms of μ since X_0 is countable. (We remark that, when all A_ν are compact, δ is necessarily finite for $\lambda_0 \neq 0$ since P must then be of finite rank.)]

COROLLARY. If all A_ν are real and n is odd, or if all A_ν are positive, then for each atom s of μ, (1) has a real or positive root, respectively.

The corollary is immediate since, under the assumptions made, (1) has real or real non-negative coefficients, respectively.

LEMMA 2. $\mathscr{E}$ is non-empty and at most countable.

Since X_0 is countable, we have $I = \mu(X_0) = \Sigma\, \mu(s_m)$ where $\{s_m\}$ is the set of atoms of μ. Hence $\{s_m\}$ is not empty and so is $\mathscr{E}$ by Lemma 1; on the other hand, the set of algebraic equations (1) with n-tuples of coefficients $\{f_\nu(s_m)\}$ $(m = 1, 2, \ldots)$ is countable, hence the same is true of $\mathscr{E}$.

If for each s_m a representative root λ_m is selected, then for any $x \in \mathscr{E}$, $x_m = \mu(s_m)x$ is in the eigenspace of (ii) corresponding to λ_m and $x = \Sigma\, x_m$ since $I = \Sigma\, \mu(s_m)$ [12, p. 156]. Hence by the corollary of Lemma 1, $\mathscr{E}_0 = \{\lambda_m\}$ can be chosen so as to satisfy the assertions of the theorem.

LEMMA 3. $\mathscr{E}$ is infinite if and only if at least one A_ν has infinitely many distinct eigenvalues.

For if each A_ν has not more than finitely many eigenvalues, it follows from condition (b), since A_ν are spectral operators, that each $\sigma(A_\nu)$ is finite. Hence X, and consequently X_0, is finite; thus $\mathscr{E}$ is finite by Lemma 1. Conversely, if $\mathscr{E}$ is finite, then the number of coefficients of all algebraic equations of order n with roots in $\mathscr{E}$ is finite, whence it follows that X_0 is finite. Now if $\sigma(A_\nu)$ were infinite for some ν, $0 \leqq \nu \leqq n-1$, A_ν would have infinitely many eigenvalues, say $\{\lambda_k\}$, and each of the disjoint sets $f_\nu^{-1}(\lambda_k)$ $(k = 1, 2, \ldots)$ would have to contain an atom of μ, since $\mu_\nu = \mu(f_\nu^{-1})$. This is contradictory and, therefore, $\sigma(A_\nu)$ is finite for all ν.

There remains to prove that any λ, not regular for (ii), is in the closure of $\mathscr{E}$ and finally, the last assertion of the theorem concerning compact A_ν. We take up these steps in order.

LEMMA 4. <u>If</u> λ_0 <u>is not regular for</u> (ii), <u>then</u> $\lambda_0 \in \bar{\mathscr{E}}$.

<u>Proof.</u> If ∞ is not regular, it is easy to see that $\infty \in \bar{\mathscr{E}}$. Thus let $\lambda_0 \neq \infty$ and assume that $\lambda_0 \notin \bar{\mathscr{E}}$. Suppose first that all spectra $\sigma(A_\nu)$ are bounded (hence compact subsets of the complex plane). There exists $\delta > 0$ such that

$$|\lambda_0^n - \sum_{\nu=0}^{n-1} f_\nu(s)\lambda_0^\nu| \geqq \delta \tag{2}$$

for all $s \in X_0$. Otherwise, by the compactness of X_0 and the continuity and boundedness of $f_\nu (0 \leqq \nu \leqq n-1)$, λ_0 would be a root of (1) for some $s_0 \in X_0$; since the set of atoms of μ is clearly dense in X_0, there would exist an atom $s \in X_0$ such that at least one root of the corresponding equation (1) were contained in a given neighborhood of λ_0, contradicting the assumption $\lambda_0 \notin \bar{\mathscr{E}}$. Hence (2) holds and, since μ is a spectral measure on X_0, it follows that

$$\lambda \rightarrow A(\lambda)^{-1} = \int \Big[\lambda^n - \sum_{\nu=0}^{n-1} f_\nu(s)\lambda^\nu\Big]^{-1} d\mu(s) \tag{3}$$

is holomorphic near λ_0. In the general case, consider the operator on $\mathscr{E}$:

$$Q = \int \Big[1 + \sum_{\nu=0}^{n-1} |f_\nu(s)|\Big]^{-1} d\mu(s) \quad ;$$

Q is clearly one-to-one, $QA_\nu = B_\nu$ $(0 \leqq \nu \leqq n-1)$ has bounded spectrum and $\lambda \rightarrow B(\lambda) = QA(\lambda)$ has the same eigenvalues as (ii). Thus if $\lambda_0 \notin \bar{\ }$, it follows from what has been shown that for some $\eta > 0$,

$$|\lambda_0^n - \sum_{\nu=0}^{n-1} f_\nu(s)\lambda_0^\nu| \geqq \eta(1 + \sum_{\nu=0}^{n-1} |f_\nu(s)|)$$

for all $s \in X_0$; hence (2) holds <u>a fortiori</u> with δ replaced by η and the proof is completed as before.

LEMMA 5. If all A_ν are compact, then 0 is the only possible accumulation point of $\mathscr{E}$, and the elements of $\mathscr{E} \sim \{0\}$ are poles of $A(\lambda)^{-1}$ of order at most n.

Proof. We assume there exists a sequence of distinct elements $\lambda_k \in \mathscr{E}$ such that $\lambda_k \to \lambda_0 \neq 0$. By Lemma 1, there exists an infinite set $\{s_k\}$ of atoms of μ such that λ_k is a root of (1) for the coefficients $\{f_\nu(s_k)\}$. Since X_0 is compact, the sequence $\{s_k\}$ can without restriction of generality be assumed to converge, $s_k \to s$ say. We consider the sequence $\{\tau_k\}$ given by

$$\tau_k = \sum_{\nu=0}^{n-1} \lambda_0^\nu f_\nu(s_k) \quad ;$$

$\{\tau_k\}$ is by necessity infinite-valued. Clearly $\tau_k \to \lambda_0^n$ as $k \to \infty$; on the other hand, $B = \sum_{\nu=0}^{n-1} \lambda_0^\nu A_\nu$ being a compact operator on $\mathscr{E}$, each $\tau_k \neq 0$ is an eigenvalue of B since $\tau_k \in \sigma(B)$. For $B = \int g\,d\mu$ where g is continuous on X_0, and this implies that $\sigma(B) = g(X_0)$ [12, Proposition 6 and Theorem 4]. Hence B has (for the same reason) the eigenvalue $\lambda_0^n \neq 0$, limit of the eigenvalues τ_k which is contradictory.

There remains to show that if $0 \neq \lambda_0 \in \mathscr{E}$, then λ_0 is a pole of order $\leqq n$. But since $\sum_{\nu=0}^{n-1} \lambda_0^\nu A_\nu$ is compact, the number of atoms s_ρ for which λ_0 is a root of (1) is necessarily finite, say r : $\rho = 1, \ldots, r$. Now

$$A(\lambda)^{-1} = \sum_{\rho=1}^{r} \frac{\mu(s_\rho)}{\lambda^n - \Sigma f_\nu(s_\rho)\lambda^\nu} + \sum_{s \neq s_\rho} \frac{\mu(s)}{\lambda^n - \Sigma f_\nu(s)\lambda^\nu}$$

where the second term on the right is holomorphic near λ_0; the first term clearly has a pole at λ_0 whose order is the greatest multiplicity of λ_0 as a root of (1) for $s = s_\rho$ $(\rho = 1, \ldots, r)$, hence $\leqq n$.

With the proof of Lemma 5 the proof of Theorem 3 is complete.

4. It is not difficult to construct operator functions (ii) where A_ν are commuting normal operators in Hilbert space.[5] We should like to indicate an example that goes beyond this classical setting.

Let $\mathscr{D}'_T$ be the space of periodic distributions (in the sense of L. Schwartz), i.e., the space of distributions one the one-dimensional torus T. Let D denote differentiation: $D\varphi = \varphi'$, and $\varphi \to g * \varphi$ convolution with some fixed real $g \in \mathscr{D}'_T$. Theorem 3 applies to the eigenvalue problem

$$\lambda^n \varphi + \sum_{\nu=1}^{n-1} \lambda^\nu D^{p\nu} \varphi + g * \varphi = 0 . \tag{4}$$

In particular, if n is odd and all p_ν even $(\nu = 1, \ldots, n-1)$, $p_\nu \geqq 2$, then (4) has infinitely many real eigenvalues.

§3. An Eigenvalue Problem of Hammerstein Type

1. By an ordered Banach space E, we understand in this section a real Banach space in which a convex cone K of vertex 0 is distinguished; K is assumed to be closed, and proper (i.e., $K \cap -K = \{0\}$). The ordering $x \leqq y$ on E is then defined by $y - x \in K$, and K is called the positive cone of E.

Let E_1, E_2 be ordered Banach spaces with respective positive cones K_1, K_2, K_2 being a total subset of E_2; we assume that there exists an injection map j of E_1 into E_2 which is continuous and positive (i.e., $jK_1 \subset K_2$). We consider a linear operator P on E_2 into E_1, and a mapping f on K_1 into K_2 having the following properties:

(a) P *is a positive, compact operator such that* jP *is not quasi-nilpotent as an endomorphism of* E_2;

(b) f *is continuous and* $f(u) \geqq ju$ *for all* $u \in K_1$.[6]

Clearly jP is a compact linear operator in E_2; since it is not quasi-nilpotent, its spectral radius r(P) is positive and hence, by a classical theorem of Krein-Rutman, an eigenvalue of jP with an eigenfunction $u_0 \geqq 0$.

THEOREM 4. *There exists a family* $\{u_c : c > 0\} \subset K_1$, *satisfying* $\|u_c\| = c$, *and a corresponding family* $\{\lambda_c\}$ *of "eigenvalues" solving the problem*

$$\lambda u = P \circ f(u) . \tag{iii}$$

Moreover, $\lambda_c \geqq r(P)$ *for all* $c > 0$.

The problem $\lambda u = P \circ f(u)$ can appropriately be called an eigenvalue problem only when $f(0) = 0$; since this assumption is not needed in our proof of Theorem 4, it has not been postulated.

2. Proof of Theorem 4. We note first that $r(P)$ is also an eigenvalue (in fact, also the greatest eigenvalue $\geqq 0$) of the adjoint $P'j'$ of jP in E'_2. There exists [10, Prop. 4] a positive eigenvector $\varphi_0 \in K'_2$ of $P'j'$ such that $\|j'\varphi_0\| = 1$ and $\langle u_0, \varphi_0\rangle = \varphi_0(u_0) > 0$.

Let $c > 0$ be fixed, $\varepsilon > 0$ an arbitrary positive number. We show that there exists $u_\varepsilon \in K_1$ and $\lambda_\varepsilon > r(P)$ satisfying

$$\lambda_\varepsilon u_\varepsilon = P \circ f_\varepsilon(u_\varepsilon) \ , \quad \|u_\varepsilon\| = c$$

where $f_\varepsilon(u) = f(u) + \varepsilon u_0$. Consider the set $\Gamma = \{u \geqq 0 : \|u\| \leqq c\}$ in K_1 ; Γ is closed and convex. For $u \in \Gamma$, we have (since $j'\varphi_0 \geqq 0$ and $\|j'\varphi_0\| = 1$)

$$\|P \circ f_\varepsilon(u)\| \geqq \langle P \circ f_\varepsilon(u), j'\varphi_0\rangle \geqq \varepsilon\langle jPu_0, \varphi_0\rangle = \varepsilon r(P)\langle u_0, \varphi_0\rangle > 0 \ ;$$

consequently, $u \to \|P \circ f_\varepsilon(u)\|^{-1}$ is continuous on Γ. The map

$$u \to c\|P \circ f_\varepsilon(u)\|^{-1} P \circ f_\varepsilon(u)$$

carries Γ onto a relatively compact subset of Γ ; hence it has, by the well-known Schauder theorem, a fixed point u_ε in Γ which is necessarily of norm c. Clearly u_ε solves the problem stated above for $\lambda_\varepsilon = c^{-1}\|P \circ f_\varepsilon(u_\varepsilon)\|$. There remains to show that $\lambda_\varepsilon > r(P)$:

$$\lambda_\varepsilon\langle ju_\varepsilon, \varphi_0\rangle = r(P)\langle f_\varepsilon(u_\varepsilon), \varphi_0\rangle \geqq r(P)\{\langle ju_\varepsilon, \varphi_0\rangle + \varepsilon\langle u_0, \varphi_0\rangle\} \ ; \tag{1}$$

here we have used that $f(u) \geqq ju$ on Γ. Since $\langle u_0, \varphi_0\rangle > 0$, we must have $\langle ju_\varepsilon, \varphi_0\rangle > 0$ and hence $\lambda_\varepsilon > r(P)$.

The remainder of the proof is now obvious: By the compactness of P, there exists a sequence $\varepsilon_n \to 0$ such that $\{\lambda_{\varepsilon_n}\}$ and $\{u_{\varepsilon_n}\}$ converge to λ_c and u_c respectively; clearly $\|u_c\| = 0$ and $\lambda_c \geqq r(P)$.

3. The condition $f(u) \geqq ju$ (see No. 1, (b)) can be weakened. The case that $f(u) \geqq \alpha ju$ for some $\alpha > 0$ is trivial; it can even be assumed that α depends on c. More generally, if there exists a function $c \to \alpha(c) > 0$ such that $\langle f(u), \varphi_0\rangle \geqq \alpha_c\langle ju, \varphi_0\rangle$ whenever $\|u\| = c$, then (1) (the only place where the assumption $f(u) \geqq ju$ was used), holds in an analogous way with the result that $\lambda_c \geqq \alpha_c r(P)$.

A concrete case of this type arises when $E_2 = \mathscr{C}(X)$, the Banach space of continuous functions on some compact space X. We assume now E_2 to contain E_1 so that j becomes the canonical imbedding, and replace condition (b) by the following:

(b') f <u>is continuous on</u> K_1 <u>into the cone of non-negative functions in</u> $\mathscr{C}(X)$. <u>There exist constants</u> $\delta > 0$, $k > 0$ <u>such that for all</u> $s \in X$,

$$ju(s) \leqq \delta \text{ implies } f[u](s) \geqq ju(s) ,$$

$$ju(s) > \delta \text{ implies } f[u](s) \geqq k .$$

Since j is continuous, $\|u\| = c$ implies $\|ju\| \leqq c_1$. The sets $A_1 = \{s : ju(s) \leqq \delta\}$, $A_2 = \{s : ju(s) > \delta\}$ are closed and open subsets of X, respectively. If χ_1, χ_2 denote the characteristic functions of A_1, A_2 respectively, then we obtain, φ_0 being a positive measure on X,

$$\langle f(u), \varphi_0 \rangle = \int f(u)\chi_1 d\varphi_0 + \int f(u)\chi_2 d\varphi_0 \geqq \int (ju)\chi_1 d\varphi_0 + kc_1^{-1}\int (ju)\chi_2 d\varphi_0$$

$$\geqq \inf\{kc_1^{-1}, 1\}\int ju(\chi_1 + \chi_2)\, d\varphi_0 = \alpha_c \langle ju, \varphi_0 \rangle$$

where $\alpha_c = \inf\{kc_1^{-1}, 1\}$, $c_1 = \sup\{\|ju\| : \|u\| = c\}$. Hence:

<u>Under conditions</u> (a), (b') <u>Theorem</u> 4 <u>continues to hold, the only modification being that</u> $\lambda_c \geqq \alpha_c r(P)$.

4. As an example for the application of Theorem 4, let us consider the boundary value problem (cf. [10])

$$\Delta u + \lambda^{-1}\tilde{f}(u) = 0 , \quad u\big|_R = 0 \tag{2}$$

on a simple region D in 3-space, with boundary R. We assume that $\tilde{f}$ is a mapping satisfying (b') on the space $E_1 = \mathscr{C}^{(1)}(D)$ of real functions continuously differentiable in D, into $E_2 = \mathscr{C}(D)$, and given by a numerical function f, Hölder continuous with respect to its seven variables t_i, u, p_i $(i = 1, 2, 3)$. (2) is then known to be equivalent with

$$u(s) = \int_D G(s,t)\, f(t, u(t), \frac{\partial u}{\partial t})\, dt \quad (s \in D) , \tag{3}$$

where G is the Green's function of the first kind associated with Δ on D. By Theorem 4, for any $c > 0$ and suitable λ_c, (2) has a solution with $\|u\| = c$ where $\|\cdot\|$ is any norm on E_2 generating the topology of uniform convergence of u and its first order derivatives. Moreover, we obtain the following perturbation result:

Let f_η, where $f_\eta(u) = u + g(u;\eta)$ with $g(0,u) = 0$, be a continuous family of mappings satisfying condition (b') for all η, $0 \leqq \eta \leqq 1$. Then for each fixed $c > 0$, there exists a family u_η, $\|u_\eta\| = c$, solving (2) for $\lambda = \lambda_\eta$, and such that

$$\lim_{\eta\to 0} \lambda_\eta = \lambda_0 , \quad \lim_{\eta\to 0} \|u - u_0\| = 0 ,$$

where λ_0 is the greatest eigenvalue of the linear problem (2), and u_0 the unique non-negative eigenfunction of G with norm c.

The simple proof of this assertion will be omitted.

5. A more concrete example is furnished by the one dimensional boundary value problem arising in connection with the longitudinal bending of a compressed rod.[7] It is assumed that a rod of unit length is fastened by two hinges A, B (through which the x-axis of a plane Cartesian system of coordinates is laid, with $A = 0$) and exposed to a force τ exerted at B in the direction of A, while A is held in place. If the form of the rod is described by the graph (x, y) of a twice differentiable function y of x, then by a change of the independent variable x to the arc length s one obtains the boundary value problem

$$\begin{aligned} \frac{d^2y}{ds^2} &= -\tau\rho(s)y(s)\sqrt{1-\left(\frac{dy}{ds}\right)^2} \qquad 0 \leqq s \leqq 1 \\ y(0) &= y(1) = 0 \end{aligned} \tag{4}$$

ρ characterizing the stiffness of the rod. To apply Theorem 4, we take $E_1 = \mathscr{C}^{(1)}[0,1]$, $E_2 = \mathscr{C}[0,1]$ to be the spaces of continuously differentiable and continuous real functions on $[0,1]$ respectively, vanishing at the endpoints. E_1 and E_2 are normed by

$$\|y\|_1 = \sup_s \left|\frac{dy}{ds}\right| , \quad \|y\|_2 = \sup_s |y(s)|$$

respectively. If G denotes the Green's function,

$$G(s,t) = \begin{cases} s(1-t) & s \leqq t \\ t(1-s) & s > t , \end{cases} \tag{5}$$

then the original problem is equivalent with the integral equation

$$y(s) = \tau \int_0^1 G(s,t)\,\rho(t)\,y(t)\,\sqrt{1-\left(\frac{dy}{dt}\right)^2}\,dt \tag{6}$$

$(0 \leqq s \leqq 1)$ under the additional condition that $\|y\|_1 < 1$.

We can apply Theorem 4 with K_1 and K_2 the cone of non-negative functions in E_1, E_2 respectively, and f given by

$$f[y](s) = \rho(s)\,y(s)\sqrt{1-\left(\frac{dy}{ds}\right)^2} \, .$$

f is, however, defined only on the unit ball of E_1. Thus[6] the result of Theorem 4 holds only for $0 < c \leqq 1$ it admits a physical interpretation only when $c < 1$.

It is evident that $\|y\|_1 = \sup\{y'(0), |y'(1)|\}$ for the positive solutions of (4). Hence if y is a solution with $\|y\|_1 = c$, the rod is forced into a position making the angle $\gamma = \text{arc tg } c/(1-c^2)$ with the x-axis at one of the endpoints. Conversely, this angle γ determines c $(0 < c < 1)$ and hence the eigenvalue τ_c, which thus appears as the force necessary to achieve the bending corresponding to the solution y_c of (4).

The present problem is a "genuine" nonlinear problem since, even for small positive solutions of (4), the form of the rod and the force τ will determine each other, whereas the cancellation of $\left(\frac{dy}{ds}\right)^2$ in (4) or (6) yields the same force for solutions of different shape which is physically incorrect.

It is moreover clear that there are no non-trivial positive solutions of (4) for small forces τ (cf. [3]). The critical force is $\tau_0 = \lambda_0^{-1}$ where λ_0 is the greatest eigenvalue of the kernel $\tilde{G}$, $\tilde{G}(s,t) = \sqrt{\rho(s)}\,G(s,t)\sqrt{\rho(t)}$; for any actual displacement of the rod a force $\tau_c > \tau_0$ is necessary. If $\rho(s) = \rho_0$ is constant, then $\tau_c > \pi^2/\rho_0$ since π^{-2} is the greatest eigenvalue of G.

Research sponsored in part by the U. S. Army Research Office.

NOTES

1. It will be seen shortly that $\kappa \in J$ if and only if $\{x_m\}$ is bounded for the pair (κ, e).
2. For recent results, see, in particular, [6] and [7].

3. ν , or its restriction to the characteristic functions of the Baire sets, is called a spectral measure on X .

4. If E is tonnelé, Σx_m converges for the given topology on E .

5. See [5] and the references there.

6. When f is defined only on a set $\{u \in K_1 : \|u\| \leqq c_0\}$ $(c_0 > 0)$, retaining the other hypotheses, Th. 4 remains valid for $0 < c \leqq c_0$.

7. This problem has been treated in detail by M. A. Krasnosel'skii [2] et al. For a brief account, see [3].

REFERENCES

1. Jentzsch, R. Ueber Integralgleichungen mit positivem Kern. Crelle's J. 141 (1912), 235-244.

2. Krasnosel'skii, M. A. Topological Methods in the Theory of Non-linear Integral Equations. Gostehizdat (Moscow 1956).

3. ___________, Investigation of the spectrum of a non-linear operator in a neighborhood of a branch point, and application to the problem of longitudinal bending of a compressed rod. A. M. S. Translations (2nd ser.) 16(1960), 418-423.

4. Ladyženskii, L. A. On non-linear equations with positive non-linearity. A. M. S. Translations (2nd ser.) 16(1960), 426-427.

5. Müller, P. H. Eine neue Methode zur Behandlung nichtlinearer Eigenwertaufgaben. Math. Zeitschr. 70(1959), 381-406.

6. Ringrose, J. R. On the Neumann series of an integral operator. Proc. London Math. Soc. 3rd ser. X, no. 37(1960), 31-52.

7. Rota, G.-C. On the eigenvalues of positive operators. Bull. A. M. S. 67(1961), 556-558.

8. Rothe, E. H. On non-negative functional transformations. Amer. J. Math. 66(1944), 245-254.

9. Schaefer, H. Ueber algebraische Integralgleichungen mit positiven Koeffizienten. Math. Ann. 137(1959), 385-391.

10. ___________. On non-linear positive operators. Pac. J. Math. 9(1959), 847-860.

11. ___________. Some spectral properties of positive linear operators. Pac. J. Math. 10(1960), 1009-1019.

12. ___________. Spectral measures in locally convex algebras. Acta Math. 107(1962), 125-173.

13. Schmeidler, W. Algebraische Integralgleichungen. Math. Nachr. 8(1952), 31-40.

14. __________. Algebraische Integralgleichungen. II. Math. Nachr. 10(1953), 247-255.

JÜRGEN MOSER

Stability and nonlinear character of ordinary differential equations

1. Introduction

We are concerned with the problem of stability of periodic solutions of ordinary nonlinear differential equations which are conservative. The customary approach to prove the stability of a periodic solution—or equilibrium, say—is to verify stability for the linearized equations and then to show that the nonlinear terms do not destroy stability. This idea is basic for many stability investigations and frequently is referred to as Liapounov's first method. One may say this method has the aim to prove the nonlinear terms to be negligible for stability.

Actually this method is applicable only if the solutions of the linearized equations die out exponentially i. e. if one has asymptotic stability, a fact which indicates dissipation of the physical situation described. For conservative systems, however, which describe frictionless motion this method becomes inapplicable. The linearized equations of a conservative system are stable only if the solutions are oscillatory. If $\omega_1, \omega_2, \ldots, \omega_n$ are the frequencies of these oscillations one has to expect instability—or resonance—from the nonlinear terms whenever these frequencies are dependent over the rationals:

$$\sum_{\nu=1}^{n} m_\nu \omega_\nu = 0 \qquad (1)$$

where the m_ν are integers which do not all vanish. In fact, by examples one can verify that in all these cases particular nonlinear terms can give rise to instability. This fact corresponds to the difficulty of the "small divisors" encountered in celestial mechanics.

On the other hand, there is a simple nonrigorous argument which suggests that the nonlinearity may have a stabilizing effect. To explain this argument we specialize to the case of $n = 2$ frequencies $\omega_1 = \omega$ and $\omega_2 = 1$ so that the condition (1) takes the form

$$\omega = \text{rational}.$$

For nonlinear systems the frequency ω will not be a constant but usually depends on some amplitude A (measuring the distance from the periodic solution). The argument is the following: In case of instability the amplitude A has to increase, hence $\omega(A)$ will vary and certainly pass through irrational values. Therefore one expects that a solution cannot penetrate a region in which ω is irrational.

This is indeed the case and one has stability on account of the nonlinearity which corresponds to the non-constant frequency amplitude relation $\omega = \omega(A)$.

It is the purpose of this talk to give a precise statement in this direction. The basic idea of this paper was stimulated by the address of Kolmogorov at the International Congress in Amsterdam, 1954 ([8], [9]). Recently, Arnold [1] gave detailed proofs of theorems concerning circle mappings which also involve small divisors. Arnold also announced some new results in the Doklady Akad. Nauk [2] which are closely related to the present work. [1])

The main results of this note have been presented at the Symposium on Nonlinear Differential Equations in Colorado Springs, 1961, [12] and the proofs will appear shortly [13].

It is a well-known method in stability theory to associate with the differential equation a mapping near a fixed point (see for instance [14] or [16]). This has the advantage that the investigation near the periodic solution is reduced to a local one and, moreover, the time variable is eliminated. For the sake of simplicity we will begin with the discussion of such mappings and relate the results to differential equations afterwards.

2. Geometrical Theorem

(a) The relevant geometrical theorem can be formulated as follows: Consider an annulus

$$1 \leq r \leq 2 \; ; \quad \theta \; (\text{mod } 2\pi)$$

in the plane and let M_0 denote the following simple mapping of the annulus onto itself:

$$\begin{cases} \theta_1 = \theta + \alpha(r) \\ r_1 = r \end{cases}$$

where $\alpha(r)$ is a monotone function of r . We refer to M_0 as twist mapping, and denote by

$$\Delta = |\alpha(2) - \alpha(1)| \tag{2}$$

the "frequency range."

The problem is concerned with the study of closed invariant curves surrounding $r = 1$ of a mapping M which is close to M_0. Obviously M_0 possesses the circles $r = \text{const.}$ as invariant curves. For a perturbed mapping

$$\begin{cases} \theta_1 = \theta + \alpha(r) + f(\theta, r) \\ r_1 = r \qquad\qquad + g(\theta, r) \end{cases} \tag{3}$$

to have invariant curves near concentric circles requires more than smallness conditions on f, g.

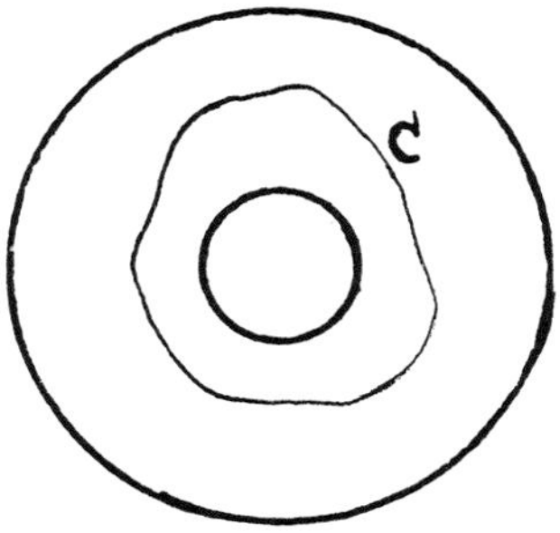

If $g > 0$, say, then r will increase under the mapping and no closed invariant curves will exist. But we will require that every closed curve $C : r = \phi(\theta)$ with small $|\phi'|$ and its image curve $MC = C_1$ intersect. This property can frequently be verified if M preserves area and one boundary circle.

We introduce the following abbreviation: If ℓ is a positive integer and $h(\theta, r)$ has partial derivatives up to order ℓ let

$$|h|_\ell = \sup \left| \frac{\partial^{\rho+\sigma} h}{\partial r^\rho \partial \theta^\sigma} \right| \quad \text{where } \rho + \sigma \leq \ell . \tag{4}$$

(b) <u>Theorem 1</u>: We assume with some constant $c \geq 1$ that

$$c^{-1}\Delta < |\alpha'| \; ; \quad |\alpha'|_\ell < c\Delta < c^2 \tag{5}$$

and that every closed curve $r = \psi(\theta)$ and its image curve intersect.

Then for $\varepsilon > 0$ there exists a $\delta = \delta(\varepsilon, c) > 0$ such that for

$$|f|_\ell + |g|_\ell < \delta \cdot \Delta \qquad (\ell = 333) \tag{6}$$

the mapping M possesses invariant curves

$$r = r_0 + \phi(\theta)$$

with

$$|\phi'| < \varepsilon \ .$$

Additional statement: Moreover, one obtains a set of such curves whose measure is at least $(1-\varepsilon)$ times the measure of the total annulus.

The proof of this theorem without the additional remark will appear shortly [13] and it is too lengthy to be presented here. However, we discuss some relevant points and applications.

(c) It is surprising that the above theorem requires bounds for high derivatives of the mapping M, even though the statement of the existence of invariant curves is of topological nature—if one ignores the differentiability of $\phi(\theta)$. But one can show by counter-examples that such invariant continuous curves $r = \phi(\theta)$ need not exist if ℓ is replaced by 0, no matter how small δ is chosen. Needless to say that $\ell = 333$ is a very generous choice which, however, was used in the proof cited.

3. Application to area preserving mappings near fixed points

(a) If one considers Hamiltonian systems of two degrees of freedom in the neighborhood of periodic solutions one is lead to area preserving mappings of the plane near a fixed point. (See G. D. Birkhoff [6], also Siegel [16]). If we assume the linearized equation to be stable, then the linear part of the mapping can be assumed to be a rotation about the fixed point. If the coordinates in the plane are given by a complex coordinate $z = x + iy$, the mapping can be written in the form

$$z_1 = ze^{i\alpha} + F(z, \bar{z}) = f(z, \bar{z}) \tag{7}$$

where F vanishes at least quadratically at the fixed point $z = 0$. Moreover, the area preserving property amounts to

$$\frac{\partial(f, \bar{f})}{\partial(z, \bar{z})} \equiv 1 \ .$$

The function $F(z, \bar{z})$ is assumed to be ℓ times differentiable near $z = 0$. For any function $G(x, y)$ we use the notation

$$G(x, y) = \mathcal{O}_\ell(|z|^s)$$

if with some $c > 1$ the inequalities

$$|z|^{\rho+\sigma} \left| \frac{\partial^{\rho+\sigma}}{\partial x^\rho \partial y^\sigma} G \right| \le c|z|^s$$

hold for $|z| < c^{-1}$ and $\rho + \sigma \le \ell$. In (7) we assume $F = \mathcal{O}_{\ell'}(|z|^2)$ with some ℓ'.

(b) The problem is to investigate whether the fixed point $z = 0$ is stable under the mapping (7). For this purpose we use

<u>Theorem 2:</u> If $\alpha, 2\alpha, \ldots, q\alpha$ are not integral multiples of 2π there exists a real analytic coordinate transformation of (x, y) into (u, v) with non-vanishing Jacobians such that the mapping (7) takes the form

$$w_1 = w e^{i(\alpha+\beta|w|^s)} + \mathcal{O}_\ell(|w|^q) \qquad (w = u+iv) \tag{8}$$

where s is an even integer with $0 < s < q-1$ and $\beta = 0$, $+1$ or -1 (independently of the choice of $q > s + 1$). Here it is assumed that in (7) one has $F = \mathcal{O}_{\ell'}(|z|^2)$ with $\ell' \ge \ell + q-2$.

This is an extension of a theorem of Birkhoff (see [5], [11]).

One can consider $\beta \ne 0$ as the general case, in which case we choose $q = s + 2$:

$$w_1 = w e^{i(\alpha+\beta|w|^s)} + \mathcal{O}_\ell(|w|^{s+2}) . \tag{9}$$

(c) Ignoring the error term $\mathcal{O}_\ell(|w|^{s+2})$ one has a twist mapping and the angle of rotation

$$\alpha + \beta|w|^s$$

depends on the radius $|w|$. We want to show that theorem 1 is applicable in this situation to give the existence of invariant curves in every neighborhood $w = 0$, thus proving the stability of $w = 0$.

For this purpose consider the annulus

$$a \le |w| \le 2a$$

for sufficiently small $a > 0$ and introduce $r = \frac{|w|}{a}$, $\theta = \arg w$. Then (8) takes the form

$$\theta_1 = \theta + \alpha + \beta a^s r^s + \mathcal{O}_\ell(a^{s+1})$$

$$r_1 = r + \qquad\qquad + \mathcal{O}_\ell(a^{s+1})$$

in $1 \leq r \leq 2$. Hence the frequency range is

$$\Delta = a^s(2^s-1) \geq a$$

and the condition (5) of theorem 1 takes the form

$$\frac{|f|_\ell + |g|_\ell}{\Delta} = \frac{\mathcal{O}_\ell(a^{s+1})}{a^s(2^s-1)} = \mathcal{O}_\ell(a) < \delta$$

is satisfied if $a < a_0$.

Moreover, the curve condition is a consequence of the area preserving character of the mapping: After the transformation of z into w the area element goes into another one $p(u, v)\,du\,dv$ with positive $p(u, v)$. Let C be any closed curve in $a < |w| < 2a$ surrounding the fixed point. Then the image curve C_1 has to intersect C since the areas enclosed by C and C_1 agree.

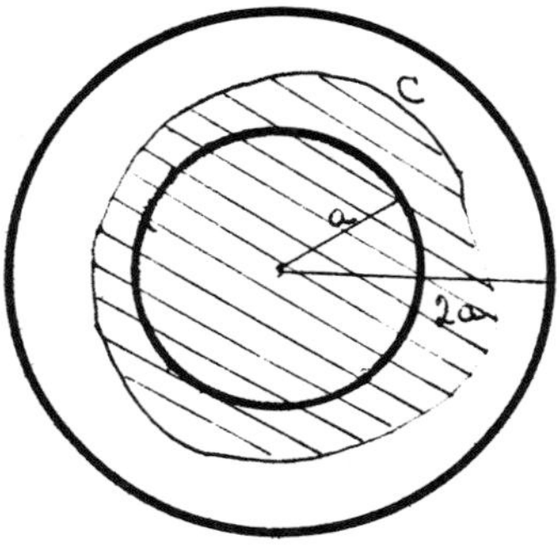

This proves that every sufficiently small circle $|w| < 2a$ contains an open invariant set containing $w = 0$, namely the interior of the invariant curve.

Theorem 3: The mapping (8) is stable if $\beta \neq 0$.

This result has several applications and its importance lies in the fact that the stability investigation is reduced to the determination of the numbers α , β . In fact, since the conditions for stability take the form of finitely many inequalities

$$\begin{cases} \dfrac{\nu\alpha}{2\pi} \neq \text{integer} & \text{for } r = 1, 2, \ldots, s+2 \\ \beta \neq 0 & \end{cases} \tag{10}$$

one has to compute α, β only approximately. This can frequently be done and we mention, in particular, the following application to the restricted 3 body problem.

4. Restricted three body problem

(a) The restricted 3 body problem deals with a special situation of the 3 body problem: We assume that 3 mass points move in a plane and one of them, called the planet P , has mass exactly zero. Therefore the other two mass points, called S (Sun) and J (Jupiter) are not affected by P and it is assumed that S and J move on circles about their common center of gravity with a frequency n_J . The problem is to study the motion of P .

Normalizing the total mass of the system to 1 we denote the mass of Jupiter by $\mu > 0$ and that of the Sun by $1 - \mu > 0$. One usually restricts oneself to the case of small values of μ (in fact in the astronomical application one has $\mu < 10^{-3}$). For $\mu = 0$, P will describe the well-known motion on a conic section. In particular, we concentrate to the case where P also describes a circular motion of frequency n_P about the center of gravity of all three mass points for $\mu = 0$.

(b) For small values of $\mu > 0$ one can find[2] periodic motions of P near the circular ones if one views the motion of P from a rotating coordinate system in which S and J are at rest. These orbits are the so-called periodic solutions of the first kind (Poincaré). To investigate the stability of these periodic solutions one has—according to the previous section—to determine the numbers $\alpha(\mu)$, $\beta(\mu)$. In fact it suffices to compute them for $\mu = 0$ since they depend analytically on μ . The calculation gives (see [10])

$$\frac{\alpha(0)}{2\pi} = \frac{n_j}{n_P - n_J} \; ; \quad \beta(0) \neq 0 \quad \text{for} \quad s = 2 \quad . \tag{11}$$

The conditions for stability (10) therefore are:

$$\frac{n_j}{n_P - n_J} \neq \frac{p}{q} \qquad \text{for } q = 1, 2, 3, 4, \text{ and all integers } p$$

or equivalently:

$$\frac{n_J}{n_P} \neq \frac{r}{s} \qquad \text{for } |r-s| \leq 4 \; . \tag{12}$$

Under these conditions the theorem of the previous section guarantees stability of the periodic solutions of the first kind for small $\mu > 0$.

A closer analysis shows that (12) can be replaced by

$$\frac{n_J}{n_P} \neq \frac{r}{s} \qquad \text{for } |r-s| \leq 3 \; . \tag{13}$$ [3]

If $\frac{n_J}{n_P} = \frac{r}{s}$ one calls for the number $|r-s|$ the "order of the resonance," if r, s are relatively prime. Therefore we have the following statement: <u>For sufficiently small $\mu > 0$ [4] a periodic solution of the first kind is stable if the order of resonance is ≥ 4</u> .

It was shown by Levi-Civita [7] that third order resonance leads to instability for sufficiently small $\mu > 0$. For second order resonance one also has instability in general (see, for instance [17]).

(c) The expression

$$\alpha + \beta |w|^s$$

has a simple physical interpretation. It represents the frequency of the oscillations about the periodic solution. This frequency depends on the amplitude $|w|$ and is not a constant if $\beta \neq 0$. Moreover, the invariant curves of Theorem 1 correspond to almost periodic solutions of the restricted three body problem.

(d) We mention another consequence of Theorem 1 which refers to the solutions of the restricted three body problem in the Hill curve surrounding the Sun—provided the energy constant is so large that the three Hill's regions are disjoint. In this case one can associate with the differential equations an annulus mapping which for $\mu = 0$ preserves concentric circles. In fact, this construction is due to Poincaré [15] (see also [4]) and led him to his fixed point theorem which was later proven by Birkhoff.

The application of Theorem 1 yields: For the solution in the Hill's curve surrounding S the major axis and excentricities can for all t be enclosed between constant bounds whose difference tends to zero as $\mu \to 0$ (actually like $\mathcal{O}(\mu^{3/8})$).

5. Generalization to several degrees of freedom

The Theorem 1 has an immediate generalization to several dimensions which is of importance for systems of several degrees of freedom. For this purpose let $\theta = (\theta^{(1)}, \ldots, \theta^{(n)})$ be a vector of angular variables (mod 2π) and $r = (r^{(1)}, r^{(2)}, \ldots, r^{(n)})$ with

$$1 \leq r^{(\nu)} \leq 2 \ .$$

The mapping M will have the form

$$\begin{aligned} \theta_1 &= \theta + \alpha(r) + f(\theta, r) \\ r_1 &= r \qquad\qquad + g(\theta, r) \end{aligned} \tag{14}$$

where we require that the Jacobian

$$\alpha' = \left(\frac{\partial \alpha^{(\nu)}}{\partial r^{(\mu)}}\right), \qquad \nu, \mu = 1, 2, \ldots, n$$

does not vanish so that the $\alpha^{(\nu)}(r)$ cover an open set in the θ space. We require

$$c^{-1} < |\det \alpha'| \quad ; \quad |\alpha'|_\ell < 0 \tag{15}$$

in place of (5). Moreover, we require that every torus

$$r^{(\nu)} = \psi^{(\nu)}(\theta) \qquad (\nu = 1, 2, \ldots, n) \tag{16}$$

and its image torus intersect each other. The norm $|f|_\ell$ will be defined as the maximum of the norm of the components $|f^{(\nu)}|_\ell$.

Theorem 4: Under the conditions (15) and the assumption that every torus (16) intersects its image[6] the mapping (14) possesses an invariant torus $r = \phi(\theta)$ provided

$$|f|_\ell + |g|_\ell \qquad (\ell = 267 + 66n)$$

is sufficiently small.

In fact, for $\varepsilon > 0$ there exists a $\delta(c, \varepsilon)$ such that for

$$|f|_\ell + |g|_\ell < \delta(c, \varepsilon)$$

the invariant tori $t = \phi(\theta)$ constructed form a set of measure $> (1-\varepsilon)$ times the measure of the total region and satisfy $|\theta'| < \varepsilon$.

This result is only partially useful for stability investigations insofar as an n-dimensional torus is not a boundary of a 2n-dimensional set if $n > 1$. However, one can obtain stability results if one neglects setsof small measure. In this sense Theorem 3 has a generalization to conservative mappings in a 2n-dimensional space near a fixed point.

6. Conclusion

(a) The general idea underlying the preceding sections can be summarized as follows (see [9]): We consider conservative systems which lie sufficiently close to "integrable" ones, i.e. systems which describe oscillatory motion with frequencies $\omega_1, \omega_2, \ldots, \omega_n$ which depend on n amplitudes (or action variables) $A_1, \ldots, A_n$. If the Jacobian

$$\det \left(\frac{d\omega_\nu}{\partial A_\mu} \right)$$

does not vanish one can find tori which are invariant under the flow (i.e. consist entirely of solutions). The solutions on these tori are almost periodic.

In particular, in the neighborhood of periodic solutions any conservative system is closely approximated by an integrable system;

this is the observation of G. D. Birkhoff [6], which was made in §3.

(b) However, there are several other limit situations which lead to systems close to integrable ones, which arise when one of the frequencies becomes very large. We mention two possible cases: The solutions of the restricted 3 body problem for which P remains close to J are called lunar orbits. By appropriate regularization of the singularity at J one can verify that the differential equations are close to integrable ones if the energy constant becomes large.[5] On this basis one can obtain bounds for lunar orbits.

Another example is the motion of a charged particle in a magnetic field where the mass of the particle is a small parameter. Such a particle performs rapid oscillation about the magnetic field lines (see, for instance, Berkowitz and Gardner [3]). For rotation symmetric configurations of magnetic fields C. S. Gardner was able to find bounds for the solutions of this problem for all times.

These remarks indicate that the above ideas may lead to a number of other applications.

This paper represents results obtained under the sponsorship of the Office of Naval Research, Contract No. Nonr-285(46). Reproduction in whole or in part is permitted for any purpose of the United States Government.

NOTES

1. The main difference between Arnold's announcement [2] and theorems 2 and 3 of the present paper is that Arnold's assumption of the irrationality of $\alpha/2\pi$ (or λ) is changed here to the exclusion of only a discrete set of numbers. It seems to be a technical difference that Arnold assumes analyticity while we require the existence of finitely many derivatives only.

2. Provided $\alpha(0)$ is not an integer in the notation of (11).

3. This statement sharpens a previous result [11] in which only an estimate for the time of escape was given while the present statement guarantees stability for <u>all</u> times.

4. The choice of μ depends on the particular periodic solution.

5. Dissertation of Charles C. Conley of M. I. T., 1961, (not yet published).

6. For canonical mappings it suffices to require this assumption only for those tori on which the associated closed 2-form vanishes identically.

REFERENCES

1. Arnold, V. I., "Small divisors, I. On circle mappings. Isvest. Akad. Nauk, Ser. mat. vol. 25, pp. 21-86, 1961.

2. Arnold, V. I., "On the stability of the equilibrium solutions of Hamilton systems of differential equations in the general elliptic case." Doklad. Akad. Nauk SSSR, vol. 137, pp. 255-257, 1961.

3. Berkowitz, J. and Gardner, C. S., "On the asymptotic series expansion of the motion of a charged particle in slowly varying fields." Communications on Pure and Applied Mathematics, vol. 12, pp. 501-512, 1959.

4. Birkhoff, G. D., "The restricted problem of three bodies, Rend. Circ. mat. Palermo, vol. 39, pp. 1-70, 1915.

5. Birkhoff, G. D., "Surface transformations and their dynamical applications, Acta math. 43, pp. 1-119, 1920.

6. Birkhoff, G. D., Dynamical Systems, New York, 1927.

7. Levi-Civita, T., Sopra alcuni criteri di instabilita, Ann. Mat. pura appl. (3) vol. 5, pp. 221-307 (1901).

8. Kolmogorov, A. N., "On the conservation of conditionally periodic motions for a small change in Hamilton's function. Doklad. Akad. Nauk, SSSR, vol. 98, pp. 527-530, 1954.

9. Kolmogorov, A. N., Théorie générale des systèmes dynamiques et mécanique classique. Proc. of the Int. Congress of Math., Amsterdam 1954, vol. 1, pp. 315-333, Amsterdam 1957.

10. Moser, J., Periodische Lösungen des restringierten Dreikörperproblemes, die sich erst nach vielen Umläufen schliessen. Math. Ann. vol. 126, pp. 325-335, 1953.

11. Moser, J., Stabilitätsverhalten kanonischer Differentialgleichungssysteme. Nachr. Akad. Wiss. Göttingen, Math.-phys. Kl., pp. 89-120, 1955.

12. Moser, J., Perturbation theory for almost periodic solutions for undamped nonlinear differential equations. Presented at the Symposium on Nonlinear Differential Equations in Colorado Springs, Aug. 1961.

13. Moser, J., On invariant curves of area-preserving mappings of an annulus. To appear in Nachr. Akad. Wiss. Göttingen, Math. phys. Kl. 1962.

14. Poincaré, H., Les méthodes nouvelles de la mécanique céleste. Gauthier Villars, Paris 1892, 1893, 1899.

15. Poincaré, H., Sur un théorème de géométrie. Rend. Circ. mat. Palermo, vol. 33, pp. 375-407, 1912.

16. Siegel, C. L., Vorlesungen über Himmelsmechanik, Springer-Verlag 1956.

17. Wintner, A., On the periodic analytic continuations of the circular orbits in the restricted problem of three bodies. Proc. of the Nat. Acad. of Sciences, vol. 22, pp. 435-439, 1936.

DAVID GILBARG

Boundary value problems for nonlinear elliptic equations in n variables

1. Let $A_k(u) \equiv A_k(u_1, \ldots, u_n)$, $k = 1, \ldots, n$ be functions defined for all real n-vectors $u = (u_1, \ldots, u_n)$. We consider the Dirichlet problem for solutions $\phi(x_1, \ldots, x_n)$ of quasi-linear elliptic equations of the form

$$\frac{\partial}{\partial x_k} A_k(u_1, \ldots, u_n) = A_{k,i}(u) \frac{\partial^2 \phi}{\partial x_i \partial x_k} = 0 \qquad (1)$$

where $u_j = \partial\phi/\partial x_j$ and $A_{k,i} = \partial A_k/\partial u_i$. (The summation convention is assumed.)

The boundary value problem for (1) is studied here under two different ellipticity hypotheses. The first is a kind of uniform ellipticity, namely,

$$\lambda_1(u)|\xi|^2 \leq A_{k,i}(u)\xi_i\xi_k \leq \lambda_2(u)|\xi|^2 \qquad (2)$$

where $\xi = (\xi_1, \ldots, \xi_n)$ is an arbitrary real n-tuple and λ_1, λ_2 are positive functions of u with uniformly bounded ratio,

$$\lambda_2/\lambda_1 \leq \gamma < \infty \qquad (3)$$

for some constant γ . By continuity λ_1 and λ_2 can approach zero or infinity only if u becomes unbounded. Under these hypotheses we shall establish the existence of solutions of the Dirichlet problem for (1), assuming sufficiently smooth, but otherwise arbitrary, boundaries and boundary values. The precise conditions are stated below.

We consider secondly the boundary value problem under the general ellipticity condition

$$A_{k,i}\xi_i\xi_k > 0 \quad \text{if} \quad |\xi| \neq 0 . \qquad (4)$$

This includes the minimal surface equation, for which the defining functions $A_k = u_k/(1+\Sigma u_i^2)^{\frac{1}{2}}$ do not satisfy (2), (3). As is well-known from the classical case of two independent variables, the solvability of the boundary value problem for the minimal surface equation and other non-uniformly elliptic equations requires in general that the boundary curve satisfy a convexity condition. A similar assumption can be expected in the case of $n > 2$ variables, and indeed, the Dirichlet problem for (1) is solved here under the hypothesis that the boundary surface is strictly convex. This result is therefore a generalization of Haar's theorem [3] on the solution of two-dimension regular variational problems.

The results presented in this paper are related to those of other authors, in particular, Morrey [6], Ladyzhenskaya and Uraltseva [4], and Stampacchia. Both Morrey (in connection with variational problems) and Ladyzhenskaya-Uraltseva have considered the boundary value problem for quasi-linear elliptic equations with self-adjoint principal part.

$$\frac{\partial}{\partial x_k} A_k(x, \phi, u) + B(x, \phi, u) = 0 \ , \quad x = (x_1, \ldots, x_n) \ , \quad u = (\phi_{x_1}, \ldots, \phi_{x_n}) \tag{5}$$

The functions A_k are required to satisfy several conditions, among them an ellipticity hypothesis of the form,

$$\mu_1(\phi)(1+|u|)^m |\xi|^2 \le A_{k,i}\xi_i\xi_k \le \mu_2(\phi)(1+|u|)^m|\xi|^2 \ , \quad |u| = (\Sigma u_i^2)^{\frac{1}{2}} \tag{6}$$

where μ_1 and μ_2 are bounded away from zero and infinity, and $m > -1$. This condition is of course more restrictive than the ellipticity condition (2), (3) assumed here.

The boundary value problem under the general ellipticity condition (4) has also been considered by Stampacchia (oral communication)*. He solves the Dirichlet problem by variational methods under essentially the same conditions as those of the present paper for the class of equations (1) derived from a variational problem (namely, those for which $A_{k,i} = A_{i,k}$).

2. Let D be a bounded region in n-space with boundary ∂D and let f be a function with Holder continuous second derivatives defined on the closure $\bar{D} = D + \partial D$. It is assumed that ∂D can be covered by a finite number of neighborhoods on which ∂D can be represented by a function with Hölder continuous second derivatives. Suppose the functions $A_k(u)$ are in C^2 and satisfy <u>either</u> of the following two conditions:

$$\text{I.}\quad \begin{array}{l} \lambda_1(u)\,|\xi|^2 \le A_{k,i}(u)\,\xi_i\xi_k \le \lambda_2(u)\,|\xi|^2 , \qquad \lambda_1 > 0 , \\ \lambda_2(u)/\lambda_1(u) \le \gamma < \infty \ . \end{array}$$

II. $A_{k,i}\,\xi_i\xi_k > 0$ for $|\xi| \neq 0$; in this case it is assumed that ∂D is strictly convex.

Under these hypotheses we establish the existence of a C^2 solution of (1) such that

$$\phi = f \quad \text{on} \quad \partial D \ . \tag{7}$$

In the following we refer to the uniformly elliptic or non-uniformly elliptic problem according as condition I or II is assumed. The method of proof is basically the same for the two problems.

3. Let S be the Banach space of vectors $u(x) = (u_1(x), \ldots, u_n(x))$ defined in D with the norm

$$\|u\| = \operatorname*{lub}_D |u(x)| + \operatorname*{lub}_D \frac{|u(x)-u(y)|}{|x-y|^{\alpha}}$$

where $0 < \alpha < 1$. The particular choice of α will be determined later.

Let an arbitrary element u of S be inserted in the coefficients $A_{k,i}(u)$ of (1) and consider the linear uniformly elliptic equation thus obtained, namely,

$$a_{ik}(x)\frac{\partial^2\psi}{\partial x_i \partial x_k} = 0 , \tag{8}$$

where

$$a_{ik}(x) = a_{ki}(x) = \frac{1}{2}[A_{k,i}(u(x)) + A_{i,k}(u(x))] \ .$$

Suppose $\psi(x)$ is the solution of this equation in D having the boundary values $\psi = f$ on ∂D . Since the coefficients a_{ik} are Hölder continuous in $\bar{D}$, such a solution exists by the Schauder theory of linear elliptic equations ([1], p. 331 ff). Setting $U = \operatorname{grad}\psi$ and noting that $U \in S$, we define by this procedure a mapping $u \to Tu = U$ of S into itself. The boundary value problem (1), (7) is solved if the equation $u = Tu$ is shown to have a solution.

Consider the family of equations

$$u - \sigma Tu = 0 , \quad 0 \le \sigma \le 1 \ . \tag{9}$$

We observe that for each σ, u coincides with grad ϕ where ϕ is the solution (1) with boundary values $\phi = \sigma f$. A simplified version of the Leray-Schauder fixed point theorem due to H. Schaefer [9] asserts that equation (9) has a solution $u(x;\sigma)$ for each $\sigma \in [0,1]$ provided that

(a) T is completely continuous in S; and

(b) the solutions of (9) are uniformly bounded in S for all $\sigma \in [0,1]$, i.e.,

$$\| u(x;\sigma) \| < \text{constant independent of } \sigma \;. \tag{10}$$

The complete continuity of T is an immediate consequence of the classical Schauder theory; (see [1], p. 331 ff). It is here that the stated regularity properties for ∂D and the boundary values are required.

4. Interior estimates. To prove (10) let us suppose first that an a priori bound,

$$|u(x)| \leq M \quad \text{for } x \in D \;, \tag{11}$$

has already been established for the magnitude of the solutions of (9), where M is independent of σ, and depends only on f, D, and γ (in case I). We defer the proof of this bound (to §6), for it is only at this point that the details differ in the solution of the uniformly and non-uniformly elliptic problems.

The bound (11) insures that the functions $A_{k,i}(u)$ are bounded for all solutions of (9) and that the minimum eigenvalue $\lambda_1(u)$ of the quadratic form $A_{k,i}(u)\xi_i\xi_k$ is uniformly bounded from below. Thus, if u varies over solutions of (9), we can set

$$\lambda_1(u) \geq \mu > 0 \quad \text{and} \quad |A_{k,i}(u)| \leq \nu < \infty \;, \tag{12}$$

for suitable constants μ, ν.

Consider the linear equation satisfied by any first derivative $v = u_j = \partial\phi/\partial x_j$, $j = 1, \ldots, n$, obtained by formally differentiating (1) with respect to x_j and then setting $a_{ik}(x) = A_{k,i}(u(x))$. This equation is

$$\frac{\partial}{\partial x_k}\left[a_{ik}(x)\frac{\partial v}{\partial x_i}\right] = \frac{\partial}{\partial x_k}\left(A_{k,i}\frac{\partial v}{\partial x_i}\right) = 0 \;. \tag{13}$$

Actually, it suffices to consider only the weak form of this equation, namely

$$\int a_{ik}\, v_{x_i}\, \zeta_{x_k}\, dx = 0 \tag{14}$$

where $\zeta(x)$ is an arbitrary smooth function with compact support in D. This can be derived in the standard way from the integral form of (1), $\int A_k \zeta_{x_k} dx = 0$, without assuming the additional regularity of v and A_k implicit in (13). However, we refer to (13) in the following for convenience.

The basic a priori Hölder estimate of DeGiorgi [2, 7] shows that solutions of (13) in any compact subregion $D' \subset D$ satisfy an inequality

$$\frac{|v(x) - v(y)|}{|x-y|^{\alpha}} \leq \frac{C}{d^{\alpha}} \operatorname{lub}_{D} |v| \tag{15}$$

where $C = C(\mu, \nu)$ and $\alpha = \alpha(\mu, \nu)$, $0 < \alpha < 1$, and d is the minimum distance between D' and ∂D. Thus the solutions of (9) satisfy in D' a Hölder inequality

$$\frac{|u(x) - u(y)|}{|x-y|^{\alpha}} \leq \frac{\text{const.}}{d^{\alpha}}, \quad x, y \in D', \tag{16}$$

where the constant is independent of σ and depends only on the bound M in (11).

5. Boundary estimates. To establish the required uniform bound for $\|u\|$ we must now show that the solutions of (9) satisfy a uniform Hölder inequality in the neighborhood of ∂D as well as in compact subdomains of D. The desired boundary estimates can be obtained as follows (see also [4], §6).

In a neighborhood N_q of each boundary point q a suitable smooth one-to-one transformation $x \longleftrightarrow y$ can be found that takes the surface element of ∂D at q into a plane element $y_n = 0$, and equation (1) into one of the form

$$\frac{\partial B_k}{\partial y_k} = 0, \quad \text{where } B_k = B_k(y_1, \ldots, y_n, \phi_{y_1}, \ldots, \phi_{y_n}). \tag{17}$$

This is easily seen by making the transformation $x \to y$ in the integral form of (1), $\int A_k \zeta_{x_k} dx = 0$, where ζ is any smooth function with compact support in N_q. This gives an equation $\int B_k \zeta_{y_k} dy = 0$ which is equivalent to (17). By subtracting f from the solution ϕ in (17), we may suppose also that ϕ vanishes on $\partial D(y_n = 0)$.

The equation satisfied by any first derivative $v = \partial\phi/\partial y_j$, $j = 1, \ldots, n$, obtained by formally differentiating (17) with respect to y_j , is

$$\frac{\partial}{\partial y_k}[b_{ik}(y)\frac{\partial v}{\partial y_i} + c_k(y)] = 0 , \tag{18}$$

where $b_{ik}(y) = \partial B_k/\partial\phi_{y_i}$, $c_k(y) = \partial B_k/\partial y_j$. Because of (12) and the regularity properties of ∂D , this equation is uniformly elliptic, and the functions c_k uniformly bounded, for all solutions of (9). Also, $v(=\phi_{y_j})$ vanishes on $y_n = 0$ for $j = 1, \ldots, n-1$. It follows from the results of Morrey [5] and Stampacchia [10] that v satisfies a uniform Hölder property up to the boundary $y_n = 0$, and that its Dirichlet integral, $D(r) = \int|\text{grad } v|^2 dy$, taken over spheres of radius r , satisfies a growth condition $D(r) \leq Cr^{n-2+2\beta}$, where β is the Hölder exponent for v , and the constants C, β depend only on the ellipticity constants and on bounds for the coefficients b_{ik} , c_k in (18).

Considering now ϕ_{y_n} , we see from (17), by expressing $\partial^2\phi/\partial y_n^2$ in terms of the other derivatives appearing in the equation, that the Dirichlet integral for ϕ_{y_n} can be bounded in terms of those for ϕ_{y_j} , $j = 1, \ldots, n-1$, and hence also satisfies a growth condition $D(r) \leq Cr^{n-2+2\beta}$ in spheres of radius r . It is well known by a lemma of Morrey (see, e.g. [5]) that this implies a uniform Hölder property, with exponent β , for ϕ_{y_n} . This proves a Hölder estimate for $\phi_{y_1}, \ldots, \phi_{y_n}$, and hence for the solutions u of (9), in a neighborhood of each boundary point, and provides the required Hölder bound for u in a boundary strip.

6. <u>The bound on $|u|$</u> . To obtain the bound (11) for $|u(x)|$ in D we first establish the same bound on ∂D . This is accomplished by the following two lemmas, corresponding respectively to the uniformly and non-uniformly elliptic problems.

It will be assumed that ∂D and f have the properties stated in §2 (although only C^2 smoothness is required for the proofs).

LEMMA 1. Let

$$L\phi = a_{ik}(x)\frac{\partial^2\phi}{\partial x_i \partial x_k} = 0 , \quad x \in D , \tag{19}$$

satisfy the uniform ellipticity condition,

$$\lambda_1(x)\,|\xi|^2 \le a_{ik}(x)\,\xi_i\xi_k \le \lambda_2(x)\,|\xi|^2 , \quad \lambda_1(x) > 0 ,$$
$$\lambda_2(x)/\lambda_1(x) \le \gamma < \infty , \tag{20}$$

and suppose that $\phi = f$ on ∂D. Then there is a constant M depending only on γ, D, and f, such that

$$\max_{\partial D} |\operatorname{grad} \phi| \le M . \tag{21}$$

<u>Proof:</u> Let $\psi = \phi - f$ and $g(x) = -Lf$. Then

$$L\psi = g(x) \text{ in } D , \text{ and } \psi = 0 \text{ on } D . \tag{22}$$

It is known (see [1], p. 343) that corresponding to each point $q \in \partial D$ a function $w(x)$ can be found with the properties:

(i) $w(x) \ge 0$ in $\overline{D}$, $w(q) = 0$;

(ii) $Lw(x) \le -|g(x)|$ in D ;

(iii) $\dfrac{\partial w}{\partial n}(q) \le k \operatorname*{lub}_D |g(x)/\lambda_1(x)|$, where k is a constant depending only on γ and D ; (n denotes the inward normal to ∂D).

In particular, if S_R is a closed sphere of radius R having only the point q in common with $\overline{D}^*$, and if r denotes radial distance from the center of S_R, then for suitable constants k and m depending only on γ, R and the dimensions of D, the function

$$w = k(R^{-m} - r^{-m}) \operatorname*{lub}_D |g/\lambda_1|$$

satisfies conditions (i)-(iii). This is easily seen by direct calculation.

It follows from (i), (ii) and (22) that

$$L(w \pm \psi) \le 0 \text{ in } D , \quad w \pm \psi \ge 0 \text{ on } \partial D .$$

The maximum principle implies that $w \pm \psi \ge 0$ throughout D, and since $w(q) \pm \psi(q) = 0$, we have

$$\left| \frac{\partial \psi}{\partial n}(q) \right| \le \frac{\partial w}{\partial n}(q) \le k \operatorname*{lub}_D |g/\lambda_1| .$$

We observe that this inequality holds uniformly in q and that $\operatorname*{lub}_D |g/\lambda_1|$ depends only on γ and f. Since $\operatorname{grad} \phi = \operatorname{grad} \psi + \operatorname{grad} f = \partial\psi/\partial n + \operatorname{grad} f$ on ∂D, we conclude the desired bound (21).

LEMMA 2. Let equation (19) be elliptic (but not necessarily uniformly), and let $\phi = f$ on ∂D. Suppose ∂D is strictly convex. Then there is a constant M depending only on f and D such that

$$\max_{\partial D} |\text{grad}\,\phi| \leq M .$$

<u>Proof:</u> The proof is based on the existence of a linear barrier function at each point $q \in \partial D$; namely, a linear function $b_q(x)$ such that: (i) $b_q(x) = 0$ at $x = q$; (ii) $b_q(x) \geq |f(x) - f(q)| = |\phi(x) - \phi(q)|$ for $x \in \partial D$; (iii) the slope of the hyperplane $z = b_q(x)$ in (x, z) space is bounded by a constant M independent of q and depending only on f and D. In other words, planes of slope $+M$ and $-M$ can be passed through each point $(q, \phi(q))$, $q \in \partial D$, in such a way that the entire boundary surface $(x, \phi(x))$, $x \in \partial D$, lies between two planes.

The existence of such planes is geometrically more or less evident under the hypotheses of strict convexity of ∂D and C^2 smoothness of the boundary values of ϕ. An analytic proof of existence is straightforward but we omit the details here.

Suppose then the linear function $b_q(x)$ satisfies conditions (i)-(iii) above. Since $L[b_q(x) \pm (\phi(x) - \phi(q)] = 0$, it follows from the maximum principle that $b_q(x) \geq |\phi(x) - \phi(q)|$ for all $x \in D$ as well as on ∂D. Thus the entire solution surface $(x, \phi(x))$, $x \in \bar{D}$, lies between the two barrier planes of slope $+M$ and $-M$ through q. Hence $|\text{grad}\,\phi| \leq M$ on ∂D, and the lemma is proved. We remark that $b_q(x)$ plays the same part in this lemma as the function $w(x)$ did in Lemma 1.

The use of linear barrier functions to bound the gradient of solutions is well known from the theory of nonlinear elliptic equations in two independent variables (see, e.g. [8]). The applicability to the present problem was pointed out to the author by J. Serrin.

Returning now to the proof of the bound (11), let $u = \text{grad}\,\phi$ be a solution of (9) which we insert in the coefficients of equation (1), thereby obtaining the linear equation (19). Lemmas 1 and 2 provide bounds for $|u(x)|$ on ∂D in the uniformly and non-uniformly elliptic cases respectively. Since the components of $u(x)$ satisfy equation (13), and therefore enjoy the (weak) maximum principle, it follows that the bound $|u(x)| \leq M$ holds also throughout $\bar{D}$ and (11) is proved. This completes the proof of the existence theorems stated in §2.

This work was supported in part by the Office of Naval Research.

FOOTNOTES

*See his abstract on p. of this volume.

** R can be chosen independent of the point q , by virtue of the assumed smoothness of ∂D .

REFERENCES

1. R. Courant and D. Hilbert, "Methods of Mathematical Physics," vol. 2; Interscience, New York, 1962.

2. E. De Giorgi, "Sulla differenziabilita e l'analiticita delle estremali degli integrali multipli regolari," Mem. Accad. Sci. Torino. Ser. 3a, vol. 3 (1957), pp. 25-43.

3. A. Haar, "Über das Plateausche Problem," Math. Ann. 97(1927), pp. 124-258.

4. O. Ladyzhenskaya and N. Uraltseva, "Quasi-linear elliptic equations and variational problems with many independent variables," Uspehi Mat. Nauk 16 No. 1(1961), 19-92; transl. in Russian Math. Surveys 16 No. 1 (1961), 17-91.

5. C. B. Morrey, "Second order elliptic equations in several variables and Holder continuity," Math. Zeit. 72(1959), pp. 146-164.

6. C. B. Morrey, "Existence and differentiability theorems for variational problems for multiple integrals," Partial Differential Equations and Continuum Mechanics, University of Wisconsin Press, 1961, pp. 241-270.

7. J. Moser, "A new proof of de Giorgi's theorem concerning the regularity problem for elliptic differential equations," Comm. Pure Appl. Math. 13(1960), pp. 457-468.

8. L. Nirenberg, "On nonlinear elliptic partial differential equations and Hölder continuity," Comm. Pure Appl. Math. 6(1953), pp. 103-156.

9. H. Schaefer, "Über die Methode der a priori Schranken," Math. Ann. 129(1955), pp. 415-416.

10. G. Stampacchia, "Problemi al contorno ellittici, con dati discontinui, dotati di soluzioni hölderiane," Ann. Mat. Pura Appl. (IV) 51(1960), 1-38.

J. E. LITTLEWOOD

I. On van der Pol's equation with large k
II. Celestial mechanics over a very long time

I. Van der Pol's equation

The equation is

$$\ddot{y} - k(1-y^2)y + y = bk\cos\phi, \phi = t + \alpha ,$$

where k is large, and $0 < b < \frac{2}{3}$: $\frac{2}{3}$ is a critical value. ϕ is the phase. For the theory see (1) , which is an account of results with indications of proofs; it is immediately followed (in Acta Mathematica, but in the next volume, 98) by a paper (of 110 pages) giving full proofs; we shall not be concerned with this.[1]

The range $0 < b < \frac{2}{3}$ consists of a set S_1 of intervals B_1, another set S_2 of intervals B_2, and very short[2] transitional intervals between the B_1, B_2. When $b \in S_1$ there is just one set[3] of stable periodic trajectories Γ of order $2n+1$ (i.e. of time-length $(2n+1)2\pi$) ; when $b \in S_2$ there are two sets of stable periodic Γ of orders $2n+1$ and $2n-1$. As b increases through the B_1 and B_2 (jumping the transitional intervals) there are successively one set of order $2n+1$; two sets of orders $2n+1$, $2n-1$; one set of order $2n-1$; two sets of orders $2n-1$; $2n-3$; and so on. n is of the order of $(\frac{2}{3} - b)k$. Adjacent B_1, B_2 are approximately equal and of length of order $1/k$, the length varying slowly with b.

For $b \in S_1 + S_2$ almost all Γ's , "normal" Γ's , converge each to some stable periodic Γ.

Any theory of the transitional intervals would be a matter of the utmost difficulty, and has not been attempted. We may, however, surmise that for a b in one of them (unless possibly in its very central core) there will be at least one set of stable, and fairly normal, periodic Γ.

When there are two stable periods very remarkable topological behavior must ensue (see the Appendix). There exists, further, in this case, an intricate fine structure of (inter alia) periodic Γ's of an infinite number of different periods[4] (and so arbitrarily long). These are highly unstable, as we describe below.

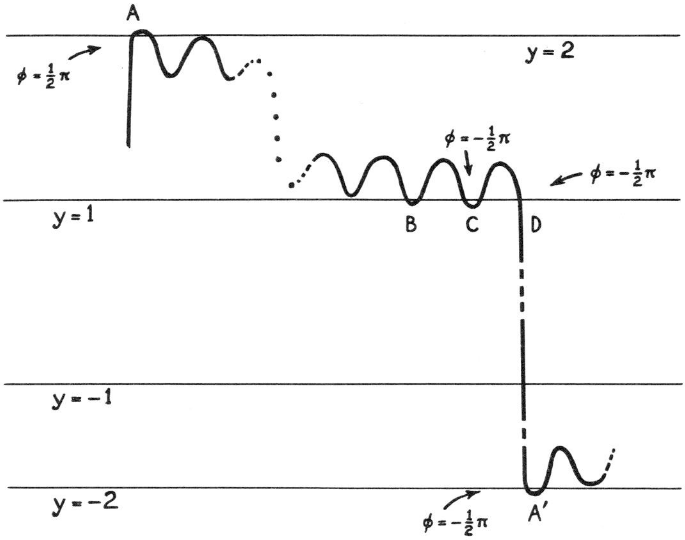

Figure 1

A schematic graph of a stable periodic Γ is given in Fig. 1. From A to A' is a half-period of the Γ, which is skew-symmetric. The upper set of waves can (for part of the range of b) make a few "dips" (depth of order $k^{-\frac{1}{2}}$) below $y = 1$, as at C, D, acquiring more penetrating power each time, and they finally penetrate into $y < 1$, at E in the figure. The negative damping then takes charge and produces a "shoot-through" from E to A', with slope of order k (and accordingly made infinite in Fig. 1). The phase ϕ is approximately $-\frac{1}{2}\pi$ at C, D, E.

The neighborhood of $y = 1$, where the damping changes sign, is naturally highly critical, and the last of the upper set of waves are distorted near their minima. The waves, e.g., at C, D have actually, in the limit as $k \to \infty$, the shape of the curves C_1, in Fig. 2 (with non-zero slopes at C, D); the last of the upper waves are approximately a succession of C_1's.

We have seen that a "normal" Γ converges to some stable periodic Γ. We have now to discuss abnormal ones. In Fig. 2 these are three curves C_1, C_2, C_3, each periodic 2π, shown by broken lines. The points D, E, D', E' in Fig. 2 have $\phi = -\frac{1}{2}\pi$. Each curve is a branch of the locus

$$\frac{1}{3}y^3 - y + \frac{2}{3} = b(1 + \sin\phi) \quad . \tag{1}$$

The upper set of waves of a normal Γ end, like those of a stable one, in a succession of approximate C_1 waves. They do <u>not</u> follow[5] $C_1, C_2, C_1, C_2, \ldots$. Motion along a C_1 (short of a shoot-through) or along a C_3, is stable. C_2, on the other hand, is an exponentially thin "tight-rope" (exponentially unstable). It is possible for a Γ (abnormal) to follow part of a C_2, and then slice or pull off it with slope of order k. The trajectory Γ, in Fig. 2 slices off C_2 at R down to C_3 and follows that; Γ_2 pulls off C_2 at S up to C_1, follows that to E, and shoots through to E', after which it follows the ensuing first upward waves S'B', ... of Fig. 1, which are approximately C_3 waves. (Note that the useful terms pull and slice refer to the upper C_2) a pull off lower C_2 counts as an "inverted" slice.)

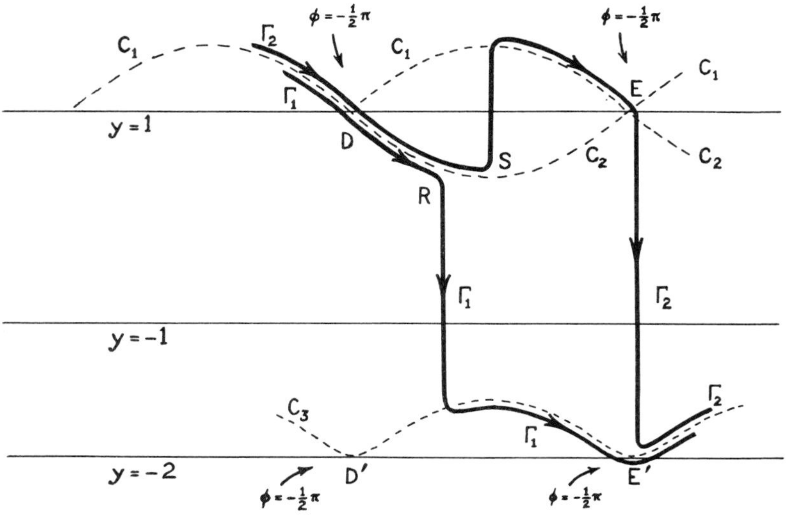

Figure 2

While they are on C_2, Γ_1 and Γ_2 are walking a tight-rope. The unstable periodic Γ behave in this way just before each of their shoot-throughs, which are many when the period of Γ is long, and they slice or pull differently each time.

I have now to report a very surprising development, an electroni calculation discovery by A. Warren,[6] under the supervision of H. P. F. Swinnerton-Dyer. His k is 5, which is a good bet for being "large For a certain b he finds two stable periods of orders 11 and 25, completely breaking the $2n \pm 1$ rule. Photographs of these are given in Fig. 3.

We can interpret the photographs in terms of the " k large" behavior we have been describing, at least up to a point. We must, of course, regard b as being in a transitional interval, and Γ_{11} as the stable Γ we said above might be expected. (Γ_{11} does not make any dips, but this is normal in part of the range of b.)

The photograph of Γ_{25} reveals that it follows part of a C_2 at A, B, and C (pulling at B and invertedly slicing at A and C). The surprise, of course, is that, its three "exponentially" dangerous walk notwithstanding, it succeeds in being stable. After this there seems no reason why, for a single b and different initial conditions, slices and pulls could not happen in a variety of ways, in which case there could be 3 or more different stable periods.

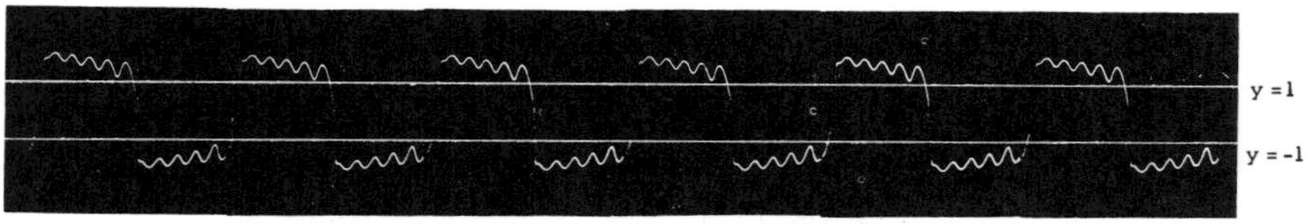

Stable trajectory - sub-harmonic order 11

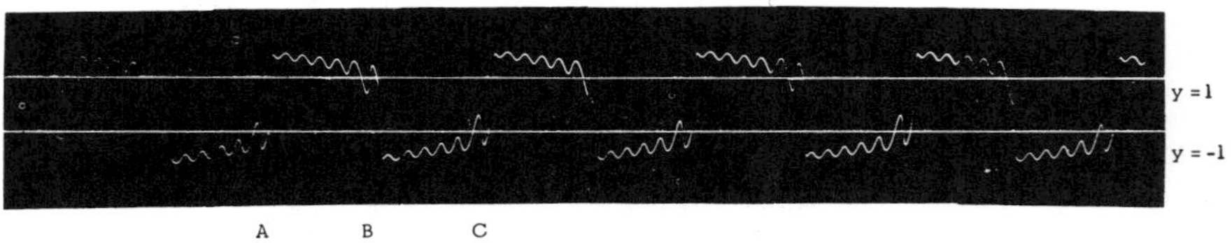

Stable trajectory - sub-harmonic order 25

Figure 3

The connection of the numbers 11 and 25 may be elucidated as follows. 11 is the "normal" stable period. Now a slice or a pull off a C_2, as opposed to a shoot-through at the beginning of the C_2, has the effect of delaying the first upward wave by time 2π. Next, Γ_{25} is not skew-symmetric, and its 25 primary periods are made up of two "normal" 11's and 3 extra from the pulls and slices.

Appendix

1. It is the case, for very general equations with a forcing term of period 2π and with positive damping for large $|y|$, that a suitable large closed curve in (y, y') space shrinks under the transformation T from $t = 0$ to $t = 2\pi$. It must then shrink, under iteration of T, to a limit set S, boundary B, say. In the van der Pol case with $b < \frac{2}{3}$, S has positive area. (By another very general topological theorem there must exist a periodic Γ_0 of period 2π. For $b < \frac{2}{3}$ this is totally unstable. Points near the representative point P_0 of Γ_0 recede from P_0 under iteration of T, and a small area around P_0 must expand, which is possible only if P_0 is an interior point of S.) If now B were a Jordan curve J, with a fixed point on it of order p, then (by a general topological theorem) all fixed points on B would have to have the same order, namely p. On the other hand, when there are stable periods of orders $2n \pm 1$ it is easy to see that B must contain all the corresponding fixed points of orders $2n \pm 1$. B is therefore nothing so simple as a J. It seems almost certain, indeed, that it is an indecomposable continuum, which is about the dirtiest thing one can say about a set of points.

In the case of a J with fixed points of order p the "rotation-number" of T is $2\pi/p$ for both the exterior and the interior of J. In the case of B two rotation-numbers (possibly the same) must exist (by general theory). What are they? Nobody knows. Electronic calculation might establish particular cases.

2. The case $b > \frac{2}{3}$. I must apologize for jumping to the conclusion [in] that there is just one stable Γ of period 2π, to which all Γ converge. [The original plan was to include an account of $b > \frac{2}{3}$, but this would have involved a considerable delay, and the problem was never actually considered.] Once one thinks of it, there will be harmonics (of sub-multiple order) when b is large enough. It is pretty certain, however, that the Γ of period 2π will always be stable, and the problem, while not at all trivial, should not be too bad.

It is known that all Γ have ultimately $|y| < A(b)$, $|y'| < A(b)k$.

II. Celestial Mechanics over a Very Long Time

§1. It is notorious that no non-trivial system of gravitating point masses is known to be stable, or bounded, over infinite time. Consider the coplanar case of Lagrange's equilateral configuration of point-masses S, J, with $m_J = \mu m_S$, and a massless planetoid P; the equilateral triangle SJP, of side 1, rotates in equilibrium with unit angular velocity about the center of mass of S and J. If $\mu < \mu_0$, the smaller root of $\mu_0(1+\mu_0)^{-2} = \frac{1}{27}$, the configuration is "stable", in the sense that a small disturbance, measured by ε, in coordinates ξ, η and velocities $\dot{\xi}$, $\dot{\eta}$ of P with respect to the obvious axes, does not double in a few revolutions; these are real periods $2\pi/\lambda_1$, $2\pi/\lambda_2$ of "normal oscillations." This is the simplest non-trivial system, because of the equilibrium state $\xi = \eta = \dot{\xi} = \dot{\eta} = 0$. We can now ask: <u>for how long, in terms of</u> ε, <u>can we say that the disturbance remains</u> $O(\varepsilon)$? Until recently it was not known that this was so even for a time like ε^{-4}. I am now reporting on two methods, which I will call A and B; these show that given any n, there is an $\varepsilon_0(n)$ such that when $\varepsilon < \varepsilon_0(n)$ the disturbance remains $O(\varepsilon)$ over a time $T_n = K_n \varepsilon^{-n}$. For (A) see my papers[7] (2), (3), (4). Method B, which has since been found by H. P. F. Swinnerton-Dyer, is unpublished.

It is possible by hard work to push the results further (this has been done for A in (3)). If λ_1/λ_2 is a normal irrational, then the disturbance remains $O(\varepsilon)$ over time[8]

$$T_e = \exp(K\varepsilon^{-1/2}|\log \varepsilon|^{-3/4}) .$$

Methods A and B are potentially very general; it does not seem beyond hope,[9] if certain difficulties can be overcome, to establish that a "Solar System" for which there is an "ε" will remain bounded over time T_n. Taking for simplicity N coplanar planets without satellites, we would suppose (i), that the mass of every planet is less than ε times the Sun's; we need not necessarily suppose the eccentricities small, but we should require (ii), that to begin with the planets keep well out of each other's way (as happens in the actual Solar System). Various <u>angles</u> would vary, but we might find that the major axes and the squares of the eccentricities vary only by $O(\varepsilon)$ over time T_n.

§2. In explaining method A, I will take as simple a model as possible.

The equations for the triangle can be transformed linearly to the form

$$(\Delta)\quad \dot{x} = \lambda y + X(x, y, z, w) \qquad \dot{y} = -\lambda x + Y(x, y, z, w)$$

$$\dot{z} = \mu w + Z(x, y, z, w)\ , \qquad \dot{w} = -\mu z + W(x, y, z, w)\ ,$$

where $X(x, y, z, w) = \sum_1^\infty \varepsilon^m X_m(x, y, z, w)$, etc. The X_m etc. are subject to

$$(H)\qquad \frac{\partial X_m}{\partial x} + \frac{\partial Y_m}{\partial y} + \frac{\partial Z_m}{\partial z} + \frac{\partial W_m}{\partial w} = 0\ ;$$

so that the equations (Δ) are Hamiltonian. They are of course autonomous (i. e. do not contain t explicitly). Further, λ/μ is supposed irrational.

Our model will be second order instead of the fourth order of (Δ). It is, however, of no use to take an autonomous pair of Hamiltonian equations, since there is then an integral H = constant which immediately settles the question.

The position, however, is completely changed if we take for our model

$$(E)\quad \dot{x} - \lambda y = X = \sum_1 \varepsilon^m X_m(x, y; t)\ ,\ \ \dot{y} + \lambda x = Y = \sum_1 \varepsilon^m Y_n(x, y; t)\ ,$$

where λ is a normal irrational, and where we suppose about X_m and Y_m only that they are polynomials in x, y (not necessarily homogeneous), with coefficients analytic and periodic 2π in t ; that they satisfy

$$(H)\quad \frac{\partial X_m}{\partial x} + \frac{\partial Y_m}{\partial y} = 0\ ;$$

and finally that, in analogy with the problem of (Δ), we have $|X_m|$, $|Y_m| < K_1^m$ for $|x|$, $|y| < K_2$ and all t , so that the convergence problems are completely trivial (as with (Δ)) and can be ignored. (We could abolish them altogether by supposing the series for X and Y to be finite.) The equation (E) covers such familiar problems as that of the disturbed pendulum.[10]

We have now

THEOREM 1. For any given n , for $\varepsilon < \varepsilon_0(n)$, and for

$$0 \le t \le T_n = K_n \varepsilon^{-n}\ ,$$

the equations (E) have an asymptotic integral

$$(a)\quad F_n(x, y; t) = \sum_0^n \varepsilon^m f_m(x, y; t) = C + O(\varepsilon^{n+1})\ ,\ \ C = a^2 + \sum_1^m a_m \varepsilon^m\ ,$$

where $f_m(x, y; t)$ depends only on m, λ, and the X_r, Y_r. We have

(b) $$f_0 = x^2 + y^2 ,$$

and for $m > 0$ the f_m are polynomials in x, y with coefficients periodic 2π in t, and they satisfy

(c) $$\int_{-\pi}^{\pi}\int_{-\pi}^{\pi} f_m(\cos\theta, \sin t; t)\, d\theta dt = 0$$

(and are then uniquely determined).

We have further, for $t \le T_n$,

(d) $$F_\nu(x, y; t) = C_\nu + O(\varepsilon^\nu) \qquad (1 \le \nu \le n) ;$$

in particular

(e) $$x^2 + y^2 = a^2 + O(\varepsilon) .$$

(a), (d) and (e) are true also in formally differentiated form.

Our main interest, of course, is (e), which shows that the "disturbance" stays $O(\varepsilon)$ over time T_n. The proof of this, however, is arrived at via (a).

We prove the results of the theorem in the first instance for $t \le 2\pi$. Suppose this done. Now $\varepsilon^{-\frac{1}{2}n}$ steps, in each of which t increases by 2π, and each with error $O(\varepsilon^n)$, give a total error $O(\varepsilon^{\frac{1}{2}n})$. The successive steps will have error $O(\varepsilon^n)$ provided the "initial $x^2 + y^2$" for each step does not get out of hand; it does not, because (a) implies (e) [and (d)]. We get (a), and (d), (e) along with it, with error $O(\varepsilon^{\frac{1}{2}n})$, over time $K_n\varepsilon^{-\frac{1}{2}n}$. Since $\varepsilon^m f_m = O(\varepsilon^\nu)$ for $\nu \le m \le \frac{1}{2}n$, we have $F_\nu = C_\nu + O(\varepsilon^\nu)$ for $\nu \le \frac{1}{2}n$, and the ν is now arbitrary along with n.

§3. From now on, then, we are concerned only with $0 \le t \le 2\pi$. We shall understand that all functions written with explicit t, as $\psi(x, y\ t)$, are periodic 2π in t and polynomial in x, y.

The first order solution of (E) is

$$x = x_0(t) + O(\varepsilon),\ y = y_0(t) + O(\varepsilon)\ ;\ \dot{x} = \dot{x}_0(t) + O(\varepsilon),\ \dot{y} = \dot{y}_0(t) + O(\varepsilon),$$

$$\text{where } x_0 = x_0(t) = a\cos\theta,\ y_0 = y_0(t) = a\sin\theta,\quad \theta = x - \lambda t . \tag{1}$$

We shall in the future use x_0, y_0 to denote these functions of t.

The corresponding first order integral is

$$F_0(x, y) = f_0(x, y) = x^2 + y^2 = a^2 + O(\varepsilon) \ . \tag{2}$$

We now proceed by induction, assuming Theorem 1 up to n−1 and deducing it for n . We have, substituting for $\dot{x}, \dot{y}$ from (E),

$$\frac{d}{dt} F_{n-1}(x, y; t) = \sum_0^{n-1} \varepsilon^m \left\{ \frac{\partial f_m}{\partial x} (\lambda y + X) + \frac{\partial f_m}{\partial y} (-\lambda x + Y) + \frac{\partial f_m}{\partial t} \right\} \ .$$

The right side, rearranged as a power series in ε , is seen to be of the form $\sum_1^{\infty} \varepsilon^m \phi_m(x, y; t)$. Thus[11]

$$\frac{d}{dt} F_{n-1}(x, y; t) = \sum^{n} \varepsilon^m \phi_m(x, y; t) + O(\varepsilon^{n+1}) \ .$$

By hypothesis this is also $O(\varepsilon^n)$ for all small ε (and for all initial conditions for x, y , and for $0 \le t \le 2\pi$); we must therefore have $\phi_m = 0$ identically for $m \le n-1$; so

$$\frac{d}{dt} F_{n-1}(x, y; t) = \varepsilon^n \phi_n(x, y; t) + O(\varepsilon^{n+1}) \ . \tag{3}$$

We are now at a key point. In what follows we switch backwards and forwards between functions of the respective forms $\psi(x, y; t)$ and[12] $\psi(x_0, y_0; t)$: the two ψ's differ by $O(\varepsilon)$ (since (1), or (e) of Theorem 1, is true under the inductive hypothesis.)

We have from (3) (the first "switch")

$$\frac{d}{dt} F_{n-1}(x, y; t) = \varepsilon^n \phi_n(x_0, y_0; t) + O(\varepsilon^{n+1}) \ . \tag{4}$$

Now we have (recalling $\theta = \alpha - \lambda t$)

$$\phi_n(x_0, y_0; t) = (\kappa + \sum_{r \neq 0} C_r e^{rti}) + \sum_m^* b_m e^{m\theta i} \left\{ \sum_p^* a_{pm} e^{pti} \right\} \ . \tag{5}$$

where the stars denote that m and p are subject to[13]

$$m \neq 0 \ , \quad -m\lambda + p \neq 0$$

For the terms with m = 0 belong in the round bracket, and since λ is irrational there are no terms with $-m\lambda + p = 0$.

We now define $f_n(x_0, y_0; t)$, and thereby, of course $f_n(x, y; t)$. We take

$$f_n(x_0, y_0; t) = - \sum_{r \neq 0} \frac{Cr}{ri} e^{rti} - Q \ ,$$

where

$$Q = \sum_m^* b_m e^{m\theta i} \left\{ \sum_p^* \frac{a_{pm}}{(-m\lambda + p)i} e^{pti} \right\} .$$

In Q we have $e^{m\theta i} = \{(x_0 \pm iy_0)/a\}^{|m|}$ according as $m \gtrless 0$, and $\sum_p^*$ is periodic 2π in t . Thus $f_n(x, y; t)$ is polynomial in x, y , with coefficients analytic and periodic 2π in t , and it is easily seen to satisfy (c) of Theorem 1 (and is uniquely determined when the earlier f_m are). Since by (5)

$$\frac{d}{dt} f_n(x_0, y_0, t) = -\left\{ \phi_n(x_0, y_0; t) - \kappa \right\}$$

we now have

$$\frac{d}{dt} F_n(x, y; t) = \frac{d}{dt} F_{n-1}(x, y; t) + \varepsilon^n \frac{d}{dt} f_n(x_0, y_0; t) + O(\varepsilon^{n+1}) ,$$

by a "switch,"

$$= \kappa\varepsilon^n + O(\varepsilon^{n+1}) . \tag{7}$$

by (4) and (6). <u>Provided $\kappa = 0$ this gives</u> $F_n(x, y; t) = C_n + O(\varepsilon^{n+1})$ as desired. [The inductive hypothesis effectively assumes that all previous κ's are 0 . But for this κ would be $\kappa_1 t^{n-1} + \ldots + \kappa_{n-1}$].

§4. It remains to prove that $\kappa = 0$. This we do by invoking Liouville's theorem of constant volume.

Consider the curve (in general a hyper-surface) S , or (in neutral coordinates X , Y)

$$F_n(X, Y; t) = C > 0 ,$$

with $t = 0, 2\pi$: the two values of t give identical curves for the same given C . S is of the form

$$X^2 + Y^2 = C + O(\varepsilon) ,$$

and is closed. It expands as C increases, because two S with different C's cannot have a point in common, and S is approximately $X^2 + Y^2 = C$ for large C . If now S(0) is

$$F_n(h, k; 0) = 1$$

in (h, k) space, and x, y are the solutions at $t = 2\pi$ of (E) with initial conditions $x = h$, $y = k$, then, as (h, k) ranges over the interior $V(0)$ of $S(0)$, (x, y) ranges over the interior $V(2\pi)$ of a closed curve $S(2\pi)$; and by Liouville's theorem the areas $|V(0)|$, $|V(2\pi)|$ are equal[14]. We have, however,

$$F(x, y; 2\pi) = 1 + \int_0^{2\pi} \left(\frac{d}{dt} F_n\right) dt = 1 + 2\pi\kappa\varepsilon^n + O(\varepsilon^{n+1}) ,$$

by (7) of § 3. If $\kappa \neq 0$, then for small ε $S(0)$ and $S(2\pi)$ have different C's and so different $|V|$'s.

This completes the proof of Theorem 1.

§5. It must be added that the application of Liouville's theorem here is abornmally simple, and I must sketch what happens when there are more pairs of variables, e.g. in the triangle problem. In this there are two integrals

$$F_n(x, y, z, w) = C_n + O(\varepsilon^{n+1}) , \quad G_n(x, y, z, w) = C'_n + O(\varepsilon^{n+1}) ;$$

two constants a and b instead of a, the first order integrals being

$$f_0 = x^2 + y^2 = a^2 + O(\varepsilon) , \quad g_0 = z^2 + w^2 = b^2 + O(\varepsilon) .$$

Finally, instead of κ there are

$$\kappa_1 = P_1(a^2, b^2) , \quad \kappa_2 = P_2(a^2, b^2) ,$$

with P_1, P_2 polynomials, so that κ_1 and κ_2 could each now take different signs for different values of b^2/a^2.

We can use for $S(0)$ the surface

$$\frac{F_n(h, k)}{\alpha} + \frac{G_n(h, k)}{\beta} = 1 ,$$

and Liouville's constant volume theorem then gives an identity with α, β as parameters; the coefficients of each $\alpha^p\beta^q$ in this must be 0. Method A unfortunately runs into difficulties at this point. It succeeds in dealing with the coplanar case of the Lagrange triangle with the help of Jacobi's integral, but as it stands it does not extend to more than 2 pairs of variables with autonomous equations. Its main achievement apart from such equations is in fact Theorem 1 and its Corollary below. Fortunately Method B is free from parallel difficulties (this is rather odd, since one would expect parallelism).

§6. Suppose the solutions of an equation (E) are in fact bounded over $t < \infty$. There are then two main topological possibilities. Let P_m be the point $\{x(2m\pi), y(2m\pi)\}$ and let T be the transformation $P_0 \rightarrow P_1$. First, there may be isolated closed curves C exactly invariant under T and its iteration; invariant curves intermediate to consecutive C will then spiral outwards and inwards to the two C's as limits (an invariant curve cannot cut itself without becoming closed and must otherwise spiral). Secondly, there may be a closed invariant curve through each point (and finally we could have a mixture of the two).

If now in the integral (a) of Theorem 1 we take $t = 0$, we get an invariant curve, to error $O(\varepsilon^n)$, through every point; i.e. we have approximately, the second possibility.

§7. THEOREM 1, COROLLARY. The results of Theorem 1 are true mutatis mutandis, if the coefficients of the polynomials X_m, Y_m are almost periodic functions of t, for simplicity finite sums of the form

$$\sum a \cos\{(m_1\lambda_1 + m_2\lambda_2 + \dots + m_k\lambda_k)t + \alpha\}, \tag{1}$$

where $\lambda, \lambda_1, \dots, \lambda_k$ are linearly independent, and the a's and α's depend on $m_1, m_2, \dots, m_k$. The polynomials $f_m(x, y; t)$ have likewise coefficients of the form (1).

This is established by substantially the same argument as that for Theorem 1.

A special case is that of the linear equation

$$\ddot{x} + x\{\lambda^2 + \varepsilon\psi(t)\} = 0, \quad \psi(t) = \sum_{r=1}^{k} a_r \cos(\lambda_r t + \alpha_r).$$

In this case, with $z = x$, we have $X = 0$, $Y = -x\psi(t)$, and the ϕ_n of §3 is $-x\dfrac{\partial f_{n-1}}{\partial y}\psi(t)$, with the result that F_n is a quadratic polynomial in x, y. The integral has the form

$$x^2\{1 + \sum_1^n \varepsilon^m a_m(t)\} + 2xy\sum_1^n \varepsilon^m b_m(t) + y^2\{1 + \sum_1^n \varepsilon^m c_m(t)\} = a^2 + O(\varepsilon^n$$

where $a_m(t)$, etc. are of the form (1). In particular x and $\dot{x}$ remain bounded over time T_n.

§8. A typical case of Theorem 1 is the disturbed pendulum. Consider a pendulum of (limiting) period $2\pi/\lambda$, with λ a normal irrational, starting with a displacement of order ε, and subject to a disturbing tangential acceleration $\varepsilon \cos t$. In the equation

$$\ddot{\theta} + \lambda^2 \sin\theta = \varepsilon \cos t$$

write

$$\ddot{\theta} = \varepsilon(a \cos t + x), \quad a = 1/(1 - \lambda^2). \quad \text{Then}$$

$$\ddot{x} + \lambda^2 x = \sum_{1}^{\infty} \varepsilon^{2m} \frac{(-)^{m+1}}{(2m+1)}, (a \cos t + x)^{2m+1}.$$

Since ε appears here in even powers only, we conclude that θ remains $O(\varepsilon)$ over $t < \exp(\kappa\, \varepsilon^{-1} |\log \varepsilon|^{-3/4})$.

§9. Method B. In its simplest form this has turned out to be a result already in "Dynamical Systems," by G. D. Birkhoff, N.Y. (1927), pp. 84, 85. (Swinnerton-Dyer and I were amateurs who had entered the subject by a side-line.) This is perhaps a case of fortunate ignorance, since the method can be developed much further.

FOOTNOTES

1. The papers are based on joint work with Miss M. L. Cartwright.

2. Almost certainly exponentially short (in terms of k), but the theory does not go into such refinements.

3. Given one periodic Γ of period $m2\pi$, there is a "set" of m of them got by translating the first by $t = 0, 2\pi, \ldots, (m-1)2\pi$.

4. Of every order of the form $2(k+r+s)n+r-s$, when k, r, s are non-negative integers, not all 0.

5. Though the two branches of (1) other than C_3 would naturally be taken to be $C_1, C_2, C_1, C_2, \ldots$ and $C_2, C_1, C_2, \ldots$, which have continuous tangents. It is perhaps instructive to mention that a reversed Γ started along a reversed C_2 wave would begin by following reversed C_2 waves, but gradually sink, and finally converge to the reversed Γ_0, which is stable.

6. Published only in a Ph. D. thesis, Cambridge, England.

7. So that incidentally there exist a second set of 25 inverted Γ_{25}'s.

8. $(3)_m$ almost completely supersedes $(2)_m$. Any intending reader should look first at the half-page Addendum in $(4)_m$, and then go on to $(3)_m$.

9. "e" for exponential. T_e and T_n occur frequently in the sequel. The index $3/4$ could be slightly improved.

10. I had reached this conclusion in $(2)_m$, but it was then a very long shot.

11. This may serve as a sop to those who would regard celestial mechanics as not quite the real thing in non-linearity.

12. We ignore convergence problems, recall.

13. See the remark after (1).

14. (i) p and m range, of course, over both positive and negative values.
    (ii) Since the curly bracket is analytic the convergence problems raised by the coefficients a_{pm} are quite trivial.

15. This is secured by condition (H) (for the triangle problem as well as in our present case). It is easily verified from $\frac{d}{dt}\frac{\partial x}{\partial h} = \frac{\partial}{\partial h}(\dot{x}) = \frac{\partial}{\partial h}(\lambda y + X)$, etc., that for the "magnification"

$$J = \frac{\partial(x, y)}{\partial(h, k)}$$

we have

$$\frac{dJ}{dt} = J\left(\frac{\partial X}{\partial x} + \frac{\partial Y}{\partial Y}\right) = 0 .$$

16. This is not essential, but we will suppose it for simplicity.

17. Swinnerton-Dyer's proof will appear in due course, with extensions.

REFERENCES

Part I

(1) J. E. Littlewood: "On non-linear differential equations of the second order. III. Acta Mathematica 97 (1957).

Part II

(2) J. E. Littlewood. "On the equilateral configuration in the restricted problem of three bodies." Proc. London Math. Soc. (3), 9(1959), 343-72.

(3) __________. "On the Lagrange configuration in Celestial Mechanics." Ibid., 525-43.

(4) __________. "Corrigendum and Addendum," Ibid., (10) (1960), 640.

LOUIS NIRENBERG

Rigidity of a class of closed surfaces

§1. Description of the problem. A closed surface in three dimensional space is said to be rigid if any other surface in 3-space that is isometric to it is congruent to it, i.e., differs from it by a rigid body motion plus (possibly) a reflection. It is easy to give examples of surfaces which are not rigid. I might remark that I know of no example of a closed surface which may be deformed continuously, preserving the metric. On the other hand, there are examples of surfaces that are infinitesimally not rigid, i.e., for which there exist non-trivial deformations that are isometric to first order in the deformation parameter; see S. Cohn-Vossen [3].

It is well known that closed convex surfaces are rigid. Perhaps the shortest proof of this, for smooth surfaces with positive Gauss curvature, is due to Herglotz. It may be found in any of the standard expository works on the subject; see N. V. Efimov [4], or H. Hopf [8]. It is not difficult, with the aid of some additional remarks to extend the proof to closed surfaces with $K \geq 0$ (see the recent expository article by K. Voss [13], p. 132). A. V. Pogorelov [12] has proved the rigidity of closed convex surfaces which need not be smooth.

To show that a surface is rigid involves proving a uniqueness theorem (modulo trivial solutions) for the nonlinear differential equations expressing the fact that a surface has a given metric. Indeed at one point in our discussion we shall prove a local uniqueness theorem for certain nonlinear differential equations.

One would naturally like to prove rigidity for some non-convex surfaces. In 1938, A. D. Alexandrov [1] introduced a class of closed surfaces of positive genus, which are characterized, essentially, by

Condition A: $$\int_{K>0} K\,dA = 4\pi .$$

Here dA represents element of area on the surface, and the integration is extended only over the region S^+ on the surface where the Gauss curvature K is positive. These surfaces are as close to being convex as a surface of positive genus can be in that Condition 1 is equivalent to the "convexity property": the tangent plane to the surface at any point in S^+ is a supporting plane for the surface. These surfaces have recently been considered by S. S. Chern and R. K. Lashof [2], as well as higher dimensional analogues called "minimal embeddings." See also N. H. Kuiper [9], [10].

Alexandrov [1] proved the rigidity of analytic surfaces satisfying Condition A, i.e., the congruence of two isometric analytically embedded surfaces satisfying the condition. I wish to present here a partial result obtained in an attempt to extend Alexandrov's theorem to the non-analytic case. Unfortunately our proof uses two restriction (condition 1 and 2) which seem unnecessary, and I hope that someone will be able to eliminate them—especially the last, which is not of an intrinsic nature since it depends on the immersion of one of the surface

We now formulate our hypotheses. Let M be a compact two-dimensional orientable Riemannian manifold of positive genus satisfying Condition A. Let S be an isometric immersion (i.e. locally 1-1 map of rank 2 of M into 3-space). We denote by M^- the part of M where $K < 0$, by S^- its immersion, and by M' the complement of M^- (by S' its immersion). We shall sometimes speak of S and M interchangeably; S^+ denotes the region where $K > 0$. For convenienc we assume the immersions to be of class C^4.

Condition 1: At every boundary point of M^-, grad $K \neq 0$

This condition implies that the boundary of M^- consists of a finite number of smooth closed curves $\Gamma_1, \ldots, \Gamma_n$ (whose images will be denoted by $\Gamma'_1, \ldots, \Gamma'_n$), and that K changes sign on crossing these. The condition, which is intrinsic, was made as a technical convenience in order to treat the differential equations arising in the proof of rigidity of S^-. It turned out to have a surprising (to me) geometric consequence, which is described in § 2.3.

Condition 2: Every component of S^- contains at most one closed asymptotic curve.

Our rigidity theorem is

THEOREM: Let S be a closed orientable surface satisfying Condition A and also Conditions 1, 2, and let $\bar{S}$ be another closed surface (immersion of M) isometric to S. Then $\bar{S}$ is congruent to S.

Condition 2 is very unsatisfactory; in addition to being non-intrinsic it is probably very difficult to verify. Knowing no counter-examples, I have tried for some time, without success, to show that S^- cannot, in fact, contain any closed asymptotic curves (it is easy to verify this for a component of S^- which is a surface of revolution); at the moment I am willing to believe that it can.

In order to make this paper fairly self-contained we shall re-derive several known results especially from [1] which is not easily available, but shall simply refer to other results which are readily found in the literature.

In §2 we shall describe the geometric properties of a surface S satisfying Conditions A and 1 and also, in §2.4, Condition 2. The rigidity theorem is proved in §3 where it is shown first, by a well-known argument, that S' itself is rigid.

It is clear from our arguments that a surface satisfying the Conditions A, 1, 2 is also infinitesimally rigid.

§2. Geometric consequences of Conditions A, 1, 2. The results of §§2.1 and 2.2 are special cases of results in Alexandrov [1], and in Kuiper [9] where more general results are obtained, and non-orientable surfaces are also treated; see also Chern Lasof [2]. In [10] Kuiper presents some interesting immersions of non-orientable manifolds.

2.1. We prove first that Condition 1 is equivalent to the "convexity property." Consider the spherical image map $\nu(x)$ defined as the non-oriented normal to S at x, i.e., $\nu(x)$ is the map into Ω: the surface of the unit sphere with diametrically opposite points identified. At a point where $K \neq 0$ the map is locally 1-1, and the element of area on Ω is given by $d\omega = |K| dA$. The total area of Ω is 2π.

A point ν_0 on Ω is called a noncritical value if the Jacobian of the mapping ν is different from zero (i.e. $K \neq 0$) at every pre-image point of ν_0. By Sard's theorem the set of points on Ω which are not noncritical has measure zero. Since S is compact every noncritical value has only a finite number of pre-images.

If ν_0 is a point on Ω then at a point x_0 on S where the scalar product $\nu_0 \cdot x$ achieves its maximum or minimum as x ranges over S the vector ν_0 is normal to S, i.e., $\nu_0 = \nu(x_0)$. At x_0 necessarily $K \geq 0$, since the tangent plane there is a plane of support, and hence x_0 belongs to S'. It follows that the spherical image map of S' covers Ω, and every noncritical value ν_0 has at least two pre-images (maximum and minimum of $\nu_0 \cdot x$). Since the area of the image is $\int_{S'} K dA$ it follows that in general

$$4\pi \leq \int_{S'} K dA = \int_{S^+} K dA .$$

Suppose now that Condition A holds. If ν_0 is a noncritical value it has at most two pre-images in S^+, otherwise a neighborhood of ν_0 would be covered more than twice and hence, as above, we would have $\int_{S^+} K\,dA > 4\pi$. The pre-images are thus the points where $x \cdot \nu_0$ takes its minimum and maximum. Since at these points the tangent planes are planes of support we have demonstrated the "convexity property" at all points in S^+ that are pre-images of noncritical points. Let, now, x_0 be any point in S^+; a neighborhood if it is mapped onto a neighborhood of $\nu(x_0)$. By Sard's theorem this neighborhood contains a noncritical value. Hence, by what we have just shown, in the neighborhood of x_0 there is a point where the tangent plane is a plane of support. Since the neighborhood is arbitrary the same holds for the tangent plane at x_0. Thus we have shown that Condition A implies the "convexity property."

Suppose, conversely, that the "convexity property" holds. Let x_0 be a point in S^+ and assume that x_1 is another point in S^+ with $\nu(x_0) = \nu(x_1)$. By hypothesis the tangent planes to S at x_0 and x_1 are support planes. These planes cannot coincide, for, since $K(x_0) > 0$, a tangent plane at a point near x_0 can be found so that x_1 lies on one side of it and points of S on the other—contradicting our hypothesis. Thus the tangent planes at x_0 and x_1 are different, and, in fact S lies between them (they are parallel). It follows that there cannot be a third point x_2 with $\nu(x_2) = \nu(x_0)$. Thus we have proved that any point on Ω has at most two pre-images in S^+. Hence

$$\int_{S^+} K\,dA \leq 4\pi ,$$

and it follows from the preceding inequality that equality must hold.

We observe that the orientability of S did not enter. If Conditions A and 1 hold we can also assert that the tangent plane at <u>any</u> point $K \geq 0$ is a plane of support. This is obvious for a point x in the closure of S^+. If x, with $K(x) = 0$, is not in the closure of S^+ then it lies in a maximal open set C in S in which $K \equiv 0$. Suppose that every neighborhood of x_0 contains points which are not flat points (i.e. not both principal curvatures vanish). Then, according to the results of [6] one can infer that there is a straight line in C through x_0 with end points on the boundary of C along which the normal ν is constant. Hence the tangent plane at x_0 is the same as that at the end points. By Condition 1. however, the end points lie in the closure of S^+ and therefore the tangent plane is a plane of support. If finally a whole neighborhood of x_0 consists of flat points then the surface contains a portion of the tangent plane at x_0. At the boundary of this portion the previous situation maintains; hence the tangent plane at x_0 is a plane of support.

2.2. From now on we assume Conditions Aand 1. We show next that each of the curves Γ'_j is a convex plane curve with ν = the normal to the plane at each point of Γ'_j. A more general result is proved in [9]; with the aid of Condition 2, however, the proof is very simple, and so we include it. Consider then one of the Γ'_j, which we call Γ. To prove that Γ is a plane curve on which ν is constant it suffices to show that the normal curvature of Γ is zero. For then Γ is a line of curvature, and it follows from Rodrigues formula that $d\nu = 0$ on Γ. Because of our "convexity property" at points in S' near Γ it then follows rather easily that Γ is necessarily convex.

Let x_0 be a point on Γ. Both principal curvatures cannot vanish at x_0 for if they do the coefficients L, M, N of the second fundamental form vanish, and hence grad $K = 0$ contradicting Condition 1 Thus the principal curvatures on Γ are $k_1 = 0$ and $k_2 \neq 0$, and we may introduce lines of curvature as local parameter curves. In terms of these local coordinates (u, v), with $(0, 0)$ corresponding to x_0, suppose that the curve $v = 0$ has zero normal curvature at $(0, 0)$. We shall show that $k_{1u} = 0$ at $(0, 0)$. Since grad $K \neq 0$ at $(0, 0)$, and since $k_1 = 0$ on Γ it follows that Γ is tangent to the u-axis at $(0, 0)$, i.e., Γ has zero normal curvature. Here subscript u denotes partial differentiation with respect to u.

If the vector function $X(u, v)$ represents the surface in a neighborhood of x_0 then, by our "convexity property" at x_0, we see that

$$(X(u, 0) - X(0, 0)) \cdot \nu_0$$

does not change sign; here ν_0 is a normal vector at x_0, and the dot denotes scalar product. If we look at the local expansion of this expression up to third order terms we have $X_u \cdot \nu_0 = X_{uu} \cdot \nu_0 = 0$ at $(0, 0)$, the last because the u-axis has zero normal curvature at x_0, and hence the third order term vanishes, i.e.

$$X_{uuu}(0, 0) \cdot \nu_0 = 0 .$$

By Rodrigues formula, however,

$$\nu_u + k_1 X_u = 0 .$$

Differentiating with respect to u, and taking scalar product with X_u, we obtain

$$X_u \cdot \nu_{uu} + k_{1u} X_u \cdot X_u = 0 \quad \text{at } (0, 0) . \tag{1}$$

On the other hand since $X_u \cdot \nu = 0$, we find, on differentiating twice, that at $(0, 0)$, where $\nu_u = 0$,

$$X_u \cdot \nu_{uu} + X_{uuu} \cdot \nu = 0 \ .$$

Having shown that $X_{uuu} \cdot \nu = 0$ at the origin we see that

$$X_u \cdot \nu_{uu} = 0 \quad \text{at } (0, 0) \ .$$

It follows from (1) that $k_{1u} = 0$ at $(0, 0)$. Q. E. D.

We now derive some further consequences.

Let D_j represent the convex interior of Γ'_j in its plane. Then from our "convexity property" it follows that the union of the point set S' and $D_1, \ldots, D_n$ forms a closed convex surface Y which is the boundary of the convex hull of S. It follows from this that the point set S' is necessarily arcwise connected. For any two points x_1, x_2 in S' can be connected by a curve lying on Y. This curve may not of course lie in S', parts of it may cross some of the D_j, but these parts can easily be deformed to the boundary of D_j, i.e., onto Γ'_j which belongs to S'.

Since the spherical image mapping ν of Y covers Ω twice it follows from Condition A that M' covers the point set S' only once, and therefore the immersion $M' \to S'$ is actually 1-1. From now on we shall usually omit the primes in describing the curves Γ'_j since they are actually imbeddings of the Γ_j.

Note furthermore that since each Γ'_j is a plane curve, with ν normal to the plane, the geodetic curvature of Γ_j equals the curvature of Γ'_j in the plane, and hence is non-negative (after appropriate orientation). It follows that the immersion of each curve Γ_j is uniquely determined from the given metric (within rigid body motion). Hence also the D_j are uniquely determined.

We thus see that if a given compact 2-dimensional Riemannian manifold satisfying Conditions A and 1 can be isometrically immersed in space then necessarily M' is connected, and its boundary curves have geodetic curvatures which do not change sign.

2.3. We now investigate the structure of a component of S^-. According to what we have shown any component is bounded by a finite number $\Gamma_1, \ldots, \Gamma_k$ of convex plane curves, and on these curves the surface is tangent to the corresponding planes. Thus a component might look like

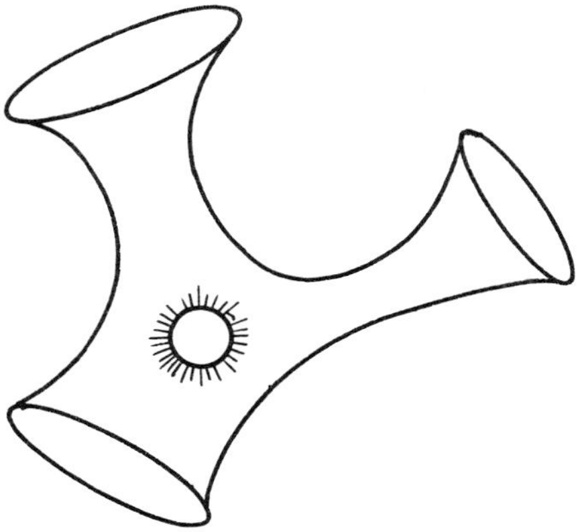

This is topologically a torus with 3 discs cut out.

It is the aim of this section to show that, because of Condition 1, such a situation cannot occur. We shall prove:

<u>Each component of</u> S^- <u>is bounded by two curves, and it is topologically a tube joining these.</u>

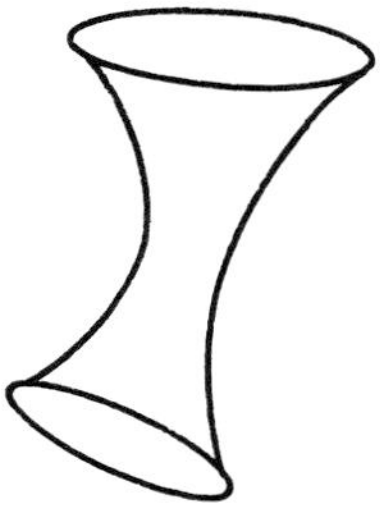

That is, if we fill in the curves by plane caps, the resulting closed surface is topologically a sphere. In order to demonstrate this we first investigate the behavior of the asymptotic curves, the curves with zero normal curvature at every point, near the boundary. As above, let us introduce the lines of curvature as local parameter curves near a point x_0 on a boundary curve Γ. With (u, v) as the local coordinates, and $(0, 0)$ corresponding to x_0, we also have, Γ, as above, corresponding to the u-axis. We may suppose that the region $v > 0$ lies in S^-. Let L, M, N be the coefficients of the second fundamental form. By our choice of coordinates $M \equiv 0$ and $N \neq 0$ (say $N > 0$), while $L = 0$ on the u-axis. Since $LN < 0$ in S^- we see that $L < 0$ for $v > 0$. We may therefore set

$$L = -v\ell(u, v) ,$$

and note that $\ell(0, 0) > 0$ for otherwise we would have $\text{grad } K = 0$

at $(0, 0)$. Thus the second fundamental form is

$$-v\ell\, du^2 + N dv^2 , \qquad \ell, N > 0 ,$$

and the asymptotic curves are therefore the integral curves of the system

$$\frac{dv}{du} = \pm \sqrt{\frac{\ell}{N}}\, v^{1/2} ,$$

and these clearly touch Γ tangentially. It is for this reason that Condition 1 was imposed—to make the behavior of the asymptotic lines near the boundary as simple as possible.

The two families of asymptotic curves are clearly distinguishable, for near an intersection of two we may mark the surface:

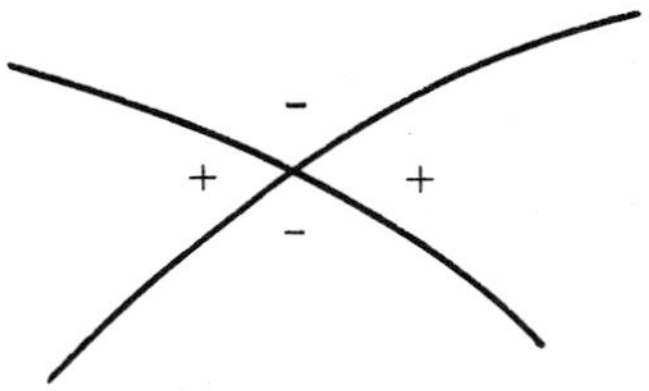

depending on whether the surface nearby rises above or below the tangent plane at the intersection. This enables us to distinguish the curves.

Consider now one of the families of asymptotic lines, which we may also imagine as well-defined on M^-: The corresponding tangents define a line field without singularity on the closure of the component of S^- under consideration. It is not necessarily a direction field, since we may not be able to assign a direction to the tangents which will be globally consistent. Suppose that the component is bounded by k curves $\Gamma_1, \ldots, \Gamma_k$. Fill these in by plane caps and extend the line field inside each cap. Since the line field is tangential at the boundary of the cap the extension must have a singularity. Make the extensions so that there is a singularity at just one point in each cap. Clearly the index of the singularity is one. Let g be the genus of the closed surface formed by the union of the component of S^- and the k planar caps. We may now apply the Poincaré theorem (here we use orientability) on the index of a field of line elements on a closed surface (this holds also for line fields as well as for direction fields, see H. Hopf [8]), according to which the sum of the indices equals $2-2g$. In our case we have k singularities, each of index one, hence

$$k = 2-2g .$$

Since $k > 0$ it follows that $g = 0$ and $k = 2$ proving our assertion.

We have thus found another necessary condition for the manifold

M satisfying Conditions A and 1 to be immersible in 3-space: that each component of the region where $K < 0$ be topologically a tube with two boundary curves.

2.4. Studying the asymptotic curves further, we claim now that we can introduce a well-defined direction on either family of asymptotic lines. On the boundary curve Γ_1 the asymptotic lines are tangent and it is possible there to introduce a well-defined direction. By suitably parametrizing the boundary curves Γ_1, Γ_2 it is possible to join points with the same parameter values on Γ_1 and Γ_2 by curves on our tube which cover the tube simply. The directions for the asymptotic line fields may then be extended in a unique way along these curves from Γ_1 without any singularity arising. We see consequently that an asymptotic curve emanating from, say, Γ_1 cannot return to Γ_1, for the direction of the asymptotic curve may be chosen in such a way that those with end points on Γ_1 are all leaving Γ_1.

Now our tube on M^- is topologically equivalent to a planar ring shaped domain and we are therefore in a position to use the Poincaré-Bendixson theorem. With its aid we shall now investigate the consequences of Condition 3. Suppose first that the interior of our tube does not contain any closed asymptotic curve. Then any asymptotic curve has its end points on Γ_1 and Γ_2 for if it doesn't end on either then by the Poincaré-Bendixson theorem it must tend to a closed asymptotic curve of the same family—but none such exists. Hence it has both end points on the boundary. By the preceding they cannot both lie on the same boundary curve, so one is on Γ_1 and the other on Γ_2. It follows that the asymptotic curves have finite length and that two curves of different families intersect at most a finite number of times.

Suppose now that there is one, and only one, closed asymptotic curve $\tilde{\Gamma}$ in the tube in M^-. Since $\tilde{\Gamma}$ can have no self intersections it is readily verified that it is homotopic to Γ_1 (i.e. goes around the tube once) and divides the tube into two disjoint parts T_1, T_2 containing respectively Γ_1 and Γ_2. Employing again the Poincaré-Bendixson theorem we find that the curves of the same family in T_1 start on Γ_1 and spiral toward $\tilde{\Gamma}$ from one side while those on the other side start on Γ_2 and spiral toward $\tilde{\Gamma}$ from the other side. Furthermore the curves of the second family go from Γ_1 to Γ_2. In T_1 (and in T_2) any curve of the second family is intersected infinitely often by a curve of the first family.

§3. <u>Rigidity.</u> Our arguments in this section are related to those in a paper by T. Minagawa and T. Rado [11].

We consider now, in addition to our surface S satisfying Conditions A, 1, 2, another surface $\bar{S}$ isometric to S, which we wish to show is congruent to S.

3.1. The first step is to show that $\bar{S}'$ is congruent to S'. As in §§2.1, 2.2 the images on $\bar{S}$ of the curves Γ_j are again plane convex curves $\bar{\Gamma}_j$ which are congruent to Γ'_j. Hence if we fill these in by discs $\bar{D}_j$ and consider the corresponding closed convex surface $\bar{Y}$, we see that $\bar{Y}$ is isometric to Y. Although Y and $\bar{Y}$ are not very smooth we may use here the uniqueness theorem for non smooth convex embeddings proved by Pogorelov [12] to conclude that Y and $\bar{Y}$, and hence S' and $\bar{S}'$ are congruent.

However the simpler proof of Herglotz (for closed surfaces including its extension to surfaces with $K \geq 0$) may also be used for the convex surfaces S' and $\bar{S}'$ with holes. This proof makes use of certain integral identities based on Green's theorem. In our case, on using Green's theorem, one obtains certain boundary integrals. Since the boundaries of both surfaces are planar curves where $K = 0$, and on which the normals are perpendicular to the planes, one verifies easily that these boundary integrals vanish (see K. P. Grotemeyer [5]). Following the remainder of Herglotz's proof (see also [12]) one finds that S' and $\bar{S}'$ are congruent.

3.2. The main problem therefore is to show that S^- and $\bar{S}^-$ are congruent. We have to prove this for any component, i.e. tube T of S^- and the corresponding tube $\bar{T}$ of $\bar{S}^-$. Since the boundaries of these tubes belong to S' and $\bar{S}'$ which, after a rigid body motion, we can assume to coincide, it follows that the boundaries of the tubes T, $\bar{T}$ coincide. We shall refer to the corresponding tube in M^- as T_0, and use the notation of §§2.3, 2.4.

We shall treat the rigidity problem as a uniqueness theorem concerning the initial value problem for a nonlinear system of hyperbolic partial differential equations, namely the system of equations satisfied by the coefficients of the second fundamental forms of T and $\bar{T}$. These coefficients together with their first order derivatives agree on Γ_1 (since it belongs to $S' = \bar{S}'$). We shall show, using a lemma on partial differential equations, that they then agree in a neighborhood of Γ_1. Then we wish to extend this neighborhood to cover all of T_0. This will prove the rigidity. Postponing the detailed discussion of the equations, as well as the local uniqueness theorem, we shall show here how to extend the uniqueness to the whole tube T_0, assuming Condition 2. All asymptotic lines on T_0 mentioned below will refer to the immersion T. We shall make use of the following lemma proved in §3.4.

LEMMA 1: <u>Let</u> B <u>be an asymptotic curve in the interior of</u> T_0 <u>and suppose that from a point</u> x_0 <u>on</u> B <u>the other asymptotic</u> A <u>(in a suitable direction) ends on</u> Γ_1. <u>Denote by</u> $A(x_0)$ <u>the closed</u>

segment of this curve connecting x_0 *to* Γ_1 . *From the continuous dependence on parameters of solutions of ordinary differential equations it follows that for points* x'_0 *on* B *near* x_0 *the corresponding segments* $A(x'_0)$ *will exist. Suppose now that the second fundamental forms of* T *and* $\bar{T}$ *agree on* $A(x_0)$ *then they agree on* $A(x'_0)$ *for* x'_0 *sufficiently close to* x_0 *on* B .

We now prove that the second fundamental forms agree on T_0 . Consider first the case that T contains no closed asymptotic line. Then, as we have remarked in §2.4 all asymptotic curves go from Γ_1 to Γ_2 . It suffices to show that the second fundamental forms agree on any asymptotic curve B going from Γ_1 to Γ_2 . Here we shall denote by B the open curve (i.e. not including end points). From any point x_0 of B a segment $A(x_0)$ of asymptotic curve of the other family will go from x_0 to Γ_1 . Let B_1 be the subset of such points x_0 on B for which the second fundamental forms of T and $\bar{T}$ agree on $A(x_0)$. By local uniqueness near Γ_1, B_1 is non empty. By continuity B_1 is closed, and by Lemma 1 it is open. Hence it is all of B . In particular the second fundamental forms agree at every point on B .

Suppose now that T_0 contains exactly one closed asymptotic curve $\tilde{\Gamma}$; according to §2.4 this divides T_0 into two disjoint parts T_1 , T_2 . Since the asymptotic curves of the other family cover T_0 it suffices to show that the second fundamental forms of T and $\bar{T}$ agree on any one of these. Let B_1 be the open part of such an asymptotic curve lying in T_1 . We may again apply the preceding argument and conclude that the second fundamental forms agree on B_1 . Thus they agree on all of T_1 . Arguing from the other side, i.e. from Γ_2 , we conclude again that the second fundamental forms agree on T_2 , and this completes the proof.

Remark: Our method of showing that the second fundamental forms agree in T_0 does not seem to be the best way to go about it. One should use the fact that the forms agree on the whole boundary of T_0 , instead of working with each boundary curve separately.

3.3. A lemma on differential equations: In proving that the second fundamental forms of T and $\bar{T}$ agree near Γ_1 we shall use a simple lemma giving uniqueness for a (slightly) singular hyperbolic initial value problem. Solutions are assumed to be of class C^1 .

LEMMA 2: *In a one-sided neighborhood,* $v \geq 0$, *of the origin in* the (u, v) plane let s, t be solutions, which vanish on $v = 0$, of the system

$$s_v - Nt_u = tO(v) + sO(1) \quad , \qquad (2)$$

$$vt_v - Ps_u + kt = tO(v) + sO(1) \quad . \qquad (3)$$

Here $k > 1/2$ is continuous, P, N are positive functions of class C^1, and $O(1), O(v)$ represent coefficients which are bounded respectively by constants and constants times v; subscripts denote partial differentiation. Then, in some neighborhood of the origin $s \equiv t \equiv 0$.

Proof: We use energy inequalities adapted to our special situation (the system is hyperbolic for $v > 0$ but becomes parabolic on $v = 0$). We shall prove $s = t = 0$ in a small characteristic triangle Δ near the origin

where the sides are characteristics, i.e. solutions of

$$\frac{dv}{du} = \pm \sqrt{\frac{v}{PN}} \quad .$$

With σ a negative number, multiply the first equation of the system by $e^{\sigma v} \frac{P}{N} s$, the second by $e^{\sigma v} t$, add, and integrate over Δ. On using Green's theorem one obtains the following, here double integrals represent integration over Δ and a single integral sign represents integration over the top boundary of Δ from left to right. There is no contribution from the lower boundary since s and t vanish there. The letter C denotes a constant depending only on the coefficients.

$$\iint e^{\sigma v}\left(\frac{P}{N} ss_v + vtt_v\right) du dv$$

$$+ \iint e^{\sigma v}\left(kt^2 - P(st)_u\right) du dv \leq C \iint e^{\sigma v}\left(vt^2 + |st| + s^2\right) du dv$$

or, using Green's theorem,

$$\frac{1}{2}\int e^{\sigma v}\left[\left(\frac{P}{N}s^2 + vt^2\right)du + 2\,Pst\,dv\right]$$

$$+\frac{1}{2}\iint e^{\sigma v}\left[(2k-1-\sigma v)t^2 - \left(\left(\frac{P}{N}\right)_v + \frac{P}{N}\sigma\right)s^2\right]du\,dv$$

$$\leq C\iint e^{\sigma v}(vt^2 + |st| + s^2)\,du\,dv \ .$$

By our choice of the characteristics the integrand in square brackets in the line integral is non-negative, it equals $du\left(\sqrt{v}t \pm \sqrt{\frac{P}{N}}\,s\right)^2$, and hence this term may be neglected. We may assume that $k - \frac{1}{2}$ is greater than a positive number ε; we have $C(vt^2 + |st| + s^2) \leq Cvt^2 + \frac{\varepsilon}{2}t^2 + \left(\frac{C^2}{2\varepsilon} + C\right)s^2$. Inserting into the preceding we find

$$\iint e^{\sigma v}\left[\left(\varepsilon - \frac{\sigma}{2}v\right)t^2 - \frac{1}{2}\left(\left(\frac{P}{N}\right)_v + \frac{P}{N}\sigma\right)s^2\right]du\,dv$$

$$\leq \iint e^{\sigma v}\left[\left(\frac{\varepsilon}{2} + Cv\right)t^2 + \left(\frac{C^2}{2\varepsilon} + C\right)s^2\right]du\,dv \ .$$

Now fix σ large negative so that

$$-\frac{\sigma}{2} \geq C\,, \quad \text{and} \quad -\frac{1}{2}\left(\frac{P}{N}\sigma + \left(\frac{P}{N}\right)_v\right) > \frac{C^2}{2\varepsilon} + C \ .$$

Then the coefficients of t^2 and s^2 on the left are greater than those on the right and we conclude that $s \equiv t \equiv 0$. Q. E. D.

This lemma can clearly be formulated in a much more general way (for instance, if we assume more regularity of the solutions, and that they vanish to sufficiently high order on $v = 0$ then the condition $k > 1/2$ can be relaxed considerably), but it is sufficient in this form for our purposes. For a treatment of hyperbolic problems which become singular on the initial line, and for further references, see G. Hellwig [7].

3.4. We consider now the differential equations satisfied by the coefficients L, M, N and $\bar{L}$, $\bar{M}$, $\bar{N}$ of the second fundamental forms of T and $\bar{T}$, in terms of local coordinates (u, v). These equations are

$$LN - M^2 = \bar{L}\bar{N} - \bar{M}^2 \tag{4}$$

and the Codazzi-Mainardi equations which take the form

$$\bar{L}_v - \bar{M}_u + a\bar{L} + b\bar{M} + c\bar{N} = 0$$
$$\bar{M}_v - \bar{N}_u + \alpha\bar{L} + \beta\bar{M} + \gamma\bar{N} = 0 , \qquad (5)$$

and are satisfied also by L, M, N; here the coefficients are given functions (the Christoffel symbols) depending on the metric.

We regard the coefficients L, M, N as known and wish to show that $\bar{L}$, $\bar{M}$, $\bar{N}$ are the same. In the neighborhood of any point in the closure of T we may introduce lines of curvature as parameter curves, so that $M = 0$. In the interior of T we have $LN < 0$; we may assume $N > 0$. Since the equations (5) hold for (L, M, N) we have

$$c = -\frac{L_v + aL}{N} , \quad \gamma = \frac{N_u - \alpha L}{N} . \qquad (6)$$

Let us first consider the equations near a point x_0 on Γ_1. As we know from §2.3, our local coordinates can be so introduced that $v = 0$ corresponds to Γ_1, and $(0, 0)$ to x_0; $L(u, 0) = 0$. We wish to apply Lemma 2 to infer that $(\bar{L}, \bar{M}, \bar{N})$ agrees with (L, M, N) knowing that there is agreement on $v = 0$. If we consider v small we know that the difference is small, so that $\bar{N}$ is also positive. Set $r = \bar{L} - L$, $s = \bar{M} - M = \bar{M}$, $t = \frac{\bar{N}}{N} - 1$, then (r, s, Nt) also satisfy (5). We may also rewrite (4) as

$$(L + r)(1 + t) - \frac{s^2}{N} = L$$

or

$$r = \frac{s^2 - LNt}{N(1+t)} .$$

Substituting this into the system (5), for (r, s, Nt), and using (6), the system takes the form

$$\left(\frac{s^2 - LNt}{N(1+t)}\right)_v - s_u + a\frac{s^2 - LNt}{N(1+t)} + bs - (L_v + aL)t = 0 ,$$

$$s_v - (tN)_u + \alpha\frac{s^2 - LNt}{N(1+t)} + \beta s + (N_u - \alpha L)t = 0 .$$

The last equation may be written as

$$s_v - Nt_u = t\left(\frac{\alpha L}{1+t} + \alpha L\right) - s\left(\beta + \frac{\alpha s}{N(1+t)}\right) ,$$

which is just of the form (2) since L vanishes on $v = 0$, while the first equation may be written in the form

$$-\frac{Lt_v}{1+t} - s_u - tL_v\left(\frac{1}{1+t} + 1\right)$$

$$= tL\left[\left(\frac{1}{1+t}\right)_v + \frac{a}{1+t} + a\right] - \left(\frac{s^2}{N(1+t)}\right)_v - \frac{as^2}{N(1+t)} - bs$$

$$= t0(v) + s0(1)$$

since $L = 0(v)$ and the derivatives of s, N and t are bounded. As in §2.3 we have $L = -v\ell(u,v)$ with $\ell(0,0) > 0$. Substituting this for L and multiplying by $(1+t)\ell^{-1}$ we find

$$vt_v - \frac{1}{\ell}s_u + t\frac{\ell + v\ell_v}{\ell}(2+t) = t\frac{s_u}{\ell} + t0(v) + s0(1)$$

$$= t0(v) + s0(1)$$

since $s_u = 0(v)$. Finally, since $t = 0(v)$ we have

$$vt_v - \frac{1}{\ell}s_u + 2t = t0(v) + s0(1)$$

which is of the form (3) with $k = 2$, $P = \ell^{-1}$.

Applying Lemma 2 we see that $s \equiv t \equiv 0$, and hence $r \equiv 0$, in a neighborhood of x_0, which implies that the second fundamental forms of $\bar{T}$ and T are the same in a neighborhood of every point of Γ_1.

Note that we have proved uniqueness for our nonlinear equations by considering them as linear hyperbolic equations (though albeit with unknown coefficients). The characteristics of the linear equations are the asymptotic curves of T.

To complete our proof of the theorem we have only to verify Lemma 1. This however follows easily from a well-known uniqueness theorem for hyperbolic equations referring to the, so-called, Goursat problem—in which one gives data on two intersecting characteristics of the system and obtains a unique solution in a small characteristic rectangle. In our situation the $(\bar{L}, \bar{M}, \bar{N})$ and (L, M, N) agree along the asymptotic curve $A(x_0)$. By what we have just shown they agree also in a neighborhood of x_1, the end point of $A(x_0)$ on Γ_1. Writing the equations for $\bar{L}, \bar{M}, \bar{N}$ locally as a linear hyperbolic system as we did above we may (working our way up along $A(x_0)$ from x_1) repeatedly apply the uniqueness theorem in the Goursat problem and infer that the $(\bar{L}, \bar{M}, \bar{N})$ agree with (L, M, N) in a whole strip about $A(x_0)$ bounded by two neighboring asymptotic lines—right up to the segment B .

This concludes the proof.

The author is a Sloan fellow.

BIBLIOGRAPHY

1. Alexandrov, A. D., On a class of closed surfaces, Recueil Math. (Moscou) 4(1938), pp. 69-77.

2. Chern, S. S., Lashof, R. K., On the total curvature of immersed manifolds I, Amer. J. Math. 79(1957), pp. 306-318; II, Michigan Math. Jrl. 5(1958), pp. 5-12.

3. Cohn-Vossen, S., Unstarre geschlossene Flächen, Math. Annalen 102(1929-30), pp. 10-29.

4. Efimov, N. V., Flächenverbiegung im Grossen, with an addition by E. Rembs and K. P. Grotemeyer, Akademie Verlag, Berlin, 1957.

5. Grotemeyer, K. P., Zur eindeutigen Bestimmung von Flächen durch die erste Fundamentalform, Math. Z. 55(1952), pp. 253-268.

6. Hartman, P., Nirenberg, L., On spherical image maps whose Jacobians do not change sign, Amer. J. Math. 81(1959), pp. 901-920.

7. Hellwig, G., Anfangswertprobleme bei partiellen Differentialgleichungen mit Singularitäten, Jrl. Rational Mechanics and Analysis, 5(1956), pp. 395-418.

8. Hopf, H., Lectures on differential geometry in the large, Stanford University, 1957.

9. Kuiper, N. H., On surfaces in Euclidean three-space, Bull. Soc. Math. Belgique, 12(1960), pp. 5-22.

10. Kuiper, N. H., Convex immersions of closed surfaces in E^3, Comment. Math. Helvetici 35(1961), pp. 85-92.

11. Minagawa, T., Rado, T., On the infinitesimal rigidity of surfaces, Osaka Math. Jrl. 4(1952), pp. 241-285.

12. Pogorelov, A. V., Die eindeutige Bestimmtheit allgemeiner konvexer Flächen (translation from the Russian), Akademie Verlag, Berlin, 1956.

13. Voss, K., Differentialgeometrie geschlossener Flächen im Euklidischen Raum I. Jahresbericht Deutschen Mat. Vereinigung 63(1960), pp. 117-136.

WOLFGANG HAHN

The present state of Lyapunov's direct method

1. STABILITY

Let the vector differential equation

$$\dot{x} = f(x, t) \qquad f(0, t) = 0 \qquad (1)$$

be given. Existence and uniqueness of the solution is taken for granted. That solution which passes through the initial point x_0 at the initial instant t_0 shall be denoted by $p(t, x_0, t_0)$. The stability theory deals mainly with the behavior, for various initial values x_0 and t_0 , of the scalar function $|p(t, x_0 t_0)|$ at large values of t where x_0 is restricted, in general, to a certain domain $|x_0| \leq h$. In the following survey, the various types of stability behavior are characterized by means of simple comparison functions, i.e., by inequalities of type $|p(t, x_0, t_0)| \leq q(t, x_0, t_0)$. The comparison functions, which are denoted uniformly by the letter q , are supposed to be continuous and positive functions of their arguments: they are distinguished by certain monotony properties:

(a) Stable: $q(|x_0|, t_0)$, $q(r, s)$ monotone decreasing to zero as $r \to 0$.

(b) Uniformly stable; $q(|x_0|)$, $q(r) \to 0$ monotone decreasing as $r \to 0$.

(c) Quasi-asymptotically stable: $q(t-t_0, x_0, t_0)$ for $|x_0| < h = h(t_0)$, $q(u, r, s) \to 0$ as $u \to \infty$.

(d) Asymptotically stable: $q_1(t-t_0, x_0, t_0) q_2(|x_0|, t_0)$, $|x_0| < h(t_0)$, $q_1(u, r) \to 0$ as $u \to \infty$, $q_2(\gamma, s) \to 0$ as $\gamma \to 0$.

When both (a) and (c) hold, the comparison function is of type $q(t-t_0, x_0, |x_0|, t_0)$ and can be decomposed into two factors provided h

is finite. In the case of asymptotic stability in the large, the decomposition cannot be considered as being self-evident.

(e) Uniformly asymptotically stable: $q_1(t-t_0) \quad q_2(|x_0|)$,
$q_1(u) \to 0$ as $u \to \infty$;
$q_2(r) \to 0$ as $r \to 0$.

The equivalence of the above given definitions with the usual $\varepsilon - \delta$ - definitions is easily shown (cf. Hahn [3]). Some more types of stability (cf., e.g. Antosiewicz [1]) can also be defined by means of suitable comparison functions. Unless the equilibrium is asymptotically stable in the large, there exists a finite "domain of attraction, i.e., the domain of those initial points x_0 from which decaying solutions originate. Generally, this domain will depend on the initial instant t_0 , and it may happen that it contracts to the origin as t_0 tends to infinite, for instance in the case of

$$\dot{x} = -x - \frac{x}{t}(1 - t^3 x^2)$$

where the domain of attraction is defined by

$$|x_0 t_0| < 1$$

2. LYAPUNOV FUNCTIONS

In order to get information about the stability behavior of the trivial solution of (1), we make use of certain functions $v(x, t)$ which may be considered as generalizing the distance of the solution $x(t)$ from the axis $t = 0$. We introduce the notation "the function $\phi(r)$ belongs to class K", i.e., $\phi(r)$ is a continuous function of the real variable r , defined and strictly monotone increasing in a certain interval $0 \le r \le r_1$ and vanishing at $r = 0$. The properties of the Lyapunov functions $v(x, t)$ which will be used in the sequel are defined by inequalities involving functions of class K :

A	$v(x, t) \le \phi_1(\|x\|)$	(v = decrescent)
B	$v(x, t) \ge \phi_2(\|x\|)$	(v = positive definite)
C	$\dot{v} \le 0$	($\dot{v}$ = negative semi-definite)
D	$\dot{v} \le -\psi_1(\|x\|)$	($\dot{v}$ = negative definite)

Here, as usual, $\dot{v}$ is the total derivative of $v(x, t) = v[x(t), t]$ with respect to t .

As already stated by Lyapunov, the existence of a function $v(x, t$ satisfying B and C is sufficient to grant stability 1(a). Properties A, B, D are sufficient to grant asymptotic stability 1(d).

3. PROBLEMS

The main problems of the theory of Lyapunov's method may be grouped as follows:

(a) Modification and weakening of the original conditions in Lyapunov's basic theorems. Significance of the single conditions 2A to 2D. The domain of attraction.

(b) Specification of the comparison functions q of 1, and ϕ, ψ of 2. Relations between these two groups.

(c) Necessity of the conditions. Converse stability theorems.

(d) Perturbed equations. The sensitivity of the stability behavior.

(e) Generalizations. Extensions of the concept of stability. Extensions of the method to functional equations of more general type.

(f) Applications of the method to practical problems.

As it is quite impossible to be complete within a one hour's lecture, I think it best to report on some interesting more or less recent results which illustrate the above given list, and I shall formulate a few little problems still unsolved which I feel are worth studying. As for details, proofs, literature, etc., I refer to Antosiewicz [1] and Hahn [3].

4. TO GROUP A

Significance of A. From B and D, not even quasi-asymptotic stability can be included. There only follows

$$\lim p(t_n, x_0, t_0) = 0$$

for a certain sequence $t_n \to \infty$. (Antosiewicz [1]).

Significance of D. (Krasovskij [7])

Let (1) be autonomous. The existence of a certain domain $|x| \leq h$ which does not contain any complete $(-\infty < t < +\infty)$ phase trajectory is necessary and sufficient for the existence of a function v with (positive or negative) definite derivative.

In the non-autonomous case, the theorem is more complicated. Let H be a subdomain of the "cylinder" $|x| \leq h$, $t \geq t_0$, and consider a monotone sequence r_k decreasing to zero. Consider the trajectory $p(t, x_0, t_0)$ for $t_0 - T_k \leq t \leq t_0 + T_k$ (T_k will be defined later), t_0 being sufficiently large in order to grant the existence of solutions at $t = t_0 - T_k$. Now, the necessary and sufficient condition for the existence of v with definite $\dot{v}$ reads: For every integer k, T_k can be chosen such that the above defined part of the trajectory is not contained in H completely provided $|x_0| > r_k$.

Weakening of D (Malkin, Antosiewicz [1]). Suppose A and B hold for a certain function v and B holds for a function v_1, and assume $\dot{v} + v_1$ to converge to zero as $t \to \infty$, this convergence being uniform with respect to x in every fixed domain $0 < h_1 \le |x| \le h$. Then asymptotic stability follows. Another modification was considere by LaSalle [8]: It is sometimes possible to conclude asymptotic stability from 2A, B, C, by studying the set $\dot{v} = 0$.

A modified criterion. (Corduneanu [2]). Consider the scalar differential equation

$$\dot{y} = g(y, t) \qquad [g(0, t) = 0] \tag{2}$$

Let $v(x, t)$ be positive definite and

$$\dot{v} \le g[v(x, t), t] \quad .$$

In accordance as 1(a) or 1(d) holds for (2), 1(a) or 1(d) holds for (1), respectively. (The criterium may also be formulated with respect to uniform stability, total stability, etc. It generalizes some criteria given formerly.)

Estimations of the domain of attraction may be found in the autonomous case if $\dot{v}$ changes the sign: if $\dot{v} > 0$, the corresponding point x cannot be within the domain. Zubov (cf. Hahn [3]) gave an exact determination of the domain in the autonomous case. This determination requires the solving of a certain partial differential equation. As for the non-autonomous case, only very few results are known.

Problem: Study in detail how the domain of attraction depends on the initial instant t_0.

5. TO GROUP B. RELATIONS BETWEEN THE COMPARISON FUNCTIONS

We shall denote the inverse functions of the comparison functions occurring in 2 by ϕ^I, etc. These functions exist and belong to class K. Sometimes, the existence of the derivatives is required. This, however, will not restrict the generality as the functions occur only in inequalities. Let A, B, D be fulfilled, i.e., consider

$$\phi_2(|x|) \le v \le \phi_1(|x|) \quad ; \quad \dot{v} \le -\psi_1(|x|)$$

one gets

$$\phi_1^I(v) \le |x| \le \phi_2^I(v) \quad ; \quad \dot{v} \le -\psi_1[\phi_1^I(v)] \quad .$$

The autonomous auxiliary differential equation $\dot{w} = -\psi_1[\phi_1^I(w)]$ has a uniform asymptotically stable equilibrium which may be reached even in a finite time whence according to 1(e) and to the inequality $v \le w$ for $v_0 = w_0$,

$$v \leq q_1(t-t_0)q_2(v_0) \quad .$$

Eliminating v_0 by x_0 we obtain

$$|p(t, x_0, t_0)| \leq \phi_2^I(q_1(t-t_0) \cdot q_2[\phi_1(|x_0|)])$$

which furnishes us with a relation connecting the comparison functions of 1(e) and the functions occurring in 2. Simultaneously, we get another proof of the well known fact that 2A, B, D are sufficient conditions even for 1(e).

Specifications. Put $\phi_2(r) = r$ (which, as a matter of fact does not restrict the generality, since v may be replaced by any $\phi(v)$, $\phi \in K$, if $\phi'(v)$ exists).

Then

$$|p(t, x_0, t_0)| \leq q_2[\phi_1(|x_0|)]q_1(t-t_0) \tag{3}$$

follows, the two factors describing the dependence on the initial instant, or initial point respectively. If the auxiliary differential equations for w is of type $\dot{w} = -aw$, we obtain an inequality

$$|p(t, x_0, t_0)| \leq \phi_1(|x_0|)e^{-a(t-t)} \quad , \quad r \leq \phi_1(r) \quad .$$

Consequently, if the order of magnitude of the space-factor in (3) is exactly $|x|$, then the time-factor decreases at least as fast as an exponential function. (It might decrease even faster.)

If in addition to 2D an inequality of type $\dot{v} \geq -\psi_2|x|$ holds, one can estimate the expression $|p(t, x_0, t_0)|$ below.

Problem: Suppose all the inequalities involved to be sharp. Study the relations between the orders of magnitude of the two factors in (3).

Problem: Let the comparison functions of (3) be known and assume the inequality to be sharp. Give sharp estimations for the comparison functions occurring in 2A to 2D.

6. TO GROUP C

The question of converse theorems may be regarded as answered completely as far as uniform asymptotic stability is involved. Massera (cf. Hahn [3]) was the first to handle the problem successfully and to solve it in a special case. With some modifications, his method proved strong enough to treat the general case. The results, obtained by various authors, differ slightly both in the requirements for the differential equations and in the properties of the Lyapunov function to be constructed. Since Massera gave, recently, a survey on the various results, I shall not do it here, but I shall restrict myself to illustrating the basic idea of the method of construction. The result will be, of course, rather poor in comparison to those mentioned above

Let us start from 1(e) and let us introduce

$$v(x, t) = \int_t^\infty w[\,|p(\tau, x, t)|\,]\,d\tau$$

$w \in K$ to be defined later. It is readily shown that

$$\dot{v} = -w(|x|)$$

whence 2D. We then introduce $\xi = |p(\tau, x, t)|^2$ as another variable of integration and get

$$\xi' = \frac{d\xi}{d\tau} = 2\sum p_i f_i$$

If we assume the functions f_i to be bounded, we have

$$\left|\frac{d\xi}{d\tau}\right| \leq c\sqrt{\xi}$$

and

$$v(x, t) = \int_{|x|^2}^{0} w(\sqrt{\xi})\frac{d\xi}{\xi'} \geq \frac{1}{c}\int_0^{|x|^2} w(\sqrt{\xi})\,\frac{d\xi}{\sqrt{\xi}}$$

whence 2B. Finally, we get for $|x| \leq h$

$$w[p(\tau, x, t)] \leq w[q_1(\tau-t)\cdot q_2(|x|)] \leq (w[g_1(\tau-t)\cdot q_2(h)]\cdot w[q_1(o)\,q_2(|x|)])^{1.}$$

and

$$v(x, t) \leq \sqrt{w[q_1(0)\cdot q_2(|x|)]}\cdot \int_t^\infty \sqrt{w(q_2(h)\cdot q_1(\tau-t))}\,d\tau$$

Therefore, 2A is fulfilled if w can be chosen to permit the integrations involved. Such a choice is possible, in view of a lemma due to Massera which is used in all the proofs. Thus, we get the function $v(x, t)$ but nothing can be said about differentiability, etc.

As far as non-uniform asymptotic stability is concerned, the question of converse theorems is still unsolved and it has not been even tackled. I advance the following problem the solution of which might be a first step:

Problem: Consider the inequality 1(d) characterizing non-uniform asymptotic stability and take specific comparison functions q_i of simple type. Find Lyapunov functions and characterize them by inequalities analogous to 2A to 2D with t-dependent functions on the right hand side.

7. TO GROUP D

Consider the equation (1) and assume the stability behavior to be known. In addition to (1), the perturbed equation

$$\dot{y} = f(y, t) + g(y, t) \qquad g(0, t) = 0 \tag{4}$$

is given, and $g(y, t)$ is assumed to be "small" in a sense which is to be defined precisely. Under which conditions has the perturbed equation the same stability behavior as the unperturbed equation?

The problem may be tackled generally by using a Lyapunov function $v(x, t)$ of the unperturbed system. The expression $\dot{v} = \text{grad } v \cdot f + \frac{\partial v}{\partial t}$ requires the additional term $\text{grad } v \cdot g$ if the derivative is formed with respect to the perturbed equation. Now, everything depends upon the question of whether the additional term may or may not affect the definiteness of $\dot{v}$. If, for instance, the equilibrium is "exponentially stable," i.e., if an inequality

$$|p(t, x_0 t_0)| \leq a \cdot |x| e^{-\alpha(t-t_0)} \qquad (\alpha > 0)$$

is valid, the procedure of construction outlined in the preceding section furnishes us with a Lyapunov function satisfying inequalities of type

$$a_1 |x|^2 \leq v(x, t) \leq a_2 |x|^2 \quad ; \quad \dot{v} \geq -a_3 |x|^2$$

and it is readily seen that the condition $g(y, t) = o(|y|)$ will guarantee the asymptotic stability of the perturbed system.

From this example it is learned that the problem is closely connected with the problems discussed in Section 5.

The case of the perturbed linear equation is of particular interest. Since uniform asymptotic stability implies exponential stability in the case of a linear equation, the main theorem reads: If the unperturbed linear term has a uniformly asymptotic equilibrium and if the perturbing term $g(y, t)$ is $o(|y|)$, then the perturbed equation still has an exponentially stable equilibrium. The theorem covers the well-known case of "stability in the first approximation." The main condition is fulfilled, for instance, for autonomous and periodic systems having an asymptotically stable equilibrium. A different condition for the linear term has been stated by Lyapunov who introduced the concept of "regularity" which is defined by means of the so-called order numbers describing the rate of increase of the solutions.

Neither the Lyapunov condition nor the condition based on exponential stability is necessary. So far, necessary and sufficient conditions for the stability in the first approximation are not known.

Problem: Is the theorem of the stability in the first approximation valid for linear equations with almost periodic coefficients?

Problem: Suppose the comparison functions l(e) are known. Give estimations for the perturbing term in order to retain the stability.

If the condition $g(0, t) = 0$ is dropped one comes to the concept of "stability under persistent perturbations" or, simply, total stability. Let $q(t, y_0, t_0)$ be the solution of the perturbed equation. If given any $\epsilon > 0$ there are two numbers $\delta > 0$ and $\eta > 0$ such that the inequalities $x_0(< \delta$, $|g(y, t)| < h$ imply $|q(t, x_0, t_0)| < \epsilon$, then the equilibrium of the unperturbed equation is said to be totally stable. If the three numbers δ, ϵ, η are known and if they correspond to the practical requirements to be put upon the system, the equilibrium may be called to be practically stable (LaSalle [8]) since rather such a performance is desired in practice than the theoretically important asymptotic stability.

The main theorem, due to Malkin (cf. Hahn [3]), reads: Uniform asymptotic stability implies total stability.

The conditions on $g(y, t)$ may be weakened. The condition "$g(y, t)$ small" may be replaced by "$g(y, t)$ small in the mean;" that is, large values of $g(y, t)$ may be admitted provided the time integral of $g(y, t)$ over any finite interval is small. Vrkoč [11] investigated, in great detail, several similar modifications. Corduneanu [2] proved criteria similar to that mentioned at the end of Section 4. The concept of total stability and its modifications are of importance in practice since in most concrete cases the perturbing terms are not known explicitly but can only be estimated.

8. TO GROUP E

(1) The general idea and the main theorems of the direct method may be extended to differential equations $\dot{x} = f(x, t)$ in linear metric spaces. The Euclidean distance has to be replaced by the norm of the space. The expression $v(x, t)$ is defined as a functional depending on the element x of the space with t being a parameter. The derivative $\dot{v}$ is obtained by Frechet's differentiation process:

$$\dot{v} = \lim \frac{1}{h}[v(x+hf, t+h) - v(x, t)]$$

K. P. Persidskij (cf. Hahn [3]) was the first to extend the direct method to differential equations with an infinite number of variables. Massera (cf. Hahn [3]) made a systematic study by checking whether the finiteness of the number of dimensions was necessary in the proofs. The basic theorems and their converses are valid; but some special theorems turned out to be restricted to Euclidean space, for instance, the theorem stating equivalence of uniform asymptotic and exponential stability in the linear case.

The trajectories of differential equations in metric spaces may be considered as special cases of dynamical (or general) systems defined as parameter-dependent mappings of the space into itself. Taking this a starting point, Zubov (cf. Hahn [3]) developed a stability theory of dynamical (and general) systems, based on ideas analogous to the direct method. But here, the aim is rather to describe the topological properties of the mapping than to get information about a particular equilibrium.

(2) The general solution of a difference differential equation or of a related functional equation depends on an arbitrary function, the "initial value," and may be interpreted as an element in a Banach space, depending on t as a parameter. Therefore, the above mentioned theory might be modified so as to cover also difference differential equations. But it is much more convenient to take a direct approach, using the specific properties of the equations in question. Of course, a norm has to be defined and Lyapunov functionals will play the role of Lyapunov functions. N. N. Krasovskij (cf. Hahn [3]) founded this part of the theory. J. K. Hale [4], quite recently, gave a complete and rigorous derivation of the main results.

(3) The extension of the direct method to difference equations is done with no difficulty. The results are such as to be expected. They have some practical interest in the theory of discontinuous control systems (Kalman and Bertram [6]).

(4) Zubov (cf. Hahn [3]) applied the direct method to partial differential equations, regarding the totality of the solutions as a dynamical or general system. In order to tackle concrete problems, this rather subtle theory has to be specialized, and again it seems to be more convenient to approach a special problem directly, as was done by Movchan [10] in dealing with the equation of the plate. He introduced a norm, fitting to that particular boundary value problem, and succeeded in deriving the well known critical loads by using Lyapunov functionals.

Problem: To extend Zubov's and Movchan's method to other concrete cases and to yield a practicable method, suitable for handling a sufficiently wide class of problems.

(5) I mention two modifications of the concept of stability, introduced in order to deal with practical problems.

Kac and Krasovskij [5] define "stable with probability" in order to discuss differential equations, depending on stochastic functions. Suitably modified theorems of Lyapunov's method are valid. Lebedev (cf. Hahn [3]) introduced "stability in a finite interval $t_0 \leq t \leq t_1$

with respect to a given domain D: $v(x, t) \leq a$ " which, roughly speaking, means that the solution being within D at $t = t_0$ will remain in D during the interval $t_0 \leq t \leq t_1$. The concept is of importance in the theory of control systems.

(6) Yoshizawa [12] developed a theory of boundedness analogous to the theory of stability. He distinguishes various types of boundedness of the solutions according to whether they are below a fixed constant k for all $t \geq t_0$, or for $t \geq t_0 + T$, whether or not the constant k and T depend on t_0 , etc. Yoshizawa states theorems analogous to the basic theorems of the direct method. The functions $v(x, t)$ involved here are defined in certain domains $|x| \geq h$, $t_0 \leq t$. Recently Yoshizawa [13] has extended the theory to cover "stability of a system" (not of a particular unperturbed solution). The distances $|x_0|$ and $|p(t, x_0, t_0)|$ are replaced by the distances $|x_0 - x_0'|$, $|p(t, x_0, t_0) - p(t, x_0', t_0)|$ respectively between any two solutions, and the behavior of the latter is studied.

9. TO GROUP F

The following survey may indicate the types or problems which can be handled by the direct method.

(1) Investigation of the stability behavior of a concrete system, by means of suitable Lyapunov functions. As no general method of construction exists, the success depends essentially on the skillfulness of the investigator.

(2) Estimations of the domain of attraction in concrete cases.

(3) Characterization of those regions in the parameter space which insure stability and

(4) Characterization of the nonlinear functions involved. In general, this type of problem will arise when dealing with control systems. (Cf. for instance, LaSalle and Lefschetz [8]). Two particular problems, named after Aizerman and Lur'e, found strongest interest and stimulated numerous papers.

(5) The category "various applications" will include the remaining applications which cannot be considered in detail here. (cf. Hahn [3]).

10. CONCLUSION

Though the revival of Lyapunov's method in the Thirties was caused by the fact that the method offered opportunities to deal with practical problems, it quickly became a domain of pure mathematics and the numbers of theoretical papers published since is much larger than the number of "applied' papers. Accordingly, the progress is not satisfying, from the point of view of practice, as it ought to be. I strongly hope that the engineers get more interested in the subject than they have thus far, for I am sure they will benefit if they do.

NOTES

1. Antosiewicz, H. A., A survey of Lyapunov's second method. Ann. Math. Studies 41, 141-166 (1958).

2. Corduneanu, C., Application des inégalitiés différentielles à la théorie de la stabilité. An. Ş ti. Univ. Iaşi, Sect. I, 6, 46-58 (1960) (in Russian, French abstract).

3. Hahn, W., Theorie und Anwendung der Direkten Methode von Lyapunov. Springer-Verlag 1959 (142 p.) (English edition in preparation).

4. Hale, J. K., Asymptotic behavior of the solutions of differential-difference equations. RIAS techn. Rep. 61-10 (1961).

5. Kac, I. Ya. and N. N. Krasovskij., On the stability of a system with random parameters. Priklad. Mat. Mech. 24, 809-823 (1960) (In Russian).

6. Kalman, R. E., and J. E. Bertram, Control analysis and design via the "second method of Lyapunov." I. II. J. basic Engin. Trans. ASME 82, 371-400 (1960).

7. Krasovskij, N. N., Some problems in the theory of motion. Moskow 1959, 211 p. (in Russian).

8. LaSalle, J. and Lefschetz, Stability by Lyapunov's direct method with applications. New York 1961, 134 p.

9. Massera, J. L., Converse theorems of Lyapunov's second method. Symposium intern. ecuav. dif. ordin, Mexico 158-163 (1961).

10. Movchan, A. A., The direct method of Lyapunov in stability problems of elastic systems. Priklad. Mat. Mech. 23, 483-493 (1959) (In Russian)

11. Vrkoč, I., Integral stability. Czechosl. math. Journ. 9(84), 71-129 (1959) (In Russian, English summary).

12. Yoshizawa, T., Lyapunov functions and boundedness of solutions. Funkcialaj Ekvacioj 2, 95-142 (1959).

13. Yoshizawa, T., Stability and boundedness of systems. Arch. rat. Mech. Anal. 6, 409-421 (1960).

M. L. CARTWRIGHT

Almost periodic solutions of equations with periodic coefficients

1.1. My starting point is an equation of the form

$$\ddot{\xi} + f(\xi, \dot{\xi})\dot{\xi} + g(\xi) = p(t) \quad , \tag{1}$$

where $p(t)$ has period 2π , and the functions, f, g, p satisfy conditions corresponding to a system damped for large values of $|\xi|, |\dot{\xi}|$. For instance the conditions

$$f(\xi, \dot{\xi}) \geq a_1 > 0 \,, \quad g(\xi)/\xi \geq a_2 > 0 \quad \text{for } |\xi| > K \,, \ |\dot{\xi}| > K \,,$$
$$\left| \int_0^t p(t)\,dt \right| < a_3 \quad \text{for all } t \tag{2}$$

in addition to the usual conditions for the existence and uniqueness of solutions will be sufficient. It is well known that the behavior of solutions can be represented by a transformation T of points $x = (\xi, \eta)$ in the ξ, η plane. For if $\dot{\xi} = \eta$ and $\xi(t, x_0)$ is a solution of (1) such that

$$\xi(0, x_0) = \xi_0 \,, \quad \eta(0, x_0) = \eta_0 \quad ,$$
$$\xi(2\pi, x_0) = \xi_1 \quad , \quad \eta(2\pi, x_0) = \eta_1 \quad , \tag{3}$$

where

$x_0 = (\xi_0, \eta_0)$ and $x_1 = (\xi_1, \eta_1)$, then $x_1 = T(x_0)$ defines homeomorphism in the ξ, η plane.

1.2. The topological properties of such homeomorphisms have been studied by many authors, [1] and in particular their minimal sets. So far as the topological properties are concerned, they apply equally well to systems of equations of the form

$$\begin{cases} \dot{\xi} = f(\xi, \eta, t) , \\ \dot{\eta} = g(\xi, \eta, t) , \end{cases} \qquad (1)$$

where f and g have period 2π in t, provided that f and g and their partial derivatives with respect to ξ, η are continuous in ξ, η, t Since all the solution of 1.1(1) are bounded as t tends to infinity, in that case $T^n(x_0)$ is bounded as $n \to \infty$ for all x_0, and in general I shall confine myself to the consideration of bounded sets.

A bounded set M such that

$$M = \overline{M} , \quad T(M) = M , \qquad (2)$$

and M is irreducible with respect to these properties is a minimal set. The orbit of a point x_0 is the set

$$O(x_0) = \bigcup_{-\infty < n < \infty} T^n(x_0) ,$$

and for all $x_0 \in M$ $\overline{O(x_0)} = M$.

It is well known that if M consists of a fixed point $f = T(f)$, or a set of periodic points $T^v(p)$, $v = 1, 2, \ldots N-1$, $T^N(p) = p$, then the corresponding solution has period 2π or $2\pi N$ as the case may be, and conversely. Actual solutions of these types are known. Certain types of u. a. p. solution are also known, and if $\xi(t, x_0)$, $\eta(t, x_0)$ are u. a. p. functions, then $\overline{O(x_0)}$ is a minimal set, but a minimal set does not necessarily represent a u. a. p. solution so far as is known, although it is possible that the differential equations imply some restriction much stronger than those which have so far been expressed topologically. The object of this paper is to discuss u. a. p. solutions and the corresponding minimal sets in relation to the bases of the u. a. p. solutions.

A good deal of work has been done on minimal sets under such transformations by Gottschalk[2] and Hedland,[3] and I use their results on isochronons sets, but so far as I know they have not considered almost periodic sets in relation to functions with bases of more than one term.

2.1. Suppose that $\xi(t, x_0)$, $\eta(t, x_0)$ are u. a. p. solutions[4] of (1.2(1). Then we may write

$$\xi(t, x_0) \sim \sum_{s=-\infty}^{\infty} A_s e^{i\Lambda_s t} , \quad \eta(t, x_0) \sim \sum_{s=-\infty}^{\infty} B_s e^{i\Lambda_s t} , \qquad (1)$$

where

$$\Lambda_s = r_0^{(s)} + \sum_{j=1}^{S(s)} r_j^{(s)} \lambda_j ,$$

$r_j^{(s)}$, $j = 0, 1, 2, \ldots S(s)$ [5] is rational, and the set of numbers $1, \lambda_1, \lambda_2, \ldots$ is rationally independent. That is to say

$$r_0 + \sum_{j=1}^{S} r_j \lambda_j \neq 0$$

for any set of rational numbers r_j, $j = 0, 1, 2, \ldots S$ except $r_0 = r_1 = \ldots = r_S = 0$. The numbers $1, \lambda_1, \lambda_2 \ldots$ are a rational base of $\xi(t, x_0)$, $\eta(t, x_0)$, and if all the $r_j^{(s)}$ are integers $p_j^{(s)}$, the base is an integral base. It should be observed that the base is not uniquely defined, and if, for instance $r_2^{(s)}/r_1^{(s)} = \alpha$ for $-\infty < s < \infty$, we may write $\lambda_1' = \lambda_1 + \alpha\lambda_2$ and omit λ_2 . A u.a.p. function for which $r_j^{(s)} = 0$ for $j \geq 1$ and all s is a limit periodic function.

2.2. Biperiodic solutions, that is solutions with a two term base $1, \lambda$ are well known; the linear equation

$$\ddot{\xi} + \lambda^2 \xi = \cos t ,$$

has a solution

$$\xi = A \cos(\lambda t + \alpha) + \frac{\cos t}{\lambda^2 - 1} ,$$

and nearly linear equations with more general biperiodic solutions have been discussed by Kryloff and Bogoliuboff[6] and other authors. These solutions may also be written in the form $\xi(t, \lambda t)$, $\eta(t, \lambda t)$, where λ is irrational and $\xi(u, v)$, $\eta(u, v)$ are continuous and have period 2π in u and v . Conversely any solution of this form has a two term base $(1, \lambda)$.

2.3. By the approximation theorem 7 there are polynomials

$$\left\{ \begin{aligned} P_k(t) &= \sum_{s=-k}^{k} c_k^{(s)} A_s e^{i\Lambda_s t} \\ Q_k(t) &= \sum_{s=-k}^{k} d_k^{(s)} B_s e^{i\Lambda_s t} \end{aligned} \right. \qquad (1)$$

such that $P_k(t)$ and $Q_k(t)$ converge uniformly to $\xi(t, x_0)$, $\eta(t, x_0)$ respectively as $k \to \infty$, and, for fixed s, $c_k^{(s)}$ and $d_k^{(s)}$ tend to 1 as $k \to \infty$. Further the numbers $c_k^{(s)}$, $d_k^{(s)}$ depend only on the numbers Λ_s and not on the coefficients A_s and B_s .

It is well known that a u.a.p. function may be expressed as

the diagonal function of a limit periodic function of a finite or infinite number of variables and that in virtue of a theorem of Kronecker the set of values of the diagonal function is everywhere dense in the set of all the values of the limit periodic function of several variables. We use a somewhat similar procedure to discuss the set

$$T^n(x_0) = (\xi(2\pi n, x_0), \eta(2\pi n, x_0)), \qquad -\infty < n < \infty .$$

For fixed k we denote by $q_j^{(k)}$ the lowest common multiple of the numbers $q_j^{(s)}$, $S(-k) \le s \le S(k)$, $j = 0, 1, \ldots S(k)$, where $r_j^{(s)} = p_j^{(s)}, q_j^{(s)}$, $p_j^{(s)}$, $q_j^{(s)}$ integers.

THEOREM 1. The set $x_k(n) = (P_k(2n\pi), Q_k(2n\pi))$, $-\infty < n < \infty$, is everywhere dense in the set M_k which is the set of points $x_k = (\xi_k, \eta_k)$ such that

$$\begin{cases} \xi_k = P_k(m, t_1, t_2, \ldots t_{S(k)}) = \sum_{s=-k}^{k} c_k^{(s)} A_s \exp\{i(2\pi r_0^{(s)} + \sum_{j=1}^{S(s)} r_j^{(s)} \lambda_j t_j)\} , \\ \eta_k = Q_k(m, t_1, t_2, \ldots t_{S(k)}) = \sum_{s=-k}^{k} d_k^{(s)} B_s \exp\{i(2m\pi r_0^{(s)} + \sum_{j=1}^{S(s)} r_j^{(s)} \lambda_j t_j)\} , \end{cases} \tag{2}$$

where $m = 0, 1, 2, \ldots q_0^{(k)} - 1$, $0 \le t_j \le 2\pi q_j^{(k)}/\lambda_j$.

2.4. The form of Kronecker's theorem required is

If $\vartheta_0, \vartheta_1, \vartheta_2, \ldots \vartheta_K$ are rationally independent, $\alpha_0, \alpha_1, \ldots \alpha_K$ are arbitrary, and U and ε are any positive numbers, then there is a number u and integers $N_1, N_2, \ldots N_K$ such that $u > U$ and

$$|u\vartheta_j - N_j - \alpha_j| < \varepsilon , \quad j = 0, 1, 2, \ldots K .$$

By our choice of $q_j^{(k)}$ we can write

$$r_j^{(s)} = p_j^{(s)}/q_j^{(k)} , \quad p_j^{(s)} \text{ an integer}, \quad j = 0, 1, 2, \ldots S(s) , \quad -k \le s \le k .$$

Hence the functions $P_k(t_0, t_1, \ldots t_{S(k)})$, $Q_k(t_0, t_1, \ldots t_{S(k)})$ have period $2\pi q_j^{(k)}/\lambda_j$, for $j = 0, 1, 2, \ldots S(k)$, where $\lambda_0 = 1$. Write $K = S(k)$, and then we may write

$$\vartheta_0 = \frac{1}{q_0^{(k)}} \quad , \quad \vartheta_j = \frac{\lambda_j}{q_j^{(k)}} \quad , \quad j = 1, 2, \ldots K ,$$

in Kronecker's theorem, and putting $\alpha_0 = m/q_0^{(k)}$, where m is an integer, and multiplying the jth relation by $q_j^{(k)}/\lambda_j$, where $\lambda_0 = 1$, we have

$$|u - N_0 q_0^{(k)} - m| < \varepsilon\, q_0^{(k)}$$

$$\left| u - \frac{N_j q_j^{(k)}}{\lambda_j} - \frac{\alpha_j q_j^{(k)}}{\lambda_j} \right| < \varepsilon \frac{q_j^{(k)}}{\lambda_j} , \quad j = 1, 2, \ldots K .$$

Hence if $n = N_0 q_0^{(k)} + m$,

$$\left| n - (N_j + \alpha_j) \frac{q_j^{(k)}}{\lambda_j} \right| < \varepsilon \left\{ q_0^{(k)} + \frac{q_j^{(k)}}{\lambda_j} \right\} , \quad j = 1, 2, \ldots K ,$$

and so

$$| n\, r_j^{(s)} \lambda_j - (N_j + \alpha_j)\, p_j^{(s)} | < \varepsilon\, r_j^{(s)} \lambda_j \left\{ q_0^{(k)} + \frac{q_j^{(k)}}{\lambda_j} \right\} , \quad j = 1, 2, \ldots K .$$

Since ε is any positive number independent of k , for every $\varepsilon' > 0$, we can choose ε so that

$$| n\, r_j^{(s)} \lambda_j - (N_j + \alpha_j) p_j^{(s)} | < \varepsilon' , \quad j = 1, \ldots K , \quad n = N_0 q_0^{(k)} + m > U .$$

This means that $x_k(n) = (P_k(2n\pi) , Q_k(2n\pi))$ takes values arbitrarily near

$$(P_k(m, \alpha_1, \alpha_2, \ldots \alpha_{S(k)}) , Q_k(m, \alpha_1, \alpha_2, \ldots \alpha_{S(k)})$$

for $n > U$, and since all values of $x_k(n)$ are included in the values of x_k we have the result stated.

2.5. It follows from the approximation theorem that as $k \to \infty$ $x_k(n) \to T^n(x_0)$ uniformly for $-\infty < n < \infty$, and the closure of $T^n(x_0)$ is the minimal set $M = \overline{O(x_0)}$ corresponding to the u.a.p. solution 2.1(1). If $x^* \in M$ but $x^* \neq T^n(x_0)$ for any n , there is a sequence $n_1, n_2, \ldots n_m \to \infty$ of integers such that $T^{n_m}(x_0) \to x^*$, and

$$\xi(t, x^*) = \lim_{m \to \infty} \xi(t + 2\pi n_m, x_0) ,$$

$$\eta(t, x^*) = \lim_{m \to \infty} \eta(t + 2\pi n_m, x_0) ,$$

where the limits are uniform for $-\infty < t < \infty$. Since the numbers ${}_{k}^{(s)}$, $d_k^{(s)}$ in 2.3(1) are the same for $\xi(t, x^*), \eta(t, x^*)$ as for $\xi(t, x_0), \eta(t, x_0)$ and the limits are uniform, by theorem 1 M is the uniform limit of the set M_k as $k \to \infty$.

At this stage it will be sufficient to observe that if ${}_{0}^{(k)} \leq N$ for all k , the set M_k and therefore the set M has at most N components, and that, unless $r_j^{(s)} = 0$, $j \geq 1$ and all s , M_k contains at least one non-degenerate continuum.

3.1. Let us now return to minimal sets. It is well known that a minimal set M cannot contain any interior points so that F(M) = M . The following classification of minimal sets in the plane is based on that of Birkhoff whose classification was somewhat ambiguous and might be interpreted as including II(a) under I.

I. Continuous minimal sets are locally connected. These include (a) fixed and periodic points, (b) sets of N closed Jordan curves,

$T^v(J)$, $v = 0, 1, 2, \ldots N-1$, $T^N(J) = J$, $T^v(J) \neq J$, $v = 1, 2, \ldots N-1$.

II. Partially discontinuous minimal sets contain non-degenerate continua but are not locally connected. This includes (a) sets of N non-degenerate continua $T^v(C)$, $v = 0, 1, 2, \ldots N-1$, $T^N(C) = C$, $T^v(C) \neq C$, $v = 0, 1, 2, \ldots N-1$, which are not locally connected (b) sets with an infinity of non-degenerate continua $T^v(C)$, $-\infty < v < \infty$, such that $T^v(C) \neq T^u(C)$, $v \neq u$ and their limit sets which may or may not contain point components.

III. Totally discontinuous minimal sets contain no non-degenerate continua and are zero dimensional. These include (a) isochronous or regularly almost periodic sets (b) non-isochronous sets. A minimal set M is isochronous if for all $x_0 \in M$ and every $\varepsilon > 0$ there is an integer $N = N(\varepsilon)$ such that $T^{nN}_{(x_0)} \in U(x_0, \varepsilon)$ for $-\infty < n < \infty$, where $U(x_0, \varepsilon)$ is the set of points whose distance from x_0 is less than ε .

Totally discontinuous sets which are not isochronous are known from a topological point of view. For instance the perfect nowhere dense set on the unit circle mentioned by Levinson, but it is not known whether any system of differential equations of the type 1.2(1) exists which has a minimal set corresponding to this topologically.

3.2. It is easy to see that a u.a.p. solution with an integral base 1 corresponds to a set of periodic points and that a u.a.p. solution with a rational base corresponds to an isochronous set which is the limit set of the periodic points $P_k(m, 0, 0, \ldots)$, $Q_k(m, 0, 0, \ldots)$.

It seems possible that the unstable discontinuous recurrent solutions discussed by Levinson [9] and others [10] are isochronous and therefore correspond to limit periodic solutions, but this point was not explicitly discussed by them. Any totally discontinous set in the plane which is stable in the sense of Lyapunov is isochronous [11], but it should be observed that asymptotically stable minimal sets can only have a finite number of components, and so if they are isochronous they are periodic.

3.3. As we saw in §2.5 a u.a.p. solution with a two term base, $1, \lambda_1$ such that

$$r_0^{(s)} = \frac{p_0^{(s)}}{N}, \quad r_1^{(s)} = \frac{p_1^{(s)}}{N}, \quad -\infty < s < \infty,$$

where $p_0^{(s)}$, $p_1^{(s)}$ are integers, corresponds to a minimal set which is the uniform limit of at most N continuous closed curves, and therefore consists of at most N continuous closed curves. Since they are locally connected, they belong to class I(b) and are Jordan curves.

Conversely, if a minimal set consists of N Jordan curves $T^{v}(J)$, $v = 0, 1, \ldots, T^N(J) = J$, then T^N has an irrational rotation number ρ, $0 < \rho < 1$, with respect to ρ, and by a suitable choice of parameter θ, the transformation T^N can be expressed on J by $\theta' = T^N(\theta) = \theta + \rho$. Hence the solution $\xi(t, x_0)$, $\eta(t, x_0)$ is of the form $\xi(t, \lambda t)$, $\eta(t, \lambda t)$, where $\lambda = 1/\rho$, $\xi(u, v)$, $\eta(u, v)$ are continuous and have period $2\pi N$ in u and v.

The solution 2.2(2) of 2.2(1) obviously corresponds to the minimal set consisting of the ellipse

$$\left(\xi - \frac{1}{\lambda^2 - 1}\right)^2 + \frac{\eta^2}{\lambda^2} = A^2 .$$

Prof. Leontovich-Androvna raised the question whether any system of equations was known with a solution corresponding to a minimal set consisting of more than one Jordan curve, but I do not know of such a system.

3.4. The remaining results are inconclusive, but suggestive. If the base is $1, \lambda_1$ and $q_0^{(k)} \leq N$ for all k, the set M consists of N continua, and unless it reduces to the case in which $r_j^{(s)} = 0$, $j \geq 2$ and all s, $q_1^{(k)} \leq N$, we have a minimal set of type II(a). If $q_0^{(k)}$ is unbounded, we shall have an infinity of non-degenerate continua and the set M is of type II(b). However, if the base contains more than two terms, the set M_k defined by 2.3(2) is the union of a continuum of closed curves which will have interior points unless the curves coincide. The corresponding minimal set cannot have interior points, and so either the curves coincide, or they coalesce as $k \to \infty$.

Suppose that the set M_k consists of $N_k = q_0^{(k)}$ Jordan curves defined by 2.3(2) by varying one particular t_j and keeping the rest constant. We may suppose that $t_j = 0$, $j > 2$, and $0 \leq t_1 \leq 2\pi q_0^k/\lambda$ Then for each fixed m there is a $t'_1 = t'_2(t_1, t_2)$ such that

$$\xi_k(m, t_1, t_2, 0, 0, \ldots) = \xi_k(m, t'_1, 0, 0, \ldots),$$

$$\eta_k(m, t_1, t_2, 0, 0, \ldots) = \eta_k(m, t'_1, 0, 0, \ldots).$$

$0 \leq t_1 \leq 2\pi q_1^{(k)}/\lambda_1$, $0 \leq t_2 \leq 2\pi q_2^{(k)}/\lambda_2$, $0 \leq t'_1 \leq 2\pi q_1^{(k)}/\lambda_1$. For fixed t_1 this defines t'_1 as a function of t_2 which is regular for all finite t_2 real or complex, and comparing the exponential terms of highest order (which we may assume to be those for $s = \pm k$) as t_2 tends to infinity along the positive and negative imaginary axes, we see that $t'_1 \sim t_1 + \alpha t_2$, where $\alpha = r_2^{(K)}, r_1^{(K)}$, $K = S(k)$, as $k_2 \to \infty$. It follows that $t_1' = t_1 + \alpha t_2$, and the terms of highest order cancel out. Repeating the process, with the terms of next highest order $s = \pm(k-1)$, say, we have $r_2^{S(k-1)}/r_1^{S(k-1)} = \alpha$, and so on. Hence we may replace λ_1, λ_2 by $\lambda'_1 = \lambda_1 + \alpha\lambda_2$, where

$$\alpha = r_2^{(s)}/r_1^{(s)}, \quad -k \leq s \leq k.$$

Now repeating the process with t_3 varying, we replace λ_1' by $\lambda_1'' = \lambda_1' + \alpha'\lambda_3$, where

$$\alpha' = r_3^{(s)}/r_1^{(s)},$$

and so on. Hence the approximating polynomials $P_k(t)$, $Q_k(t)$ can be expressed in terms of a two term base, and therefore the solution $\xi(t, x_0)$, $\eta(t, x_0)$.

So far I have not been able to exclude the possibility of curves which coalesce as $k \to \infty$ also.

4. Returning to the original equation we observe that the only examples of minimal sets corresponding to actual solutions of 1.1(1) or 1.2(1) which are known are (1) fixed points corresponding to solutions with period 2π, (ii) points with period N corresponding to subharmonic solutions with period $2\pi N$, (iii) possibly isochronous sets corresponding to limit periodic functions, (iv) biperiodic functions of the form $\xi(t, \lambda t)$, where λ is irrational and $\xi(u, v)$ is continuous and has period 2π in u and v. In the last three cases the left hand side of 1.1(1) exerts an important influence and the period $2\pi N$ in (ii)

the irrational number in (iv), and the peculiar behavior of the solutions in (iii) might be regarded as contributions of the free oscillation to the solutions. Although the behavior of solutions of non-linear equations is equite different from that of solutions of linear equations in that the free and forced oscillation cannot be separately identified, the question arises whether any recurrent solutions can exist other than those which are in a modified form a combination of a free and forced oscillation. Are they all u. a. p. solutions for which $r_j^{(s)} = 0$, $j \geq 2$, and either $r_0^{(s)} = p_0^{(s)}$ an integer for all s, or $r_1^{(s)} = 0$ for all s? For the longer period in case (ii) and the periods which tend to infinity in case (iii) are due to something like a free oscillation, and in case (iv) if there is a two term base $1, \lambda$, the frequency λ is accounted for by what corresponds to the free oscillation, and there is nothing left to account for a longer period than 2π in respect of the first term of the base.

If these conjectures are correct, it means that the topological properties which we have used so far do not adequately represent the properties of solutions, and we may ask whether further properties of the solutions can be expressed topologically.

Similar problems arise for systems of n equations in n unknowns with period 2π. A linear system of $2n$ equations in $2n$ variables may obviously have a u. a. p. solution with a base of n terms, $1, \lambda_1, \lambda_2, \ldots \lambda_{n-1}$, but not more, and some of the methods used can be applied in n dimensions.

FOOTNOTES

1. See 12, 2, 3, 7, 4.
2. 8 and 9.
3. 9.
4. The notation used is based on that in 1, in particular pp. 34-37.
5. It is convenient to assume that $S(k) \geq S(s)$, $-k \leq s \leq k$.
6. See 11 and 3.
7. See 1, pp. 29 and 46-51.
8. Theorem 444 in 10.
9. See 13.
10. See 14 and 15.
11. See 6.

ADDENDUM

A system of two equations whose coefficients have period 2π can be constructed so that the corresponding transformation has a minimal set consisting of n Jordan curves by the following method. It is based on one suggested to me by Dr. J. Moser for the case $n = 2$.

Consider the circle

$$x^2 + y^2 - x = -\frac{1}{8} .$$

By putting

$$x + iy = z = w^n , \quad w = \rho e^{i\phi} ,$$

we obtain a curve

$$H(\rho, \phi) \equiv \rho^{2n} - \rho^n \cos n\phi = -\frac{1}{8} \tag{1}$$

which consists of n Jordan curves. If $n \geq 2$, they are obviously integral curves of the system

$$\begin{cases} \dot{\rho} = -\dfrac{\omega}{\rho}\dfrac{\partial H}{\partial \phi} = -n\omega\rho^{n-1}\sin n\phi , \\ \dot{\phi} = \dfrac{\omega}{\rho}\dfrac{\partial H}{\partial \rho} = 2n\omega\rho^{2n-2} - n\omega\rho^{n-2}\cos n\phi \end{cases} \tag{2}$$

for all positive constants ω and satisfy (1) at $t = 0$. Put

$$\xi = \rho\cos\theta , \quad \eta = \rho\sin\theta , \quad \theta = \phi + t ,$$

so that the whole system represented by (2) rotates through an angle t in time t. Then the new system is

$$\dot{\rho} = -n\omega\rho^{n-1}\sin n(\theta - t) ,$$

$$\dot{\theta} = 1 - n\omega\rho^{n-2}\cos n(\theta - t) + 2n\omega\rho^{2n-2} ,$$

and

$$\dot{\xi} = -n\omega\rho^{n-1}\sin((n-1)\theta - nt) - 2n\omega\rho^{2n-1}\sin\theta - \rho\sin\theta ,$$

$$\dot{\eta} = -n\omega\rho^{n-1}\cos((n-1)\theta - nt) + 2n\omega\rho^{2n-1}\cos\theta + \rho\cos\theta .$$

Since $\rho^{2n-2} = (\xi^2 + \eta^2)^{n-1}$, and the terms in $\rho^{n-1}\cos(n-1)\theta$, $\rho^{n-1}\sin(n-1)\theta$ can be expressed as homogeneous polynomials of degree $n-1$ in ξ and η, putting $nt = \tau$ we have

$$\frac{d\xi}{d\tau} = -\eta + \omega F(\xi, \eta, \tau),$$
$$\frac{d\eta}{d\tau} = \xi + \omega G(\xi, \eta, \tau), \tag{3}$$

where F and G are polynomials in $\xi, \eta, \cos\tau$, $\sin\tau$ and therefore have period 2π in τ.

The set of solutions of (3) starting at $\tau = 0$ on the loop L of (1) for which $|\phi| \leq \frac{1}{4}\pi/n$ rotates as τ increases until at $\tau = 2k\pi/n$ they form the loop $T^k(L)$ for which $|\phi - 2k\pi/n| \leq \frac{1}{4}\pi/n$, $k = 1, 2, \ldots, n$, so that $T^n(L) = L$. The speed of each point relative to the rotating loop is

$$(\dot{\rho}^2 + \rho^2\dot{\phi}^2)^{\frac{1}{2}} = \omega\left\{\frac{1}{\rho^2}\left(\frac{\partial H}{\partial \phi}\right)^2 + \left(\frac{\partial H}{\partial \rho}\right)^2\right\},$$

and so by a suitable choice of ω we can ensure that the rotation number of T^n with respect to L is irrational. Hence the whole set of loops $T^k(L)$, $k = 0, 1, 2, \ldots, n-1$ is a minimal set for the transformation T defined by the system (3).

REFERENCES

1. A. S. Besicovitch, Almost Periodic Functions, Cambridge, 1932.
2. G. D. Birkhoff, Surface transformations and their dynamical applications, Acta Math. 43(1920) 1-119.
3. M. L. Cartwright, Forced Oscillations in Nonlinear Systems, Contributions to the theory of nonlinear oscillations, ed. S. Lefschetz, Annals of Mathematics Studies, 20 (Princeton 1950).
4. M. L. Cartwright, Some decomposition theorems for certain invariant continua and their minimal sets, Fundamenta Mathematicae 58(1960).
5. M. L. Cartwright, Almost periodic solutions of certain second order differential equations, Rendiconti del Seminario Matematico e Fisico di Milano (1961).
6. M. L. Cartwright, Almost periodic solutions of systems of two periodic equations, Proceedings of the international symposium on nonlinear oscillations (Kiev 1961), unpublished.
7. M. L. Cartwright and J. E. Littlewood, Some fixed point theorems, Annals of Mathematics 54(1951), 1-37.

8. W. H. Gottschalk, Minimal sets: An introduction to topological dynamics, Bulletin of the American Mathematical Society 64(1958) 336-351.

9. W. H. Gottschalk and G. A. Hedlund, Topological Dynamics, American Mathematical Society Colloquium Publications 36(1955).

10. G. H. Hardy and E. M. Wright, An introduction to the theory of numbers (Oxford 1938).

11. N. Kryloff and N. Bogoluiboff, Methods of nonlinear mechanics applied to the study of stationery oscillations, Monograph in Russian with a summary in French (Kiev 1934).

12. N. Levinson, Transformation theory of nonlinear differential equations of the second order, Annals of Mathematics, 45(1944), 723-737.

13. N. Levinson, A second order equation with singular solutions, Annals of Mathematics 50(1949), 127-153.

14. J. E. Littlewood, On non-linear differential equations of the second order: II $\ddot{y} - k(1-y^2)\dot{y} + y = b\mu k \cos(\mu t + \alpha)$ for large k, and its generalizations, Acta Math. 97(1957) 267- 08.

15. J. E. Littlewood, On non-linear differential equations of the second order: IV The general equation $\ddot{y} + k f(y)\dot{y} + g(y) = bk\, p(\phi)$, $\phi = t + \alpha$, Acta Math. 98(1957), 1-110.

PHILIP HARTMAN

On uniqueness and differentiability of solutions of ordinary differential equations

In [5], questions concerning the uniqueness and differentiability of the solution $y = \eta(t, t_0, v_0)$ of the initial value problem

$$y' = f(t, y) \quad , \qquad y(t_0) = v \tag{P}$$

were considered.

The main results obtained there allow, for example, an extension to n dimensions of the theorems of [4] and [7], Part VII, dealing with geodesics of a 2-dimensional Reimann metric having a continuous curvature. They do not, however, give an extension of the theorem of [7], Part VII, in which it is only assumed that the curvature is bounded.

The object of this note is to obtain generalizations of the uniqueness theorems of [5] (which, in particular, yield such an extension, Theorem 3.2 below) and to give a new, simple proof of the main result of [5]. The proofs will depend on approximating f suitably and using a priori estimates. This type of proof for a uniqueness theorem seems novel. The proof is based in part on inequalities related to those of D.C. Lewis [10], particular cases of which have recently been used ([11], [6]) to obtain results on stability in the large.

1. Differentiability theorem. In [5], the differential equation

$$y' = f(t, y) \tag{1.1}$$

was viewed as an equation for "differentials"

$$dy - f(t, y)\,dt = 0$$

rather than one for "derivatives." This involves some concepts which will now be recalled.

All variables and functions below are real-valued. t is a real variable and differentiation with respect to t is denoted by a prime; $y = (y^1, \ldots, y^d)$ is a Euclidean d-vector. Tensor summation convention will be followed for repeated indices i, j, k (not n) over the range $1, \ldots, d$.

A differential 1-form in $dy^1, \ldots, dy^d$, say $\omega = p_j(y)\,dy^j$ with coefficients $p_j(y) = p_j(y^1, \ldots, y^d)$ defined on a y-domain E is said to be of class C^m if its coefficients are of class C^m. Similarly, a sequence of differential forms is said to be uniformly convergent or uniformly bounded if the corresponding sequences of coefficients are uniformly convergent or uniformly bounded.

A continuous $\omega = p_j(y)\,dy^j$ on an open y-set E is said to possess a continuous [or bounded] exterior derivative $d\omega$ if there exists a differential 2-form $d\omega = p_{jk}(y)\,dy^j dy^k$ having coefficients $p_{jk} = -p_{jk}$ continuous [or bounded, measurable] on E and satisfying Stokes' formula

$$\int_C \omega = \iint_S d\omega \tag{1.2}$$

for every piece S of a 2-dimensional C^1 surface in E bounded by a piecewise C^1 Jordan curve C in E.

This definition of a continuous exterior derivative is due to E. Cartan [1], pp. 65-71. The concept of a bounded (not necessarily continuous) exterior derivative will, in general, not be very useful unless y is a 2-dimensional vector (for otherwise, the existence of the 2-dimensional integral in (1.2) for arbitrary S requires more assumptions on $d\omega$).

The main theorem of [5] can be stated as follows:

THEOREM 1.1. *Let* $f(t, y)$ *be continuous on an open* (t, y)*-set* E. *A necessary and sufficient condition that the initial value problem* (P) *have a unique solution* $y = \eta(t, t_0, v)$ *which is of class* C^1 *in all of its variables is that every point of* E *have an open neighborhood* $E^0 \subset E$ *on which there exists a* $d \times d$, *continuous, non-singular matrix* $A(t, y)$ *such that the* d *differential* 1*-forms*

$$\omega = A(dy - f\,dt) \tag{1.3}$$

in $dt, dy^1, \ldots, dy^d$ *on* E^0 *have continuous exterior derivatives*.

One advantage of the necessary and sufficient condition in Theorem 1.1 over the standard "sufficient" conditions in Lindelof's theorem is the fact that it is invariant under C^1 changes of the variables (t, y). For applications of this theorem to problems in geodesics, extremals, and total differential equations, see [5].

If $f = (f^1, \ldots, f^d)$ and $A = (a_{ij})$, then ω in (1.3) means

$\omega = (\omega_1, \ldots, \omega_d)$, where

$$\omega_i = a_{ij} dy^j - (a_{ij} f^j) dt . \tag{1.4}$$

Correspondingly, when $d\omega = (d\omega_1, \ldots, d\omega_d)$ exists, then $d\omega_i$ is a differential 2-form of the type

$$d\omega_i = \alpha_{ijk} dy^j dy^k + \beta_{ij} dt\, dy^j , \tag{1.5}$$

where $\alpha_{ijk} = -\alpha_{ikj}$ and β_{ij} are continuous functions of (t, y). The matrix $F(t, y) = (f_{ij}(t, y))$ defined by

$$f_{ij} = \beta_{ij} - 2\alpha_{ijk} f^k \tag{1.6}$$

plays an important role below.

COROLLARY 1.1. Let f be as in Theorem 1.1 and let A exist on $E^0 \subset E$ with the stated properties. Then the Jacobian matrix $\partial\eta/\partial v$ of $\eta(t, t_0, v)$ with respect to v has the property that $A^0 \partial\eta/\partial v$ is continuously differentiable with respect to t and

$$(A^0 \partial\eta/\partial v)' = F^0 \partial\eta/\partial v , \quad \partial\eta/\partial v = I \text{ at } t = t_0 , \tag{1.7}$$

where the superscript 0 signifies that the argument is $(t, y) = (t, \eta)$ and $\eta = \eta(t, t_0, v)$.

2. S- and L- Lipschitz continuity. The proofs in [5] involve direct applications of Stokes' theorem (1.2). Below the proof involves a priori estimates and suitable approximations for $f(t, y)$. This type of proof is possible by virtue of two lemmas: the first (Lemma 1.1) is the basic fact on the existence of continuous exterior derivatives and the second (Lemma 4.1) is a uniqueness theorem for (P).

LEMMA 2.1. Let $\omega = p_j(y) dy^j$ be continuous on an open y-set E . Then ω has a continuous exterior derivative $d\omega = p_{jk}(y) dy^j dy^k$ if and only if, for every open set E^0 with compact closure $\bar{E}^0 \subset E$, there exists a sequence of differential 1-forms $\omega^1, \omega^2, \ldots$ of class C^1 such that $\omega^n \to \omega$, $n \to \infty$, uniformly on E^0 and $d\omega^1, d\omega^2, \ldots$ is uniformly convergent on E^0 (in which case $d\omega = \lim d\omega^n$, $n \to \infty$).

Cf. Gillis [3]. The "if" portion is trivial. The "only if" portion involving the existence of $\omega^1, \omega^2, \ldots$ has a simple proof due to H. Cartan [2], pp. 62-63, in which the coefficients of $\omega^n = p_j^{(n)}(y) dy^j$

are obtained by smoothing those of ω with a convolution, $\alpha^n = p_{jk}^{(n)}(y)\,dy^j dy^k$ is obtained similarly from $d\omega$ and, finally, it is shown that $\alpha^n = d\omega^n$ by "applying" the convolution to (1.2).

This lemma can be used to show that in verifying that a continuous $p_{jk}dy^j dy^k$ is the exterior derivative of $\omega = p_j dy^j$, it is sufficien to verify the Stokes' formula (1.2) only for small rectangles S in 2-planes parallel to coordinate 2-planes.

The notion of the existence of a continuous exterior derivative $d\omega$ generalizes the concept that ω is of class C^1. Lemma 1.1 suggests the following generalization for a ω to be uniformly Lipschitz continuous: A continuous $\omega = p_j(y)\,dy^j$ on an open set E will be said to be S-Lipschitz continuous on E if there exists a sequence of differential 1-forms $\omega^1, \omega^2, \ldots$ of class C^1 on E such that $\omega^n \to \omega$, $n \to \infty$, uniformly on E and $d\omega^1, d\omega^2, \ldots$ is uniformly bounded on E.

Actually, a slightly more general concept can be used for uniqueness: Let $A(t, y)$ be a continuous $d \times d$ matrix, $f(t, y)$ a continuous d-vector on an open set E. The differential 1-forms (1.3) will be said to be L- Lipschitz continuous on E if there exists a sequence of differential 1-forms $\omega^1, \omega^2, \ldots$ of class C^1 on E of the form

$$\omega^n = A_n(t, y)\,[dy - f_n(t, y)\,dt] \tag{2.1}$$

such that $\omega^n \to \omega$, $n \to \infty$, uniformly on E and that there exist constants c, c_0 satisfying

$$c_0 I \leqq A_n^* F_n \leqq cI \quad \text{for } n = 1, 2, \ldots \tag{2.2}$$

on E.

Here the matrix $F_n = F_n(x, y)$ belongs to (2.1) and is defined by the analogues of formulae (1.4)-(1.6). B^* denotes the transpose of the (real) matrix B. $B \leqq C$ is an abbreviation for inequalities for the quadratic forms $\xi . B\xi \leqq \xi . C\xi$ for all real d-vectors ξ, where B and C are not necessarily symmetric; so that $B \leqq C$ is equivalent to $B^H \leqq C^H$ if B^H is the Hermitian part $\frac{1}{2}(B + B^*)$ of B.

The differential 1-forms (1.3) will be said to be upper L-Lipschitz continuous on E if (2.2) is replaced by

$$A_n^* F_n \leqq cI \quad \text{for } n = 1, 2, \ldots \; . \tag{2.3}$$

In order to get an idea of the significance of these definitions, recall that if y is a scalar and $f(t, y)$ is non-increasing in y, then (P) has a unique solution to the right $(t \geqq t_0)$. When y is a vector, this condition on f can be replaced by

$$[f(t, y_2) - f(t, y_1)] \cdot (y_2 - y_1) \leqq 0 \; . \tag{2.4}$$

In this case, (1.3) is upper L- Lipschitz continuous with $A = I$ and $c = 0$; in fact, when $f \in C^1$, $A = I$ and F is the Jacobian matrix $\partial f/\partial y$, then (2.4) is equivalent to $F \leqq 0$.

Another way to understanding the meaning of the definition of L-Lipschitz continuity is to suppose that (1.3) is of class C^1. A simple calculation shows that the matrix F has the elements

$$f_{ij} = \partial a_{ij}/\partial t + a_{ik} \partial f^k/\partial y^j + f^k \partial a_{ij}/\partial y^k \tag{2.5}$$

and $B = A^* F$ has elements of the form

$$b_{ij} = \frac{1}{2} \partial g_{ij}/\partial t + g_{ik} f^k{}_{,j} + h_{ij} ,$$

where $h_{ij} = -h_{ji}$, $G \equiv A^*A = (g_{ij})$, and $f^k{}_{,j}$ are the components of the covariant derivative of the contravariant vector f with respect to the metric $ds^2 = g_{ij}(t, y)\, dy^i dy^j$ for fixed t. Inequalities concerning the quadratic form $\xi \cdot B\xi$ occur in [10] (and, in cases where A, f do not depend on t, in [9], [11], [6]). Actually the inequalities in [10] are of the type $B \leqq \beta G$, $B \geqq \alpha G$ and are used in this form below, Lemma 5.1.

3. Uniqueness theorems. The following uniqueness theorem generalizes those of [5].

THEOREM 3.1. *Let* $f = f(t, y)$ *be continuous on an open* (t, y)-*set* E *and let there exist a continuous, non-singular matrix* $A = A(t, y)$ *on* E *such that the differential* 1-*forms* (1.3) *are* S- *Lipschitz continuous or, more generally,* (1.3) *is* L- *Lipschitz continuous on* E. *Then* (P) *has a unique solution* $y = \eta(t, t_0, v)$ *for all* $(t_0, v) \in E$. *Furthermore,* $\eta(t, t_0, v)$ *is uniformly Lipschitz continuous with respect to* (t, t_0, v) *on compact subsets of its domain of definition.*

This will be proved in Section 5 below.

It is possible to formulate a one-sided uniqueness theorem for (P) analogous to the one in the scalar case of (P) in which $f(t, y)$ is non-increasing with respect to y.

COROLLARY 3.1. *Let* f, A *be as in Theorem* 3.1 *except that* (1.3) *is only assumed to be upper* L- *Lipschitz continuous on* E. *Then* (P) *has a unique solution* $y = \eta(t, t_0, v)$ *to the right* $(t \geqq t_0)$ *of* t_0 *for all* $(t_0, v) \in E$. *Furthermore,*

$$|\eta(t, t_1, v_1) - \eta(t, t_2, v_2)| \leqq \text{Const.}\ |\eta(t^*, t_1, v_1) - \eta(t^*, t_2, v_2)| \tag{3.1}$$

holds for $t \geqq t^* \geqq \max(t_1, t_2)$ on compact subsets of the domain of definition of $\eta(t, t_0, v)$.

The proof[1] of Corollary 3.1 will be clear from the proofs of Theorem 3.1 and Corollary 4.1 below.

When the open set E is replaced by a product set of the form

$$R: \quad 0 \leqq t \leqq a, \quad |y| \leqq b, \tag{3.2}$$

one can obtain a generalization of Nagumo's uniqueness criterion.

COROLLARY 3.2. Let $f(t, y)$ be continuous on R in (3.2). On R, let there exist a continuous, non-singular matrix $A(t, y)$ such that (1.3) is upper L- Lipschitz continuous on every

$$R_\epsilon: \ (0 <) \ \epsilon \leqq t \leqq a, \quad |y| \leqq b \tag{3.3}$$

with (2.3) replaced by

$$F_n A_n^{-1} \leqq I/t, \quad \text{i.e.,} \quad A_n^* F_n \leqq A_n^* A_n / t \quad \text{on } R_\epsilon. \tag{3.4}$$

Then (P) has a unique solution when $t_0 = 0$, $|v| < b$.

The approximations (2.1) for (1.3) on R_ϵ can depend on ϵ. As in Nagumo's theorem, Corollary 3.2 fails if I in (3.4) is replaced by cI with $c > I$.

Theorem 3.1 implies the following theorem on the geodesics of a d-dimensional (non-singular) Riemann metric

$$ds^2 = g_{ik}(y)\, dy^i dy^k, \tag{3.5}$$

i.e., on solutions of the initial value problems

$$y^{i''} + \Gamma^i_{jk} y^{j'} y^{k'} = 0, \quad y(0) = y, \quad y'(0) = y_0'. \tag{3.6}$$

THEOREM 3.2. Let $G(y) = (g_{ik}(y))$ be a symmetric, non-singular, $d \times d$ matrix of class C^1 on a y-domain such that (3.5) has a bounded curvature. Then (3.6) has a unique solution $y = \eta(t, y_0, y_0')$ and $\eta(t, y_0, y_0')$, $\eta'(t, y_0, y_0')$ are uniformly Lipschitz continuous (in all variables) on compact subsets of their domain of existence.

In this theorem, it is to be understood that a non-singular, C^1 metric (3.5) has a bounded curvature if each of the differential 1-forms

$$\omega^i_j = \Gamma^i_{jk}(y)\,dy^k \ , \tag{3.7}$$

$i, j = 1, \ldots, d$, is S- Lipschitz continuous. This is certainly the case if there exists a sequence of C^2 metrics

$$ds^2 = g^{(n)}_{ik}(y)\,dy^i dy^k \ , \tag{3.8}$$

$n = 1, 2, \ldots$, on E such that $g^{(n)}_{ik}$ and their first order partial derivatives tend to g_{ik} and their partials, respectively, uniformly on compact subsets of E and, in addition, the sequence of Riemann curvature tensors belonging to (3.8) for $n = 1, 2, \ldots$, is uniformly bounded.

The deduction of Theorem 3.2 from Theorem 3.1 is the same as the deduction of (i) in [5] from (I) in [5]. In (i) in [5], it is required that (3.5) have a bounded curvature in the sense that (3.7) have a bounded exterior derivative. But as noted in Section 1, the latter concept is, in general, useful only in dimension 2.

4. A uniqueness lemma. The assertion which will permit the deduction of "uniqueness" from a priori estimates is the following:

LEMMA 4.1. *Let* $f(t, y)$ *be continuous on an open* (t, y)*-set* E *such that there exists a function* $y = \eta(t, t_0, v)$ *defined for* $a < t$, $t_0 < b$, $(t_0, v) \in E$ *with the properties* (i) $y = \eta(t, t_0, v)$ *is a solution of* (P); (ii) $\eta(t, t_0, v)$ *is uniformly Lipschitz continuous with respect to* v; *finally,* (iii) *two arcs* $y = \eta(t, t_1, v_1)$, $y = \eta(t, t_2, v_2)$ *in* (t, y)*-space either have no point in common or are identical. Then* $y = \eta(t, t_0, v)$ *is the only solution of* (P) *for* $a < t, t_0 < b$.

This lemma is due to van Kampen [8] in the autonomous case where f does not depend on t . A variant of his proof also gives Lemma 4.1.

It can be mentioned that the scalar equation $y' = 3y^{2/3}$ with the family of solutions $y = (v^{1/3} + t - t_0)$ shows that the lemma is false if condition (ii) is omitted or if Lipschitz continuity is replaced by Hölder continuity. It is also easy to see that (i)-(iii) do not imply that f is locally uniformly Lipschitz continuous with respect to y .

Proof of Lemma 4.1. Condition (ii) means that there exists a constant K such that

$$|\eta(t, t_0, v_1) - \eta(t, t_0, v_2)| \leqq K|v_1 - v_2| \ . \tag{4.1}$$

Since the assumption (iii) implies that any point of an arc $y = \eta(t, t_0, v)$ can be used as an initial point to determine it, (4.1) implies that

$$|\eta(t, t_1, v_1) - \eta(t, t_2, v_2)| \leqq K|\eta(t^*, t_1, v_1) - \eta(t^*, t_2, v_2)| \qquad (4.2)$$

for $a < t, t^* < b$, $(t_1, v_1) \in E$, $(t_2, v_2) \in E$.

Let $y = y(t)$ be a solution of (P), say, on $a < t < b$. It will be shown that $y(t) = \eta(t, t_0, v)$, i.e., that $\tau(t) = 0$ if

$$\tau(t) = \eta(t, t_0, v) - y(t) \qquad (4.3)$$

To this end, let t be fixed and put

$$\sigma(s) = \eta(t, s, y(s)) - y(t) \quad . \qquad (4.4)$$

Then $\sigma(t) = 0$ and $\sigma(t_0) = \tau(t)$. Thus in order to show $\tau(t) = 0$, it is sufficient to show that $\sigma(s)$ is a constant (for fixed t).

By the definition (4.4),

$$\sigma(s+h) - \sigma(s) = \eta(t, s+h, y(s+h)) - \eta(t, s, y(s)) \quad . \qquad (4.5)$$

Applying (4.2) with $t^* = s+h$,

$$|\sigma(s+h) - \sigma(s)| \leqq K|y(s+h) - \eta(s+h, s, y(s))| \quad .$$

Since the two functions $y(t)$, $\eta(t, s, y(s))$ of t have the common value $y(s)$ at $t = s$ and hence a common derivative at $t = s$, it follows that $d\sigma/ds$ exists and is 0. Thus $\sigma(s)$ is a constant as was to be proved.

Note that the proof of Lemma 4.1 has as a consequence the following one-sided uniqueness theorem.

COROLLARY 4.1. Let $f(t, y)$ be continuous on an open (t, y)-set E such that there exists a function $y = \eta(t, t_0, v)$ defined for $a < t_0 \leqq t < b$, $(t_0, v) \in E$ with the properties (i) $y = \eta(t, t_0, v)$ is a solution of (P) and (ii) the inequality (4.2) holds for $t \geqq t^* \geqq \max(t_1, t_2)$. Then, for $(t_0, v) \in E$ and $a < t_0 < b$, $y = \eta(t, t_0, v)$ is the only solution of (P) for $t_0 \leqq t < b$.

5. Proof of Theorem 3.1. Consider first the case that f, A are of class C^1. The desired a priori inequalities follow from the following variant of a result of Lewis [10].

LEMMA 5.1. Let $f(t, y)$, $A(t, y)$ be of class C^1 on an open set E^0 such that

$$\alpha(t)A^*A \leqq A^*F \leqq \beta(t)A^*A , \qquad (5.1)$$

where $\alpha(t), \beta(t)$ are continuous functions. Let $y = \eta(t) = \eta(t, t_0, v)$ be a solution of (P) for t near t_0 and $J(t, y)$ the Jacobian matrix $(\partial f/\partial y)$, $J^0 = J(t, \eta(t))$. Then any solution $y(t)$ of the equations of variation

$$y' = J^0 y \qquad (5.2)$$

satisfies, for $t \geqq t_0$,

$$|A(t_0, t)y(t_0)| \exp \int_{t_0}^{t} \alpha(s)\,ds \leqq |A^0 y(t)| \leqq |A(t_0, v)y(t_0)| \exp \int_{t_0}^{t} \beta(s)\,ds \qquad (5.3)$$

where $A^0 = A(t, \eta(t))$.

In fact, a simple calculation involving (2.5) shows that (5.2) implies

$$(A^0 y)' = F^0 y , \qquad (5.4)$$

hence

$$(A^0 y)\cdot(A^0 y)' = y\cdot(A^0)^* F^0 y . \qquad (5.5)$$

Clearly, (5.3) follows from (5.1).

Note that if

$$-cI \leqq A^*F \leqq cI , \qquad c \geqq 0 , \qquad (5.6)$$

$$mI \geqq A^*A \geqq I/m , \qquad m > 1 , \qquad (5.7)$$

then $-cmA^*A \leqq A^*F \leqq cmA^*A$. In this case, (5.3) and corresponding inequalities for $t \leqq t_0$ show that

$$|A^0 y(t)| \leqq |A(x_0, v)y(x_0)| \exp cm|t - t_0| \qquad (5.8)$$

when $\eta(t)$, hence $y(t)$ is defined. In particular, (5.6)-(5.7) imply

$$|\partial\eta(t, t_0, v)/\partial v^k| \leqq m \exp cm|t-t_0| . \qquad (5.9)$$

Estimates for $\partial\eta/\partial t_0$ follow from the standard relation

$$\partial\eta(t, t_0, v)/\partial t_0 = -f^k(t_0, v)\,\partial\eta(t, t_0, v)/\partial v^k . \qquad (5.10)$$

Let $(t_1, v_1) \in E$ and E^0 an open convex neighborhood of (t_1, v_1) with compact closure $\bar{E}^0$ in E. Peano's existence theorem makes it clear that there exist a convex neighborhood $E_0 \subset E^0$ of (t_1, v_1) and numbers a, b $(> t_1 > a)$ depending only on a bound for $|f|$ in $\bar{E}^0$ such that the solution $y = \eta(t, t_0, v)$ of (P) for $(t_0, v_0) \in \bar{E}_0$ exists and $(t, \eta(t, t_0, v)) \in E^0$ for $a \leqq t \leqq b$.

It follows from (5.9), (5.10) and the convexity of E_0 that there then exist a constant K, depending only on c, m and a bound for $|f|$ on $\bar{E}^0$ such that

$$|\eta(t, t_0, v_0) - \eta(t, t^0, v^0)| \leqq K(|t_0 - t^0| + |v_0 + v^0|) \tag{5.11}$$

for $a \leqq t \leqq b$ and (t_0, v_0), (t^0, v^0) in $\bar{E}_0$. Finally, if $t_0 = t^0$ and if t_0 and t are interchanged

$$|v_0 - v^0| \leqq K|\eta(t, t_0, v_0) - \eta(t, t_0, v^0)| \tag{5.12}$$

provided that $a \leqq t \leqq b$ and $(t, \eta(t, t_0, v_0))$, $(t, \eta(t, t_0, v^0))$ are in E_0. The last proviso holds if the interval $a \leqq t \leqq b$ is small enough and (t_0, v_0), (t_0, v^0) belong to a suitably small neighborhood E_{00} if (t_1, v_1).

It must be emphasized that E_0, E_{00} a, b and K depend only on E^0, a bound for $|f|$ on $\bar{E}^0$ and c, m.

Return now to the case that f, A are not of class C^1 but satisfy the assumption of Theorem 3.1. There exists a sequence of C^1 differential 1-forms $\omega^1, \omega^2, \ldots$ of the form (2.1) such that $\omega^n \to \omega$, $n \to \infty$, uniformly on E and (2.2) holds, say, with $c_0 = -c \leqq 0$. Since the assertion of the theorem is local, (t, y) can be restricted to a convex neighborhood E^0 of a point (t_1, v_1) of E. Thus, it can be supposed that A_n is non-singular and that there is a constant m, independent of n, such that

$$mI \geqq A_n^* A_n \geqq I/m, \quad m > 1. \tag{5.13}$$

It can also be supposed that $|f_1|, |f_2|, \ldots$ have a common bound. Thus, there is an E_0, E_{00}, a, b $(> t_1 > a)$ and a K, independent of n, such that the solution $y = \eta_n(t, t_0, v)$ of

$$y' = f_n(t, y), \quad y(t_0) = v, \tag{5.14}$$

with $(t_0, v) \in \bar{E}_0$ exists and $(t, \eta_n) \in E^0$ for $a \leqq x \leqq b$ and that (5.11), (5.12) hold for $\eta = \eta_n$.

In particular, $\eta_1(t, t_0, v)$, $\eta_2(t, t_0, v), \ldots$ is uniformly bounded and equicontinuous for $a \leqq t \leqq b$, $(t_0, v) \in \bar{E}_0$. Thus there exists a subsequence which after a renumbering can be taken to be the full sequence such that

$$\eta(t, t_0, v) = \lim_{n \to \infty} \eta_n(t, t_0, v) \tag{5.15}$$

exists uniformly for $a \leqq t \leqq b$, $(t_0, v) \in \bar{E}_0$ and is a solution of (P). Since (5.11), (5.12) hold for $\eta = \eta_n$, they hold for η. Condition (5.12) shows that no two distinct arcs $y = \eta(t, t_0, v)$ pass through the same point (t, y) if (t, t_0, v) is sufficiently near to (t_1, t_1, v_1). Hence Lemma 4.1 implies that $y = \eta(t, t_1, v_1)$ is the only solution (for t near t_1) of $y' = f$, $y(t_1) = v_1$. Since (t_1, v_1) is an arbitrary point of E, the theorem is proved.

6. Proof of Theorem 1.1. Consider "sufficiency," so that on an open subset E^0 of E, there is a continuous, non-singular A such that (1.3) has a continuous exterior derivative. By Lemma 1.1, there is a sequence of C^1 differential 1-forms $\omega^1, \omega^2, \ldots$ such that

$$\omega^n \to \omega, \quad d\omega^n \to d\omega \text{ as } n \to \infty \tag{6.1}$$

uniformly on compact subsets of E^0. Since $\det A \neq 0$, it follows that if E^0 is replaced by a smaller set, it can be supposed that ω^n is of the form (2.1) with $\det A_n \neq 0$ and that (6.1) holds uniformly on E^0. Relations (6.1) imply

$$f_n \to f, \quad A_n \to A, \quad F_n \to F \text{ as } n \to \infty \tag{6.2}$$

uniformly on E^0.

Arguing as in the last section, it follows that if E^0 is sufficiently small, it can be supposed that (5.15) holds uniformly on a fixed t-interval for $(t_0, v) \in E^0$. Since $y = \partial\eta_n/\partial v^k$ satisfies

$$(A_n(t, \eta_n)y)' = F_n(t, \eta_n)y, \quad y^j(t_0) = \delta^j_k, \tag{6.3}$$

the last two parts of (6.2) and $\det A_n \neq 0$ show that $\lim \partial\eta_n/\partial v^k$ tends uniformly to a limit which is the solution of (5.4). Hence $\partial\eta/\partial v^k$ exists and is this limit. In the standard way, it can be shown that $\partial\eta/\partial t_0$ exists and is given by (5.10). Thus $y = \eta(t, t_0, v)$ is of class C^1. Since very point of E is contained in an E^0 with properties above, the "sufficiency" half of Theorem 1.1 is proved.

The proof of "necessity" is the same as in [5] and will be omitted.

7. Proof of Corollary 3.2. It follows from Theorem 3.1 that (P) has a unique solution $y = \eta(t, t_0, v)$ for $t \geqq t_0 > 0$, $|v| < b$. Let

$0 < \theta < 1$ and, in what follows, consider only $|v| \leqq \theta b < b$. It will be verified that there are constants C, a_0 such that

$$|\eta(t, \epsilon, v_1) - \eta(t, \epsilon, v_2)| \leqq C|v_1 - v_2|/\epsilon \quad \text{for } 0 < \epsilon \leqq t \leqq a_0 . \tag{7.1}$$

Let this be granted for a moment. Suppose, if possible, that $y' = f(t, y)$, $y(0) = v$ has two solutions $y = y_1(t), y_2(t)$. Then uniqueness for $t_0 > 0$ implies that $y_i(t) = \eta(t, \epsilon, y_i(\epsilon))$ for $i = 1, 2$ and $\epsilon \leqq t \leqq a_0$. Hence (7.1) gives

$$|y_1(t) - y_2(t)| \leqq C|y_1(\epsilon) - y_2(\epsilon)|/\epsilon \quad \text{for } \epsilon \leqq t \leqq a_0 . \tag{7.2}$$

It is clear that the right side of (7.2) tends to 0 as $\epsilon \to +0$. Hence $y_1(t) = y_2(t)$ for $0 \leqq t \leqq a_0$ and the corollary follows.

Thus it suffices to verify (7.1). Arguing as in Section 5, it follows that there exists a positive a_0 $(\leqq a)$, independent of n, such that the solution $y = \eta_n(t, t_0, v)$ of (5.1) exists for $t_0 = \epsilon$, $\epsilon \leqq t \leqq a_0$, $|v| \leqq \theta b$. By (3.4) and Lemma 5.1,

$$|\partial \eta_n(t, \epsilon, v)/\partial v^k| \leqq m^2/\epsilon \quad \text{for } \epsilon \leqq t \leqq a , \quad k = 1, \ldots, d ,$$

when (5.15) holds on R_ϵ. Hence (7.1) holds with $C = m^2 a$ for $\eta = \eta_n$ and, consequently, for η. This completes the proof.

8. <u>An application</u>. A curious consequence of Theorem 1.1 is the following: Let $f(t, y, y')$ be a continuous d-vector function defined on an open (t, y, y')-set E. Let $(i_1, \ldots, i_d)$ be any set of d integers, $0 \leqq i_j \leqq d$, such that no integer except 0 occurs more than once. Let $t = y^0$ and suppose that $f^k(y^0, y, y')$ has continuous partial derivatives with respect to each of its variables except possible y^{i_k}, $k = 1, \ldots, d$. Then the initial value problem

$$y'' = f(t, y, y') , \quad y(t_0) = y_0 , \quad y'(t_0) = y_0' , \tag{8.1}$$

where $y_0'^{i_k} \neq 0$ if $i_k \neq 0$, has a unique solution $y = \eta(t, y_0, y_0')$ and $\eta(t, y_0, y_0')$, $\eta'(t, y_0, y_0')$ are of class C^1.

The case that $i_j = 0$ for $j = 1, \ldots, d$ is the classical case where f has continuous partial derivatives with respect to the components of y, y'. The components of y can be renumbered so that $(i_1, \ldots,$ becomes $(0, \ldots, 0, i_{j+1}, \ldots, i_d)$ where $0 \leqq j \leqq d$ and $i_{j+1}, \ldots, i_d$ are distinct integers on the range $1 \leqq i_k \leqq d$. Write (8.1) as a first order system

$$y' = z , \quad z' = f(t, y, z) \quad \text{or} \quad dy - z\,dt = 0 , \quad dz - f\,dt = 0 .$$

Let A be the $2d \times 2d$ matrix of the form

$$A = \begin{pmatrix} I_{d+j} & 0 \\ Q & P \end{pmatrix}$$

where I_{d+j} is the $(d+j) \times (d+j)$ unit matrix, P is the $(d-j) \times (d-j)$ diagonal matrix $\operatorname{diag}(z^{i_{j+1}}, \ldots, z^{i_d})$, $Q = (q_{mn})$ is the $(d-j) \times (d+j)$ matrix with at most one element different from 0 on each row: $q_{mn} = -f^m$ or 0 according as $n = i_m$ or $n \neq i_m$ for $m = j+1, \ldots, d$ and $n = 1, \ldots, d+j$. Thus $A(dy-zdt, dz-fdt)$ is the 2d-vector with the first d components $dy^i - z^i dt$, the next set of j components of the form $dz^k - f^k dt$ for $k = 1, \ldots, j$, and the last set of $d-j$ components $z^{i_k} dz^k - f^k dy^{i_k}$, $k = j+1, \ldots, d$. By the assumption on f^k, each of these linear differential 1-forms in the variables t, y, z has a continuous exterior derivative. Also $\det A$ is the product $z^{i_{j+1}} \ldots z^{i_d} \neq 0$ if $z^{i_k} \neq 0$ for $k = j+1, \ldots, d$. Since the conditions specified on the initial value problem (8.1) is $y_0'^{i_k} \neq 0$ if $k = i_{j+1}, \ldots, i_d$, the assertion follows from Theorem 1.1.

This research was supported by the Air Force Office of Scientific Research.

NOTE

1. The argument leading to Corollary 3.1 was employed in the proof of (II) in [6] where, however, in the subsidiary statement (*), [6], p. 490, "uniqueness" rather than "uniqueness to the right" is erroneously claimed. This does not affect the proof of (II).

REFERENCES

1. E. Cartan, Leçons sur les invariants intégraux, Paris, 1922.
2. H. Cartan, Algebraic topology, mimeographed notes, Harvard, 1949.
3. P. Gillis, "Sur les formes différentielles et la formule de Stokes," Académie Royale de Belgique, Memoires, vol. 20(1943).
4. P. Hartman, "On the local uniqueness of geodesics," American Journal of Mathematics, vol. 72(1950), pp. 723-730.

5. ___________, "On the exterior derivatives and solutions of ordinary differential equations," Transactions of the American Mathematical Society, vol. 91(1959), pp. 277-293.

6. ___________, "On stability in the large for systems of ordinary differential equations," Canadian Journal of Mathematics, vol. 13 (1961), pp. 480-492.

7. ___________ and A. Wintner, "On the third fundamental form of a surface," American Journal of Mathematics, vol. 75(1953), pp. 298-334.

8. E. R. van Kampen, "Remarks on systems of ordinary differential equations," ibid., vol. 59(1937), pp. 144-152.

9. D. C. Lewis, "Metric properties of differential equations," ibid., vol. 71(1949), pp. 294-312.

10. ___________, "Differential equations referred to variable metric," ibid., vol. 73(1951), pp. 48-58.

11. L. Markus and H. Yamabe, "Global stability criteria for differential systems," Osaka Mathematics Journal, vol. 22, 1960), pp. 305-317.

E. H. ROTHE

Some remarks on critical point theory in Hilbert space

1. Introduction

The theory of critical points of real valued functions as created by M. Morse[1] was first developed for functions defined in a finite dimensional space. In order to treat problems in analysis (e.g. problems concerning geodesics) Morse extended his theory to infinite dimensional spaces by approximating functions defined in such spaces by functions defined in finite dimensional spaces.

Similarly the results concerning critical points in Hilbert space contained in [9] were obtained from results known in Euclidean n-space by a rather laborious limit process.

It is desirable to make such a passage to the limit in the dimension superfluous by developing a theory of critical points and critical levels directly in Hilbert space.

The purpose of the present paper is to offer some remarks towards establishing such a theory. The author hopes to be able to present a more complete and systematic treatment in the future.

We will deal with the following situation: in the Hilbert space E let $V = V_R$ be the ball with positive radius R and with the origin as center. Let $i(x)$ be a real valued function defined for x in some neighborhood of V which contains a ball V_1 concentric to V of radius $R_1 > R$. $i(x)$ is supposed to have a gradient $g(x)$ [2] of the form

$$g(x) = x + G(x) \tag{1.1}$$

where $G(x)$ is completely continuous. It is assumed that on the boundary S of V the gradient field in exteriorly directed (regular boundary condition[3]). A critical point of i is a point x for which $g(x) = 0$, and a critical level of i is a real number c such that $i(x) = c$ for at least one critical point x. It is assumed that there are at most a finite number of critical levels.

For any real number e let

$$i_e = \{x \in V \mid i(x) < e\}\ ,\quad \bar{i}_e = \{x \in V \mid i(x) \leqq e\}\ .$$

We will use singular homology theory[4] and denote as usual by $H_q(B,A)$ the q-th relative homology group (with respect to a given coefficient group) . Let now c be a critical level and let a, b be two numbers such that $a < c < b$ and such that c is the only critical level in the closed interval $[a, b]$. As in the finite dimensional case one wants to define the critical group at level c as $H_q(\bar{i}_b, \bar{i}_a)$, and the q-th Morse number as the rank of this group. Consequently the first task will be to prove that (up to isomorphisms) this group is independent of a and b . This is done in section 4.

In section 5 we consider the case that the set σ of critical points at level c consists of a finite number of points. Then the critical group at level c is isomorphic to $H_q(i_c \cup \sigma, i_c)$ as in the finite dimensional case.

In section 6 we consider the case of a non degenerate critical point.[5] Here we make the two additional assumptions that the Hilbert space E is separable and that the G(x) in (1.1) has a Fréchet differential L(x, k)[6] which is completely continuous in k . Under these assumptions the second Fréchet differential of i (in a proper coordinate system) takes the form:

$$-\sum_{1}^{r} x_i^2 + \sum_{r+1}^{\infty} x_i^2 \qquad 7$$

where r is a finite positive integer or 0 in which latter case the symbol Σ_1^0 is supposed to mean 0 . In extension of a result classical in the finite dimensional case, it is proved that the q-th Morse number equals 0 for $q \neq r$, and 1 for $q = r$.[8]

Sections 2 and 3 are of a preliminary character assembling tools needed later on. Section 2 refers to topology, and section 3 to differential equations in Hilbert space.

In part the proofs follow closely the method used by E. Pitcher in [6] with modifications necessary for dealing with a space which is not locally compact.

2. Topological Tools

For convenience and easy reference some definitions and lemmas are recalled in this section. For proofs and details we refer to [3].

A couple B, A of sets is called a pair and denoted by (B, A) if $B \supset A$. If (B, A) and (D, C) are two pairs, and B and D are subsets of topological spaces then a map

$$\phi : (B, A) \to (D, C) \qquad (2.1)$$

always means a continuous map for which $\phi(B) \subset D$ and $\phi(A) \subset C$.

Let $\mathscr{A}$ be an admissible category [3, p. 5] which in particular contains the Hilbert space E (in its strong topology), all pairs (B, A) for which $B \subset E$, and all maps (2.1) if also $D \subset E$.

DEFINITION 2.1. Let I denote the closed unit interval $0 \leqq s \leqq 1$. Two maps $\phi_i : (B, A) \to (D, C)$ $(i = 1, 2)$ are called homotopic if there exists a map

$$\phi = \phi(x, s) : \quad (B \times I,\ A \times I) \to (D, C)$$

such that $\phi(x, 0) = \phi_1(x)$, $\phi(x, 1) = \phi_2(x)$, $(x \in B)$.

DEFINITION 2.2. Let (B, A) and (D, C) be two pairs for which $B \supset D$ and $A \supset C$. Let

$$\delta : \quad (B \times I,\ A \times I) \to (B, A) \quad .$$

We say δ deforms (B, A) (over itself) into (D, C) if $(x, 0) = x$ (for $x \in B$), and if $(B \times 1,\ A \times 1) \to (C, D)$.

DEFINITION 2.3. The pairs (B, A) and (D, C) are called homotopically equivalent (in symbols: $(B, A) \sim (D, C)$) if there exist maps

$$f : (B, A) \to (D, C) \quad , \qquad g : (D, C) \to (B, A)$$

such that fg is homotopic to the identity map of (B, A) and gf is homotopic to the identity map of (D, C).

LEMMA 2.1. Suppose that there exists a δ which deforms (B, A) into (D, C) in such a way that (D, C) is deformed over itself, i.e., that

$$(D \times I,\ C \times I) \to (D, C) \quad . \tag{2.2}$$

Then $(B, A) \sim (D, C)$.

LEMMA 2.2. Denote as usual the q-dimensional relative homology group of B modulo A by $H_q(B, A)$, and let the symbol $\approx$ denote the isomorphism of groups. Then: if $(B, A) \sim (D, C)$ then

$$H_q(B, A) \approx H_q(D, C) \quad . \tag{2.3}$$

COROLLARY to lemmas 2.1 and 2.2. If there exists a δ satisfying the assumptions of lemma 2.1, then (2.3) holds.

DEFINITION 2.4. (B, A) is called homotopically trivial if $(B, A) \sim (A, A)$.

LEMMA 2.3. If (B, A) is homotopically trivial then $H_q(B, A) = 0$ for all q.

LEMMA 2.4. Let $C \supset B \supset A$ with a non vacuous A. If then $H_q(B, A) = 0$ for all q, then $H_q(C, A) \approx H_q(C, B)$ for all q.

This is a special case of [3, Th. 10.4].

3. Elementary theorems on differential equations in Hilbert space

The theorems listed in this section are all classical in finite dimensional spaces and it is quite trivial to see that familiar proof methods carry over to the Hilbert space case. Consequently no proofs will be given. Of course one has to be clear about the definition of

$$\int_{t_0}^{t} F(\tau)\, d\tau$$

where $F(t)$ is a function of the real variable t with values in the Hilbert space E. Since we will have to deal only with continuous integrands it will be sufficient to adopt the Riemann integral definition as given by Graves in [2].

THEOREM 3.1. Let x_0 be a point of the Hilbert space E. Let α, β, L be positive constants. Let $f(x, t)$ be defined in

$$\| x - x_0 \| \leqq \beta , \quad | t - t_0 | \leqq \alpha , \quad (x \in E) \tag{3.1}$$

with values in E.

We assume that $f(x, t)$ is continuous in (3.1) and in this set also satisfies the Lipschitz condition

$$\| f(x_1, t) - f(x_2, t) \| < L .$$

Then:

a) there exists a positive $\overline{M}$ such that in (3.1)

$$\| f(x, t) \| < \overline{M} .$$

b) if $\alpha' = \min(\alpha, \beta/\overline{M})$ then there exists in $|t-t_0| < \alpha'$ one and only one $x = x(t)$ such that

$$\frac{dx}{dt} = f(x, t) ,$$

$$x(t_0) = x_0 , \quad \| x(t) - x_0 \| \leqq \beta . \tag{3.2}$$

c) this unique solution depends continuously on x_0.

THEOREM 3.2. Let Δ be an open set of E and let J be the interval $T_1 < t < T_2$. Let $\vartheta = \Delta \times J$. Let $f(x, t)$ satisfy the assumptions of Theorem 3.1 in some neighborhood of every point (x_0, t_0) of ϑ. Moreover we suppose that $F(x, t)$ is bounded in ϑ.

Let now $x = x(t)$ be a solution of the differential equation (3.2) which is defined in an open subinterval (a, b) of J and which is such that $x(t) \in \Delta$ for t in (a, b).

Then:

a) the limits $\alpha = \lim_{t \to a^+} x(t)$, and $\beta = \lim_{t \to b^-} x(t)$ exist.

b) if we make the additional assumption that $f(x, t)$ is bounded in the closure of ϑ and if we extend the definition of $x(t)$ by setting $x(a) = \alpha$ and $x(b) = \beta$, then the left derivative of $x(t)$ at $t = b$ exists and satisfies the differential equation, and the corresponding assertions hold at $x = a$.

THEOREM 3.3. Let Δ be a bounded domain in E. Let J be the interval $T_1 < t < T_2$ the values $T_2 = \infty$, $T_1 = -\infty$ being admitted, and let $\vartheta = \Delta \times J$. We assume that the assumptions of theorem 3.1 are satisfied in some neighborhood of every point $(x_0, t_0) \in \vartheta$. Moreover we assume that $f(x, t)$ is bounded in every subset ϑ_1 of ϑ which has a positive distance from the boundary $\dot{\vartheta}$ of ϑ. Let then $x = x(t)$ be the solution of (3.2) satisfying $x(t_0) = x_0$ with $(x_0, t_0) \in \vartheta$, and denote by $\rho(t)$ the distance of the point $(x(t), t)$ from $\dot{\vartheta}$. Then, either $x(t)$ can be continued to all $t > t_0$ such that it is a solution for all these t and such that $(x(t), t)$ remains in ϑ, or there exists a finite $b > t_0$ such that $x(t)$ can be continued as a solution to all t in the interval $t_0 < t < b$ with $(x(t), t) \in \dot{\vartheta}$ and such that $\lim_{t \to b^-} \inf \rho(t) = 0$.

The corresponding assertions hold concerning the continuation of $x(t)$ to t-values less than t_0.

We now consider the case that the right member of (3.2) does not depend on t, i.e. we consider the "autonomous" differential equation

$$\frac{dx}{dt} = f(x) \quad . \tag{3.3}$$

We use the same notations and assumptions as in theorem 3.3, J now always denoting the interval $-\infty < t < \infty$. If $x(t)$ is the solution discussed in theorem 3.3 we refer to the curve $(x(t), t)$ situated in $E \times J$ as an integral curve. The point set in E obtained by projecting the integral curve naturally into E is called a characteristic unless it reduces to a single point. If $x(t_0)$ is a point on a characteristic then the subset of the characteristic for which $t \geqq t_0$ is called a half characteristic.

LEMMA 3.1. Through the point x_0 of Δ passes a characteristic if and only if $f(x_0) \neq 0$. If there is a characteristic through x_0 it is uniquely determined.

The following theorem is a corollary to theorem 3.2:

THEOREM 3.4. If the half characteristic defined by $x(t)$, $t = t_0$ lies in a subset of Δ which has a positive distance from the boundary of Δ then $x(t)$ is defined for <u>all</u> $t \geqq t_0$.

4. The groups attached to a critical level

We start by recalling the definition of a Fréchet differential and of the gradient.

DEFINITION 4.1. Let E and F be two Hilbert spaces (which may coincide). Let $x_0 \in E$, let N be a neighborhood of x_0, and let $f(x)$ be a map of N into F. Suppose there exists a linear bounded map $l : E \to F$:

$$l(h) = d(x_0, h) \tag{4.1}$$

of the following property: if $R(x, h)$ is defined by the equation

$$f(x_0 + h) - f(x) = l(h) + R(x_0, h) \quad , \quad (x_0 + h \in N) \tag{4.2}$$

then

$$\lim_{h \to 0} \frac{R(x_0, h)}{\|h\|} = 0 \ , \text{ and } R(x, 0) = 0 \ . \tag{4.3}$$

Then f is called differentiable at $x = x_0$, and (4.1) is called the differential of f at x_0.

DEFINITION 4.2. If $F = E^1$, the real line, then (3.1) is a linear bounded functional. Consequently there exists a unique element $g(x_0)$ of E such that

$$l(h) = d(x_0, h) = (g(x_0), h) \tag{4.4}$$

where the usual notation for the scalar product is used at the right. $g(x_0)$ is called the gradient of f at $x = x_0$. If the gradient $g = \text{grad } f$ exists for every $x \in N$, then g maps N into E.

LEMMA 4.1. Let the scalar (i.e. real valued function) $i(x)$ be defined in N, and let the differential

$$di = l(x, h)$$

exist for x in N. Moreover let $x = x(t)$ be such that dx/dt exists (t real). Then

$$\frac{di}{dt} = (l(x(t), \frac{dx}{dt}) = (g(x), \frac{dx}{dt})$$

where $g(x) = \text{grad } i$.

This is well known.

Using definitions and notations given in section 1 (see in particular the paragraph containing equation (1.1)) we state our main assumptions:

A) If $I(x)$ is defined by

$$i(x) = (x, x)/2 + I(x)$$

then $G(x) = \text{grad } I(x)$ exists (this is equivalent to the assumption that $g(x) = \text{grad } i(x)$ exists and equals $x + G(x)$).

Note that A) implies the continuity of $i(x)$ and $I(x)$.

B) $G(x)$ is completely continuous in V, i.e. it is continuous and maps V onto a subset of E whose closure is compact. Moreover $G(x)$ satisfies locally a Lipschitz condition.

C) $i(x)$ has at most a finite number of critical levels $(x \in V)$.

D) For x on the boundary S of V we have $(x, g(x)) > 0$.

Note that D) implies that there are no critical points on S.

E) $G(x)$ is bounded in V.

Obviously E) implies and is implied by the existence of a positive constant M such that

$$\|g(x)\| < M \quad \text{for } x \in V. \qquad (4.5)$$

[9]

LEMMA 4.2. If $g(x)$ is of the form (1.1) with completely continuous G then a) the set of solutions of $g(x) = 0$ is compact, and b) the image of a closed set under g is closed.

This is well known.

LEMMA 4.3. The set Γ of critical points is compact.

This is a corollary to the first part of the preceding lemma.

LEMMA 4.4. For any $x_0 \in V$ denote by $\delta(x)$ the distance of x from the set Γ of critical points. Then there exists a $\gamma_0 \in \Gamma$ such that

$$|i(x_0) - i(\gamma_0)| \leqq M\delta(x_0) \qquad (4.6)$$

where M is the constant appearing in (4.5).

PROOF. We note first that on account of lemma 4.3 there exists a $\gamma_0 \in \Gamma$ such that

$$\delta(x_0) = \| x_0 - \gamma_0 \| \quad . \tag{4.7}$$

If now $d(x, h) = (g(x), h)$ is the differential of $i(x)$ then by the mean value theorem[10]

$$i(x+h) - i(x) = \int_0^1 (g(x+th), h)\, dt \ .$$

Using the Schwarz inequality and (4.5), we obtain the inequality

$$| i(x+h) - i(x) | \leqq M \| h \| \quad .$$

If we set $x = \gamma_0$ and $h = x_0 - \gamma_0$, we obtain (4.6).

LEMMA 4.5. Let

$$\Gamma_\mu = \{ x \in V \mid \delta(x) \geqq \mu M^{-1} \} \ ;$$

then Γ_μ is closed.

This follows from the well known fact that $\delta(x)$ is continuous (see e.g. [1, p. 57, Satz 6]).

LEMMA 4.6. There exists a positive $m = m(\mu)$ such that

$$\| g(x) \| > m \qquad \text{for} \qquad x \in \Gamma_\mu \tag{4.8}$$

PROOF. By the preceding lemma Γ_μ is closed. Therefore by lemma 4.2 $g(\Gamma_\mu)$ is closed. But the latter set does not contain the 0-point since Γ_μ contains no critical point.

LEMMA 4.7. Let the numbers μ, μ', ρ be such that

$$0 < \mu' < \mu \ \text{ and } \ 0 < \rho < \frac{\mu - \mu'}{M} \quad . \tag{4.9}$$

Then the ρ-neighborhood $N_\rho = N_\rho(\Gamma_\mu)$ of Γ_μ is contained in $\Gamma_{\mu'}$. Here the ρ neighborhood of any set S is defined as the union of all balls of radius ρ whose center is a point of S.

PROOF. Let x be a point of N_ρ. Then by definition of N_ρ and of Γ_μ there exists an $x_0 \in \Gamma_\mu$ such that

$$\| x - x_0 \| < \rho \ , \quad \| x_0 - \gamma \| \geqq \mu/M \ \text{ for all } \gamma \in \Gamma \ .$$

From these inequalities together with (4.9) we see that

$$\|x-\gamma\| \geqq \|x_0-\gamma\| - \|x-x_0\| \geqq \mu M^{-1} - \rho > \mu M^{-1} - (\mu-\mu')M^{-1} = \mu' M^{-1}$$

By definition of $\Gamma_{\mu'}$, this inequality shows that $x \in \Gamma_{\mu'}$.

LEMMA 4.8. Let the real numbers $a < b$ be such that the closed interval $[a, b]$ contains no critical level. Let μ be a positive number which is less than the distance of $[a, b]$ from the set of critical levels. By assumption C) such a number exists. Finally let

$$S_{a,b} = \{x \in V \mid a \leqq i(x) \leqq b\} .$$

Then $S_{a,b} \subset \Gamma_\mu$.

PROOF. We will prove: if $x \notin \Gamma_\mu$ then $x \notin S_{a,b}$. If $x \notin \Gamma_\mu$ then $\delta(x) < \mu M^{-1}$ where δ is defined as in lemma 4.4; moreover this lemma implies the existence of a $\gamma_0 \in \Gamma$ such that

$$|i(x) - i(\gamma_0) \leqq M\delta(x) . \tag{4.10}$$

Now γ_0 is a critical point and, therefore, $i(\gamma_0)$ is a critical level, and (4.10) shows on account of the definition of μ that $i(x) \not\subset [a, b]$ which is equivalent to $x \notin S_{a,b}$.

We now consider the differential equation

$$\frac{dx}{dt} = -g(x) \tag{4.11}$$

where $g = \text{grad } i$.

LEMMA 4.9. a) No characteristic goes through a critical point, and through each ordinary (that is non-critical) point x_0 of V goes exactly one characteristic of (4.11). b) If this characteristic is given by

$$x = x(t) \quad , \quad x(0) = x_0 \tag{4.12}$$

then for no positive t the point $x(t)$ is outside V. c) $x(t)$ is defined for all positive t.

PROOF. a) follows from lemma 3.1. b) if there were a positive t_1 such that $x(t_1) > R$, the radius of V, there would be a $\bar{t}$ in the interval $(0, t_1)$ such that

$$\|x(\bar{t})\| = R \text{ and } \|x(t)\| > R \text{ for } \bar{t} < t \leqq t_1 . \tag{4.13}$$

On the other hand the differential of (x, x) is $2(x, h)$ and therefore by lemma 4.1 and by (4.11)

$$\frac{d(x,x)}{dt} = 2(x, \frac{dx}{dt}) = -2(x, g(x)) .$$

Consequently

$$\|x(t)\|^2 - R^2 = \|x(t)\|^2 - \|x(\bar{t})\|^2 = -2 \int_{\bar{t}}^{t} (x(\tau), g(x(\tau))\, d\tau . \qquad (4.14)$$

Now by assumption D) the integrand is positive for $\tau = \bar{t}$ and will therefore still be positive for $\bar{t} \leqq \tau \leqq t$ if t is near enough to $\bar{t}$. But for such $t \geqq \bar{t}$, (4.14) shows that $\|x(t)\|^2 < R^2$ in contradiction to (4.13). This proves b), and c) follows now from theorem 3.4 (with $\Delta = V_1$).

LEMMA 4.10. Let $x = x(t)$ be the characteristic (4.12). Then

$$i(x(t)) < i(x_0) \quad \text{for} \quad t > 0 .$$

PROOF. From lemma 4.1 and from (4.11) we see that the derivative with respect to t of $i(x(t))$ equals $-\|g(x(t))\|^2$. Therefore

$$i(x(t)) - i(x_0) = -\int_0^t \|g(x(\tau))\|^2 d\tau < 0 \qquad (4.15)$$

since by lemma 4.9 no critical point lies on the characteristic.

DEFINITION 4.3. For $T > 0$ let I_T be the interval $0 \leqq t \leqq T$. We define a map $\Delta_T : V \times I_T \to V$ as follows:

if x_0 is an ordinary point of V then $\Delta(x_0, t) = x(t)$ where $x(t)$ is the characteristic (4.12);

if x_0 is a critical point of V then $(x_0, t) = x_0$.

LEMMA 4.11. a) for $0 \leqq t_1 < t_2 \leqq T$ we have

$i(\Delta(x_0, t_2)) < i(\Delta(x_0, t_1))$ if x_0 is ordinary

$i(\Delta(x_0, t_2)) = i(\Delta(x_0, t_1))$, if x_0 is critical.

b) if a, b satisfy the assumptions of lemma 4.8 then there exists a positive constant m such that for $x_0 \in S_{a,b}$ and for each t in the interval $0 \leqq t \leqq T$ at least one of the two following inequalities holds

I) $i(\Delta(x_0, t)) \leqq i(x_0) - tm^2$

II) $i(\Delta(x_0, t)) \leqq a$.

PROOF. The first part of a) follows from lemma 4.10, and the second part directly from the definition of Δ_T. To prove b) let us assume that for some positive $t \leqq T$, II) is false. We then have to show for this t the inequality I) holds. From the assumption made and from a) we have

$$b \geqq i(x_0) > i(x(t)) > a$$

and, again from a), we see that this inequality still holds if t is replaced by any τ in the interval $0 < \tau \leqq t$. By definition of $S_{a,b}$ this implies that for these τ, $x(\tau) \in S_{a,b}$. Therefore, by lemma 4.8, $x(\tau) \in \Gamma_\mu$, and by lemma 4.6 there exists a constant m such that $g(x(\tau)) > m > 0$ for $0 \leqq \tau \leqq t$. From this inequality and (4.15) we obtain I).

LEMMA 4.12. If $T \geqq (b-a)/m^2$ then

$$i(\Delta(x_0, T)) \leqq a \quad \text{for all } x_0 \in S_{a,b} . \tag{4.12}$$

This follows immediately from the b)-part of the preceding lemma since for a T satisfying the assumed inequality the right member of inequality I) is $\leqq a$ for $t = T$.

THEOREM 4.1. If a, b satisfy the assumptions of lemma 4.8 then

$$H_q(\bar{i}_b, \bar{i}_a) = 0 \quad \text{for all } q . \tag{4.16}$$

(For the notation see (1.2) and lemma 2.2).

PROOF. We consider $\Delta_T(x_0, t)$ with a T satisfying the assumption of lemma 4.12. By theorem 3.1 part c), $\Delta_T(x_0, t)$ is continuous in (x_0, t). This together with the results of the two preceding lemmas shows that Δ_T satisfies the assumptions made about δ in lemma 2.1 if we set $\bar{i}_b = B$, $\bar{i}_a = A = D = C$. Therefore we see from that lemma that $(\bar{i}_b, \bar{i}_a) \sim (\bar{i}_a, \bar{i}_a)$, and (4.16) follows from lemma 2.3.

THEOREM 4.2. Let c be a critical value and let a, b, α, β be such that

$$b \geqq \beta > c > \alpha \geqq a . \tag{4.17}$$

Moreover suppose that c is the only critical level in the closed interval $[a, b]$. Then

$$(\bar{i}_b, \bar{i}_a) \sim (\bar{i}_\beta, \bar{i}_\alpha) , \tag{4.18}$$

and $H_q(\bar{i}_\beta, i_\alpha)$ remains constant (up to isomorphisms) as α and β vary in the intervals indicated by (4.17).

PROOF. Since there is no critical level on $[a, \alpha]$, we may apply lemma 4.12 to this interval to assert the existence of a T such that

$\Delta_T(x_0, T) \subset i_a$ if x_0 in i_α. As in the proof of theorem 4.1 we see that Δ_T satisfies the assumptions on δ in lemma 2.1 if $i_\beta = B = D$, $i_\alpha = A$, $i_a = C$. Consequently

$$(\bar{i}_\beta, \bar{i}_\alpha) \sim (\bar{i}_\beta, \bar{i}_a) .$$

Since the corresponding argument shows that

$$(\bar{i}_b, \bar{i}_a) \sim (\bar{i}_\beta, \bar{i}_a) ,$$

(4.18) follows. The remaining part of the theorem follows now from lemma 2.2.

Theorem 4.2 shows that the following definition is unique up to isomorphisms:

DEFINITION 4.4. Let c be a critical level and let a, b be two numbers such that $a < c < b$ and such that c is the only critical level in $[a, b]$. Then $H_q(\bar{i}_b, \bar{i}_a)$ is called the critical group of dimension q at the level c. The rank M_q of this group is called the q-th Morse number of the level c.

5. The case of a finite critical set

Let c be a critical level and let σ be the critical set at level c, that is the set of critical points γ such that $i(\gamma) = c$. The main goal is the proof of theorem 5.2. As in the finite dimensional case [6, p. 13] deformations will be used which are based on characteristics of the differential equation

$$\frac{dx}{dt} = \frac{-g(x)}{\|g(x)\|^2} \qquad (g = \text{grad } i) \tag{5.1}$$

The following lemma serves to ensure the applicability of the existence theorem 3.1 in the neighborhood of an ordinary point.

LEMMA 5.1. Let the numbers μ, μ', and ρ have the same properties as in lemma 4.7. Moreover let m' be such that

$$\|g(x)\| > m' > 0 \qquad \text{for } x \in \Gamma_{\mu'} \tag{5.2}$$

the existence of such an m' being assured by lemma 4.6. Finally let x_0 be a point of Γ_μ and let

$$f(x) = -g(x)/\|g(x)\|^2 . \tag{5.3}$$

Then for

$$\|x - x_0\| < \rho \tag{5.4}$$

a) $\|f(x)\| < 1/m'$

b) for ρ small enough, $f(x)$ satisfies a Lipschitz condition in the neighborhood (5.4) of x_0.

c) if $G(x)$ satisfies a uniform Lipschitz condition on a subset S of Γ_μ then $f(x)$ satisfies a uniform Lipschitz condition on S in the sense that there exists a constant L such that

$$\|f(x_0+h) - f(x_0)\| < L\|h\| \quad \text{for all } x_0 \in S, \quad \|h\| < \rho.$$

PROOF. a) follows from (5.2) since by (5.1) $\|f(x)\| = 1/\|g(x)\|$.

To prove b) we note that by assumption $G(x)$, and therefore also $g(x)$, satisfies a Lipschitz condition in (5.4) for ρ small enough, i.e. there exists an L_0 (depending on x_0 and ρ), such that

$$\|g(x_0+h) - g(x_0)\| < \|h\| L_0, \quad \|h\| < \rho. \tag{5.5}$$

Now from the definition (5.3) of $f(x)$ we obtain $f(x_0+h) - f(x_0)$

$$= \frac{\|g(x_0)\|^2 [g(x_0+h) - g(x_0)] - g(x_0)(g(x_0+h) + g(x_0), g(x_0+h) - g(x_0))}{\|g(x_0)\|^2 \|g(x_0+h)\|^2}.$$

Using the Schwarz inequality we see from (4.5), (5.2) and (5.5) and from a) that the norm of the right member of the above equality is bounded by $(M^2 L_0 + M M^2 L_0)\|h\|(m')^{-4}$. This proves b); it also proves c) since the factor of $\|h\|$ does not depend on x_0 if L_0 does not depend on x_0.

LEMMA 5.2. a) Through every ordinary point x_0 of V goes one and only one characteristic of (5.1).

b) let the characteristic of a) be given by

$$x = x(t), \quad x(0) = x_0 \tag{5.6}$$

and let μ, μ', ρ have the same meaning as in the preceding lemma. We assume that for some $\bar{t} > 0$

$$\bar{x} = x(\bar{t}) \in \Gamma_\mu \tag{5.7}$$

Then

$$\|x(t) - \bar{x}\| < \rho \tag{5.8}$$

for

$$|t - \bar{t}| < \alpha' = m'\rho. \tag{5.9}$$

PROOF. a) since the set Γ of critical points is closed the ordinary point x_0 has a positive distance μM^{-1} from Γ. We set $\mu' = \rho = (3M)^{-1}\mu$. Application of lemma 5.1 shows then that the right member of (5.1) satisfies the assumptions of the existence theorem 3.1 with $\beta = \rho$ (cf. (3.1)).

b) we apply lemma 5.1 to the point $\bar{x}$ in Γ_μ (instead of x_0). We conclude that in $\|x - \bar{x}\| < \rho$ the assumptions of theorem 3.1 are satisfied and that in this domain $\|f(x)\| < 1/m'$. Theorem 3.1 asserts then that for the interval

$$|t - \bar{t}| < \alpha' = \frac{\rho}{(m')^{-1}} \tag{5.10}$$

there exists a unique solution $\bar{x}(t)$ of (5.1) with

$$\text{a)} \quad \bar{x}(\bar{t}) = x \qquad \text{and} \qquad \text{b)} \quad \|\bar{x}(t) = \bar{x}\| < \rho . \tag{5.11}$$

Now by the uniqueness theorem we see from (5.7) and (5.11a) that $\bar{x}(t) = x(t)$. Thus (5.8) becomes identical with (5.11b).

REMARK to lemma 5.2. We will have to deal with the differential equation

$$\frac{dx}{dt} = K(x_0) f(x) \quad , \quad K(x_0) > 0 \tag{5.12}$$

where $K(x_0)$ is continuous and $f(x)$ is defined by (5.3). Since the substitution

$$s = tK(x_0) \quad , \quad x(t) = y(s)$$

transforms (5.12) into $dy/ds = f(y)$ it is easy to see how our results concerning (5.1) should be modified in order to apply to (5.12). In particular: if $x = x(t)$ is the characteristic of (5.12) satisfying the initial condition $x(0) = x_0$, then the results of lemma 5.2 apply to $x(t)$ if the α' in (5.9) is replaced by

$$\alpha'' = \rho m'/K(x_0) . \tag{5.13}$$

Moreover from $x(t) = y(s) = y(tK(x_0))$ it is clear that the result of theorem 3.1 c)(continuous dependence of $x(t)$ on x_0) is still true.

LEMMA 5.3. Let c be a critical level, and let b be such that $c < b$ and that c is the only critical level in the closed interval $[c, b]$. Moreover let the point x_0 be such that

$$c < i(x_0) \leqq b . \tag{5.14}$$

Let $K(x_0) = i(x_0) - c$, and let $x = x(t)$ be the characteristic of (5.12) determined by $x(0) = x_0$.

Then: a) $d\,i/dt = -K(x_0)$

b) $c < i(x(t)) \leqq b$ for $0 \leqq t < 1$

c) $\lim_{t \to 1^-} i(x(t)) = c$

PROOF. a) By lemma 4.1 and by (4.12)

$$\frac{di}{dt} = (g(x), \frac{dx}{dt}) = -K(x_0)$$

b) and c): from a) we have

$$i(x(t)) - i(x_0) = -(i(x_0)-c)t \; .$$

From this and (5.14) the assertions follow.

REMARK. We see from lemma 5.3 b) that the right member of (5.12) is defined for $x = x(t)$ if $0 \leqq t < 1$. However since the denominator in (5.3) is zero if x is a critical point it is not clear what happens for $t \to 1^-$.

The following theorem answers this question in a special case.

THEOREM 5.1. If the critical set σ at level c consists of a finite number of points, then, with the assumptions and notations of lemma 5.3, $\lim_{t \to 1^-} x(t)$ exists.

PROOF. We distinguish two cases.

a) $\lim_{t \to 1^-} \inf \rho(t) > 0$ where $\rho(t)$ is the distance of $x(t)$ from σ. It follows then from lemma 5.3 that the set of points $x(t)$ $(0 \leqq t < 1)$ has a positive distance from σ. But by lemma 5.3, $x(t) \subset S_{c,b}$, and $S_{c,b}$ has a positive distance from $\Gamma' = \Gamma - \sigma$; indeed if we modify lemma 4.4 by replacing Γ by Γ', and $\delta(x)$ by $\delta'(x)$, the distance of x from Γ', the modified inequality (4.6) shows that $\delta'(x)$ for $x \in S_{c,b}$ is bounded away from 0 since for these x, $i(x)$ has positive distance from the set of critical levels other than c. It follows that in our case a)

$$x(t) \in \Gamma_{\mu} \qquad 0 \leqq t < 1$$

for some positive μ. Consequently lemma 4.6 is applicable and (4.8) holds for $x = x(t)$; therefore

$$\| f(x(t)) \| = 1/\| g(x(t)) \| < 1/m \quad \text{for } 0 \leqq t < 1 \; .$$

From this inequality and from (5.12) we see that for $0 < t' < t'' < 1$

$$\|x(t'')-x(t')\| = \left\| \int_{t'}^{t''} K(x_0) f(x(t))\,dt \right\| < (t''-t')K(x_0)/m \ .$$

Using Cauchy's convergence criterion we see that the last inequality implies the existence of $\lim\limits_{t \to 1^-} x(t)$.

b) $\liminf\limits_{t \to 1^-} \rho(t) = 0$. Since σ consists of a finite number of points there must for at least one of them, say x_1 , exist a sequence t_n monotonically converging to 1^- such that

$$\lim_{n \to \infty} x(t_n) = x_1 \ . \tag{5.14}$$

We will prove

$$\lim_{t \to 1^-} x(t) = x_1 \ .$$

If (5.15) were not true there would exist a positive ε and a sequence t'_n monotonically converging to 1^- such that

$$\|x(t'_n) - x_1\| \geqq \varepsilon > 0 \ . \tag{5.16}$$

For any positive r let us denote by $V(r)$ the ball with center x_1 and radius r , and by $S(r)$ its boundary. Let now $\beta > 0$ be such that

1) $g(x)$ satisfies a uniform Lipschitz condition in $V(5\beta)$.
2) $0 < 5\beta < \varepsilon$.
3) $V(5\beta)$ has a distance $> \beta$ from $\Gamma - x_1$.

We now set $\mu = 3\beta M$, $\mu' = M\beta$, $\rho = \beta$. Then μ , μ' , ρ satisfy the assumptions in lemma 4.7. Moreover by definition of Γ_μ (lemma 4.5)

$$S(3\beta) \subset \Gamma_\mu \ , \quad V(5\beta) - V(\beta) \subset \Gamma_{\mu'} \ .$$

Now by (5.14) there exists an N_0 such that

$$\|x(t_n) - x_1\| < \beta \qquad \text{for } n > N_0 \ . \tag{5.17}$$

On the other hand since $t'_n \to 1^-$ there exists to $n > N_0$ an $n' = n'(n)$ such that

$$t_{N_0} < t_n < t'_{n'} < 1 \ , \tag{5.18}$$

and by (5.16)

$$\|x(t'_{n'}) - x_1\| \geqq 5\beta \ . \tag{5.19}$$

By continuity of $x(t)$ we see from (5.17) and (5.19) that there exists a t^*_n with

$$t_n < t^*_n < t'_{n'} \tag{5.20}$$

such that $\|x(t^*_n) - x_1\| = 3\beta$, i.e. such that

$$x(t^*_n) \in S(3\beta) \ . \tag{5.21}$$

Moreover we see from (5.18) and (5.20) that $x(t^*_n) \to 1^-$.

Now by 1), $g(x)$ satisfies a uniform Lipschitz condition on the subset $S(3\beta)$ of Γ_μ, and by lemma 5.2 with $\beta = \rho$, and by the remark to this lemma we can make the following assertion: if for any $\bar{t} > 0$, $\bar{x} = x(\bar{t}) \in S(3\beta)$ then (cf. (5.8) and (5.13))

$$\|x(t) - \bar{x}\| < \beta \tag{5.22}$$

if

$$|t - \bar{t}| < \eta = \beta m'/K(x_0) \ . \tag{5.23}$$

Now we choose an n so big that $0 < 1 - t^*_n < \eta$. By (5.21) we may take this t^*_n as $\bar{t}$. Then (5.23) is true for all t satisfying

$$t^*_n = \bar{t} < t < 1 \ . \tag{5.24}$$

Consequently (5.22) is also true for these t. But because of (5.14) there is an $n = n_1$ such that (5.24) is true for $t = t_{n_1}$ and such that $\|(x(t_{n_1}) - x_1\| < \beta$. This however contradicts (5.22) with $t = t_{n_1}$ for $\|x(t_{n_1}) - \bar{x}\| \geqq \|\bar{x} - x_1\| - \|x(t_{n_1}) - x_1\| \geqq 3\beta - \beta = 2\beta$.

LEMMA 5.4. Let $a < c < b$, and let c be the only critical level in $[a, b]$. If the critical set σ at level c consists of a finite number of points then

$$(\bar{i}_b, \bar{i}_a) \sim (\bar{i}_c, \bar{i}_a) \ . \tag{5.25}$$

PROOF. We define a deformation $\delta = \delta(x, t)$ of the pair forming the left member of (5.25) into the pair forming the right member as follows: if $x_0 \in \bar{i}_c$ we set $\delta(x_0, t) = x_0$ $(0 \leqq t \leqq 1)$. If $x_0 \in \bar{i}_b - \bar{i}_c$, let $x(t)$ be the characteristic defined in lemma 5.3 for $0 \leqq t < 1$.

By theorem 5.1 we may define $x(1) = \lim_{t \to 1^-} x(t)$. Then $\delta(x_0, t) = x(t)$ for $0 \leqq t \leqq 1$. It is clear from lemma 5.3 that δ satisfies the assumptions of lemma 2.1 which proves (5.25).

LEMMA 5.5. Let c be a critical level, let $a < c$ such that there is no critical level in $[a, c)$. Let $x_0 \in i_c - \bar{i}_a$, and $K(x_0) = i(x_0) - a$. If $x(t)$ is the characteristic of (5.12) determined by $x(0) = x_0$ then

a) $di/dt = -K(x_0)$

b) $a \leqq i(x(t)) \leqq i(x_0) < c$ for $0 \leqq t \leqq 1$

c) $i(x(1)) = a$

We omit the proof since it is essentially the same as the proof of lemma 5.3.

LEMMA 5.6. Let a, c be as in lemma 5.5. Then the pair $(i_c, \bar{i}_a)$ is homotopically trivial (cf. definition 2.4).

PROOF. For $0 \leqq t \leqq 1$ let $\delta(x_0, t) = x_0$ if $x_0 \in \bar{i}_a$, and $\delta(x_0, t) = x(t)$ where $x(t)$ is the characteristic of lemma 5.5 if $x_0 \in i_c - \bar{i}_a$. This lemma shows that δ deforms the pair $(i_c, \bar{i}_a)$ into the pair $(\bar{i}_a, \bar{i}_a)$ and satisfies the assumptions of lemma 2.1. Therefore these pairs are homotopically equivalent which proves the lemma.

THEOREM 5.2. If the critical set σ at level c consists of a finite number of points then the critical groups at level c are isomorphic to the homology groups of the pair $(i_c \cup \sigma, i_c)$.

PROOF. Let a, b be as in lemma 5.4. Assume first that c is not the level of an absolute minimum. Then a can be chosen such that $\bar{i}_a$ is not empty. Then (5.25) is true. Therefore

$$H_q(\bar{i}_b, \bar{i}_a) \approx H_q(\bar{i}_c, \bar{i}_a) . \tag{5.26}$$

Now $\bar{i}_c \supset i_c \supset \bar{i}_a$ and by lemmas 5.6 and 2.3 the homology groups of the pair $(i_c, \bar{i}_a)$ are all 0. Therefore by lemma 2.4, $H_q(\bar{i}_c, \bar{i}_a) \approx H(\bar{i}_c, i_c)$ which together with (5.26) proves that $H_q(\bar{i}_b, \bar{i}_a) \approx H_q(\bar{i}_c, i_c)$. This proves our theorem, since $\bar{i}_c = i_c \cup \sigma$ if c is not the minimum level. If c is the minimum level then i_c and $\bar{i}_a$ for a c are empty, and $H_q(\bar{i}_b, \bar{i}_a) \approx H(\bar{i}_b) \approx H(\bar{i}_c)$ $\approx H_q(\bar{i}_c, i_c)$.

6. Non degenerate critical points

Let σ_0 be a critical point in V. In addition to assumptions A) - E) of section 4, we will assume:

F) $g(x) = \text{grad } i$ has in some neighborhood of σ_0 a differential $l(x, k)$, and the differential $L(x, k)$ of $G(x) = g(x) - x$ is completely continuous in k and continuous in x.

We first recall a few known facts in the form of lemmas. For more details we refer to [8].

LEMMA 6.1. The second differential $d_2(x, h, k)$ of i at the point x exists and

$$d_2(x, h, k) = (l(x, k), h) .$$

(Proof in [8, p. 78]).

DEFINITION 6.1. A bilinear form $q(h, k)$ is degenerate if there exists a $k_0 \neq 0$ such that $q(h, k_0) = 0$ for all h. If there is no such k_0, then $q(h, k)$ is called non degenerate. A linear operator $l(k)$ is called singular if there exists a $k_0 \neq 0$ such that $l(k_0) = 0$. If there exists no such k_0, then $l(k)$ is called non singular.

LEMMA 6.2. The differential $l(x, k)$ of $g(x)$ (as linear operator in k) is non singular if and only if $d_2(x, h, k)$ (as bilinear form in h and k) is non degenerate. (Proof in [8, p. 78]).

DEFINITION 6.2. The critical point σ_0 is non degenerate if the bilinear form $d(\sigma_0, h, k)$ is non degenerate.

LEMMA 6.3. Let σ_0 be a non degenerate critical point. Then σ_0 is isolated. Moreover if l and L are defined as in assumption F) and in lemma 6.2 then $l(\sigma_0, k) = k + L(\sigma_0, k)$. $l(\sigma_0, k)$ is non singular, and the completely continuous operator $L(\sigma_0, k)$ is symmetric. (Proof: [8, lemma 3.1 and lemma 4.1]).

LEMMA 6.4. Let the Hilbert space E be separable. Otherwise we make the same assumptions as in lemma 6.3. Then there exists a non negative (finite) integer r, and an orthogonal base $\{\beta_i\}$ of E such that

$$d_2(\sigma_0, k, k) = - \sum_1^r x_i^2 + \sum_{r+1}^{\infty} x_i^2$$

where $x_i = (k, \beta_i)$, and where the symbol Σ_1^r denotes 0 if $r = 0$. (Proof in [8, theorem 4.1]).

If σ_0 is a critical point then the Taylor expansion at σ_0 starts with the second differential d_2. We make the following assumption concerning the remainder term:

If $\rho(k) = \rho(\sigma_0, k)$ is defined by

$$i(\sigma_0+k) - i(\sigma_0) = d_2(\sigma_0, k, k) + \rho(k) \tag{6.1}$$

then $\operatorname{grad} \rho(k)$ exists and

$$\lim_{k \to 0} \frac{\rho(k)}{\|k\|^2} = 0 , \qquad \lim_{k \to 0} \frac{\operatorname{grad} \rho(k)}{\|k\|} = 0 . \tag{6.2}$$

THEOREM 6.1. We assume that (6.2) is true and that all hypotheses of lemma 6.4 are satisfied. For r positive let E^r be the r dimensional subspace of E defined by $x_{r+1} = x_{r+2} = \dots = 0$, and for any neighborhood N of σ_0 denote by N^r the projection of N into E^r. Then for N a small enough ball

$$((i_c \cup \sigma_0) \cap N, i_c \cap N) \sim (N^r, N^r - \sigma_0) . \tag{6.3}$$

PROOF.[11] As an abbreviation we will write $d_2(k)$ for $d_2(\sigma_0, k, k)$ and define

$$q(k) = d_2(k)/\|k\|^2 . \tag{6.4}$$

The proof will consist in defining a deformation satisfying the assumptions of lemma 2.1 by use of the differential equation

$$\frac{dx}{dt} = - \operatorname{grad} q(x) . \tag{6.5}$$

We will need the following two lemmas:

LEMMA 6.5. If $x = x(t)$ is a characteristic of (6.5) then $\|x(t)\|$ is constant.

PROOF.

$$\frac{d(x, x)}{dt} = 2(x, \frac{dx}{dt}) = -2(x, \operatorname{grad} q) . \tag{6.6}$$

Now $q(x)$ is homogeneous of degree 0. Therefore the scalar product at the right member of (6.6) equals 0 by Euler's lemma on homogeneous forms.[12]

LEMMA 6.6. If $x = x(t)$ is a characteristic of (6.5) then

$$\frac{dq(x(t))}{dt} = - \frac{4(1 - q^2)}{\|x\|^2} \tag{6.7}$$

PROOF. By lemma 4.1

$$\frac{dq(x(t))}{dt} = (\text{grad } q, \frac{dx}{dt}) = - \|\text{grad } q\|^2 . \qquad (6.8)$$

But a straightforward computation which uses the fact that grad $\|x\|^2 = 2x$, that $\|\text{grad } d_2\|^2 = 4\|x\|^2$, and that by Eulers lemma $(x, \text{grad } d_2) = 2d_2$ shows that the right members of (6.8) and (6.7) agree.

Let now ε be a positive number <1. Then by (6.2) there exists a positive α such that

$$\frac{|\rho(k)|}{\|k\|^2} < \varepsilon , \quad \frac{\|\text{grad } \rho\|}{\|k\|} < \varepsilon \quad \text{for} \quad \|k\| < \alpha . \qquad (6.9)$$

From now on we suppose that N is a ball around σ_0 with radius $< \alpha$. For simplicity of notation let us suppose that σ_0 is the origin of E and that $c = i(\sigma_2) = 0$.

We will now show that the characteristics of (6.5) may be used to define a deformation $\delta(x_0, t)$ of $A = i_c \cap N$ into E^r. If $x_0 \in A \cap E^r$ we set $\delta(x_0, t) = x_0$. Let then $x_0 \in A$ and $\notin E^r$. Then by (6.1) and (6.9)

$$q(x_0) = \frac{i(x_0)}{\|x_0\|^2} - \frac{\rho(x_0)}{\|x_0\|^2} < \frac{|\rho(x_0)|}{\|x_0\|^2} < \varepsilon . \qquad (6.10)$$

Let now $x = x(t)$ be the characteristic of (6.5) determined by $x(0) = x_0$. It follows easily from the two lemmas above and from (6.10) that there exist numbers

$$t_0 < 0 < t_1 < t_2$$

such that

$$q(x(t_0)) = \varepsilon , \; q(x(t_1)) = -\varepsilon , \; q(x(t_2)) = -1 . \qquad (6.11)$$

and such that moreover $q(x(t))$ is monotone decreasing in $[t_0, t_2]$. Since $q(x) = -1$ if and only if $x \in E^r$, (6.11) obviously shows that $\delta(x_0, t) = x(t)$ defines a deformation of A into E^r.

To prove that this deformation satisfies the assumptions of lemma 2.1 it will be sufficient to show that $x(t) \subset A$ for $0 < t \leqq t_2$. Since $x(t) \in N$ by lemma 6.5 it remains to prove that

$$i(x(t)) < 0$$

for $0 < t \leqq t_2$. We first show that (6.12) holds in $[t_1, t_2]$: from (6.11) and from the fact that $q(x(t))$ is monotone decreasing it follows that in this interval $-\varepsilon \geqq q(x(t)) \geqq -1$. But then by (6.1) and (6.9)

$$\frac{i(x(t))}{\|x\|^2} \leqq q(x(t)) + \frac{|\rho(x(t))|}{\|x\|^2} < -\varepsilon + \varepsilon = 0 .$$

This proves (6.12) for the interval $[t_1, t_2]$.

To prove (6.12) for the interval $0 < t < t_1$ we note first that by (6.1), (6.9) and (6.11)

$$\frac{i(x(t_0))}{\|x\|^2} \geqq q(x(t_0)) - \frac{|\rho(x(t_0))|}{\|x\|^2} > -\varepsilon + \varepsilon = 0 .$$

Thus $i(x(t_0)) > 0$ while $i(x(0)) < 0$, and $i(x(t_1)) < 0$. Therefore (6.12) will be true in $(0, t_1)$ if we can show that $i(x(t))$ is monotone decreasing in (t_0, t_1). Now $q(x(t))$ is monotone decreasing and since by lemma 6.5, $\|x(t)\|$ is constant, $d_2 = \|x\|^2$ q is also decreasing. Therefore it will be sufficient to show that

$$(\operatorname{grad} i(x(t)), \operatorname{grad} d_2(x(t))) > 0 \quad \text{for } t_0 \leqq t \leqq t_1 . \tag{6.13}$$

By (6.1), the left member equals

$$\begin{aligned}(\operatorname{grad} d_2, \operatorname{grad} d_2) &+ (\operatorname{grad} \rho , \operatorname{grad} d_2) \\ &= 4\|x\|^2 + (\operatorname{grad} \rho , \operatorname{grad} d_2) \\ &\geqq 4\|x\|^2 - \|\operatorname{grad} \rho\| \, 2 \, \|x\|\end{aligned}$$

and this is > 0 by (6.9).

THEOREM 6.2. Under the assumptions of theorem 6.1 we have for $r = 0$

$$H_q((i_c \cup \sigma_0) \cap N , i_c \cap N) \approx \begin{cases} 0 & \text{for } q \neq r \\ \tilde{G} & \text{for } q = r \end{cases} \tag{6.14}$$

where $\tilde{G}$ is the coefficient group.

PROOF. We consider first the case $r > 0$. By lemma 2.2 the homology groups of the two pairs in (6.3) are isomorphic. But the homology groups of the pair forming the right member of (6.3) is well known to be isomorphic to the groups at the right member of (6.14) (see e.g. [3, p. 45]).

If $r = 0$ then it follows easily from (6.1) and (6.9) that $i(x) > 0$ in N, i.e., that $i_c \cap N$ is empty, and the left member of (6.14) reduces to $H_q(\sigma_0)$ which proves (6.14) for $r = 0$.

REMARK. Theorem 6.2 holds if N is a ball with center σ_0 and radius $< \alpha$ where α is such that (6.9) holds with an $\varepsilon < 1$. (6.14) shows that the homology-groups at its left member are independent of N. Therefore we may define these groups as the critical groups at the non degenerate critical point σ_0, and the ranks of these groups as the type numbers at σ_0.

It is then clear that theorem (6.2) implies the assertions made at the end of the introduction.

NOTES

1. See [4] and the literature listed there. Numbers in brackets refer to the bibliography at the end of the paper.
2. See definition 4.2.
3. For a more precise formulation see assumption D) in section 4. In the finite dimensional case this is condition IIIβ in [5, p. 562]. The treatment of more general boundary condition like e.g. IIIγ in [5] is left to further investigation.
4. [3].
5. Definition 6.2
6. Definition 4.1.
7. [8, theorem 4.1].
8. This was proved in [9, theorem 7.2] by going to the limit in the dimension.
9. For a sufficient condition on G and its differential for such an M to exist see [9, corollary to theorem 2.1].
10. [2].
11. [10, p. 37]. The proof of theorem 6.1 is an adaptation of the proof given in [10] for the finite dimensional case.
12. For a proof of Euler's lemma in Hilbert space see [7, p. 272].

BIBLIOGRAPHY

1. P. Alexandroff and H. Hopf, Topologie, Berlin, 1935.
2. L. M. Graves, Riemann integration and Taylor's theorem in general analysis, Trans. Am. Math. Soc. 29(1927), 163-177.
3. S. Eilenberg and N. Steenrod, Foundations of algebraic topology, Princeton, 1952.
4. M. Morse, Calculus of variations in the large, Am. Math. Soc. 1934.

5. M. Morse and G. B. Van Schaack, The critical point theory under general boundary conditions, Annals of Mathematics 35 (1934), 545-571.

6. E. Pitcher, Inequalities of critical point theory, Bull. Am. Math. Soc. 64(1958), 1-30.

7. E. H. Rothe, Completely continuous scalars and variational methods, Ann. of Math. 49(1948), 265-278.

8. ___________, Critical points and gradient fields of scalars in Hilbert space, Acta Math. 85(1951), 73-98.

9. ___________, Leray-Schauder index and Morse type numbers in Hilbert space, Ann. of Math. 55(1952), 433-467.

10. H. Seifert and W. Threlfall, Variationsrechnung in Grossen, Leipzig und Berlin, 1938.

W. T. KOITER

Elastic stability and post-buckling behavior

Summary

The shortcomings of the classical linear theory of elastic stability have been recognized for many years. Approximate solutions of the inherently nonlinear problem of stability and post-buckling behavior have been discussed by many writers. In this paper emphasis will be put on a general theory which is rigorous in an asymptotic sense for the initial stage of post-buckling behavior. This general theory is also capable of assessing the effect of small initial imperfections in the structure. Some results are discussed of an application of the theory to the buckling of thin cylindrical shells in axial compression.

1. Introduction

The theory of elastic stability and the behavior after buckling is one of the oldest branches of the theory of elasticity. Euler's famous investigation of the stability of an initially straight bar under compressive end loads included a complete discussion of the behavior of the bar after buckling has occurred. Euler's theory has remained for nearly two centuries the only complete investigation of post-buckling behavior. It seems that even now only one second example of a complete post-buckling analysis is available for the problem of a thin circular plate under uniform edge thrust [1].

Although the problem of elastic stability belongs inherently to the domain of the nonlinear theory of elasticity, important results may already be obtained from a linearized theory. This classical linear theory of elastic stability is valid for infinitesimal deflections from a supposedly known configuration of equilibrium, the so-called fundamental state. Its general formulation is due to Southwell [2], Biezeno and Hencky [3], and Trefftz [4, 5]. A modern account of these theories is given in the book by Green and Adkins [6].

The behavior after buckling is of particular importance for thin-walled plate and shell structures. The needs of the aircraft designer have stimulated a vast literature in this essentially nonlinear field. The books by Vol'mir [7] and by Mushtari and Galimov [8] give excellent accounts of the present state of knowledge. Nearly all investigations have been based on more or less ad-hoc approximate methods. It may therefore be worthwhile to emphasize that a general nonlinear theory of post-buckling behavior is also available in the case of small finite deflections from the fundamental state. It is the purpose of the present paper to discuss this general theory on the basis of earlier work by the author [9].

We assume that the external loads on the structure are specified as the product of a unit load system and a single load parameter λ . Furthermore, we assume that the loads are conservative. A potential energy then exists for the mechanical system consisting of the elastic structure and its external loads. Two basic types of singular behavior due to loss of stability may now occur, characterized by a limit point (fig. 1) or a bifurcation point (fig. 2), in the terminology of Poincaré.

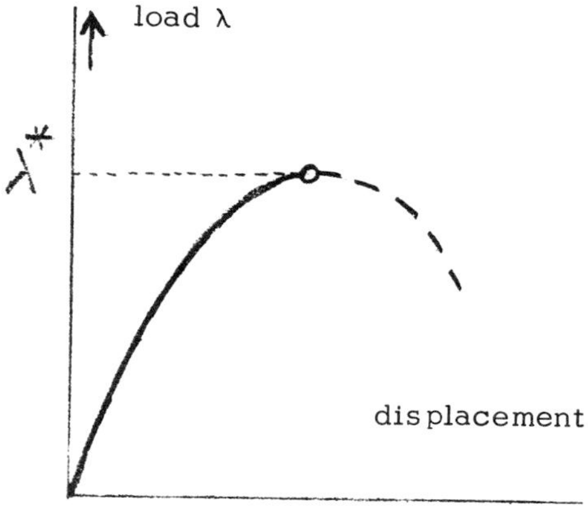

Fig. 1
Limit point

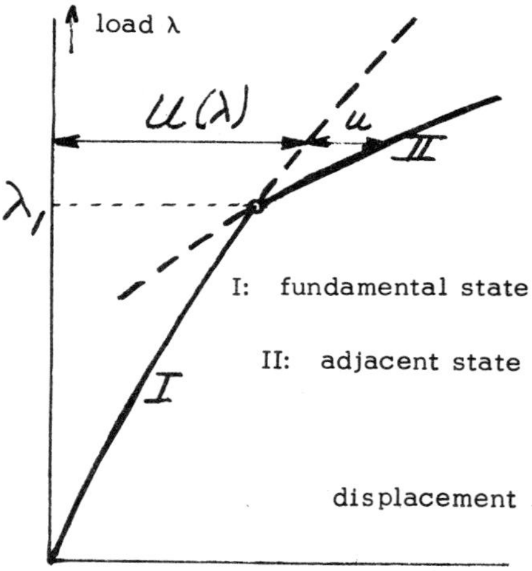

Fig. 2
Bifurcation point

We shall focus our attention on bifurcation problems.[1] They are characterized by the existence of a fundamental state of equilibrium I. The displacement vector $\underline{U}(\lambda)$ in this fundamental state is a single-valued and continuously differentiable function of the load parameter λ in the range under consideration and it vanishes for $\lambda = 0$. Equilibrium in this fundamental state is stable for $0 \leqq \lambda < \lambda_1$, where λ_1 is the critical value of the load parameter, and unstable for $\lambda > \lambda_1$. An alternative solution of the equations of equilibrium branches off from the fundamental state at the critical point, the so-called adjacent state of equilibrium II. According to Poincaré's theory for systems specified by a finite number of generalized coordinates, an exchange of stability occurs between the branches I and II at the bifurcation point. This state of affairs has also been indicated in fig. 2, where a drawn curve indicates a stable branch and a dotted curve an unstable branch. It should be noted, however, that the exchange of stability is the general case only in a formal mathematical sense. Exceptions to this rule are quite frequent in the theory of elastic stability.

Critical points may be characterized by two equivalent criteria. In the first approach it is observed that in such a critical point infinitesimally adjacent configurations of equilibrium exist at the same load. This definition has been employed by Southwell [2] and Biezeno-Hencky [3] in their derivation of the linear and homogeneous equations of neutral equilibrium at a critical point. The second characterization is obtained from the energy criterion of stability. The potential energy has a proper minimum in the stable part of the fundamental branch but only a stationary value in the unstable part. The critical point may now be characterized by a positive semi-definite second variation of the energy. This variational formulation, due to Trefftz [4, 5] in its general form, yields of course the same linear equations of neutral equilibrium.

In spite of the fact that all critical points are characterized by similar linear eigenvalue problems, the actual behavior of the structure at loads in the vicinity of the critical load may vary widely. For example, flat plates can support loads in their plane far in excess of the critical load. On the other hand, some shell structures fail already at loads which are only a fraction of the critical load predicted by the linear theory of stability. These divergent phenomena can only be understood if all adjacent configurations of equilibrium are investigated at loads slightly above or below the critical load. Such an investigation requires of course a higher approximation than is implied in the linear eigenvalue problem which characterizes the critical point.

The most important result of the more refined general analysis is that the initial stage of post-buckling behavior is specified completely by the stability or instability of equilibrium at the critical load itself. In other words, the initial stage of behavior after buckling is governed by the answer to the question whether the critical bifurcation point still belongs to the stable part of the fundamental branch I or should be reckoned to its unstable part. Since the second variation is semi-definite at the bifurcation point, this question can of course only be answered by an investigation of higher order variations of the potential energy.

2. The energy criterion of stability

The transition from the fundamental state I to an arbitrary kinematically admissible configuration II is described by the kinematically admissible displacement field $\underline{u}$. The energy increment

$$P_{II} - P_{I} = P[\underline{u}] \tag{1}$$

is a functional of the kinematically admissible displacement field $\underline{u}$. For a three-dimensional elastic body this functional is the sum of a volume integral, extended over the volume of the body, and a surface integral over its boundary.[2] The integrands are functions of the Cartesian displacement components u_i and their first partial derivatives $u_{i,j}$ $(i, j = 1, 2, 3)$. A non-vanishing neighborhood $\{g, g'\}$ of the fundamental state I is specified by the class of kinematically admissible displacement fields $\underline{u}$ for which

$$|u_i| \leqq g \ , \quad |u_{i,j}| \leqq g' \ , \tag{2}$$

where g and g' are positive constants. A necessary and sufficient condition for the stability of the fundamental state is the existence of some non-vanishing neighborhood $\{g, g'\}$ in which the energy increment (1) is always non-negative, i.e.

$$P[\underline{u}] \geqq 0 \ . \tag{3}$$

We assume that the integrands of $P[\underline{u}]$ have continuous partial derivatives with respect to their arguments u_i and $u_{i,j}$, up to any order required in the analysis. The functional (1) then admits a Taylor-expansion of the form

$$P[\underline{u}] = P_2[\underline{u}] + P_3[\underline{u}] + \ldots + P_m^*[\underline{u}] \ , \tag{4}$$

where the first-order term is absent because the fundamental state is an equilibrium configuration. The asterisk attached to the last term in (4) indicates that it is the m-th order remainder in the expansion. Furthermore we assume that any geometric conditions which restrict the class of kinematically admissible displacement fields $\underline{u}$, are linear and homogeneous. A necessary condition for stability then follows immediately from (3) and (4) in the form

$$P_2[\underline{u}] \geqq 0 \ . \tag{5}$$

A rigorous derivation of sufficient conditions for stability is a far more difficult problem. We introduce an auxiliary quadratic functional $T_2[\underline{u}]$, consisting of the sum of a volume integral over the volume of the body and surface integral over its boundary.[3] The integrands of $T_2[\underline{u}]$ are homogeneous positive definite quadratic forms in u_i and $u_{i,j}$ with continuously differentiable coefficients. Apart from these restrictions $T_2[\underline{u}]$ is completely arbitrary. We now assume that the minimum problem

$$\omega_1 = \text{Min.} \frac{P_2[\underline{u}]}{T_2[\underline{u}]} \tag{6}$$

has a solution. Once this assumption has been made, it may be proved that a positive solution $\omega_1 > 0$ is a sufficient condition for stability of the fundamental state [9, ch. 2]. The necessary condition (5) may now of course also be rephrased in the form $\omega_1 \geqq 0$. No decision on stability has been obtained as yet in the critical case of neutral equilibrium $\omega_1 = 0$, which characterizes the bifurcation point in our bifurcation problem.

3. Stability at a critical point

For the investigation of stability at a critical point, characterized by a solution $\omega_1 = 0$ of minimum problem (6), we need the general form of the displacement field $\underline{u}$ for which (6) vanishes. This general minimizing displacement field is also the general solution of the equations of neutral equilibrium. Let $\underline{u}_h$ $(h = 1, 2, \ldots n)$ denote a complete set of linearly independent solutions of the equations of neutral equilibrium. The general form of the minimizing displacement field is then

$$\underline{u} = a_h \underline{u}_h , \tag{7}$$

where we adopt the summation convention to imply summation from 1 to n for a repeated subscript. Without any loss in generality we may assume that the so-called buckling modes $\underline{u}_h$ are orthogonal and normalized with respect to the auxiliary quadratic functional $T_2[\underline{u}]$

$$T_{11}[\underline{u}_h, \underline{u}_j] = 0 \text{ for } h \neq j \; ; \; T_2[\underline{u}_h] = 1 . \tag{8}$$

The bilinear functional $T_{11}[\underline{u}, \underline{v}]$ is here defined by the identity in the displacement fields $\underline{u}$ and $\underline{v}$

$$T_2[\underline{u}+\underline{v}] = T_2[\underline{u}] + T_{11}[\underline{u}, \underline{v}] + T_2[\underline{v}] . \tag{9}$$

We now assume also that the minimum problem

$$\omega_{(n+1)} = \text{Min.} \frac{P_2[\underline{v}]}{T_2[\underline{v}]} \tag{10}$$

under the side conditions

$$T_{11}[\underline{u}_h, \underline{v}] = 0 \ , \quad h = 1, 2, \ldots n \tag{11}$$

has a solution. Since (7) is the general solution of the equations of neutral equilibrium, this solution must be positive, $\omega_{(n+1)} > 0$. The inequality

$$P_2[\underline{v}] \geqq \omega_{(n+1)} T_2[\underline{v}] \ , \tag{12}$$

under the restrictions (11) on $\underline{v}$, is essential for the derivation of a sufficient condition for stability at the critical point.

We first derive some additional necessary conditions for stability in a critical case of neutral equilibrium. These are obtained immediately by substituting from (7) into (4). In order that equilibrium may be stable, we must obviously have

$$P_3[a_h \underline{u}_h] = 0 \ , \tag{13}$$

$$P_4[a_h \underline{u}_h] \geqq 0 \ , \tag{14}$$

for arbitrary values of the amplitudes a_h.

The derivation of a sufficient condition for stability at a critical point is based on the fact that any kinematically admissible displacement field $\underline{u}$ may be written in the form

$$\underline{u} = a_h \underline{u}_h + \underline{v} \ , \text{ where } T_{11}[\underline{u}_h, \underline{v}] = 0 \ , \quad h = 1, 2, \ldots, n. \tag{15}$$

We now investigate the minimum of the energy increment functional

$$P[\underline{u}] = P[a_h \underline{u}_h + \underline{v}] \tag{16}$$

as a functional of $\underline{v}$, orthogonal to all buckling modes, for fixed values of the amplitudes a_h. The results of this analysis are expressed by two inequalities [9, ch. 2].

$$P[a\underline{u}_h + \underline{v}] \geqq P_4[a_h \underline{u}_h] - \frac{1}{1-\varepsilon} P_2[a_h a_i \underline{v}_{hi}] + O(a^5) \ , \tag{17}$$

$$\underset{(a_h = \text{const.})}{\text{Min}} P[a \underline{u}_h + \underline{v}] \leqq P_4[a_h \underline{u}_h] - \frac{1}{1+\varepsilon} P_2[a_h a_i \underline{v}_{hi}] + O(a^5), \tag{18}$$

where $O(a^5)$ stands for a term which tends to zero as a homogeneous polynomial of the fifth degree in the amplitudes a_h, and ε is a positive constant which may be made as small as we please by choosing the neighborhood $\{g, g'\}$ sufficiently small. The displacement field $\underline{v}_{hi}$ $(h, i = 1, 2, \ldots, n)$ is the unique solution of the non-homogeneous quadratic minimum problem

$$P_2[\underline{v}] + \frac{1}{2} P_{111}[\underline{u}_h, \underline{u}_i, \underline{v}] = \text{Min.} \tag{19}$$

under the side conditions

$$T_{11}[\underline{u}_h, \underline{v}] = 0, \; h = 1, 2, \ldots, n. \tag{20}$$

The trilinear functional in the second term of (19) is defined by the aggregate of all terms in the expansion of $P[\underline{u}_h + \underline{u}_i + \underline{v}]$ which are linear in all three constituents $\underline{u}_h$, $\underline{u}_i$, $\underline{v}$[3]. The conditions for a minimum of (19) are equivalent to a set of non-homogeneous linear differential equations and boundary conditions for the components of $\underline{v}_{hi}$.

Since ε in (17) and (18) may be selected as small as we please, stability conditions may be formulated in terms of the quartic form in the amplitudes of the buckling modes[4]

$$P_4[a_h \underline{u}_h] - P_2[a_h a_i \underline{v}_{hi}] = A_{hijk} a_h a_i a_j a_k . \tag{21}$$

A further necessary condition for stability, sharper than (14), is that (21) must be non-negative for arbitrary values of a_h. A sufficient condition is that the quartic form (21) is positive definite.

It is perhaps somewhat surprising that the necessary and sufficient conditions based on the quartic form (21) are apparently not to be found in the mathematical literature on the calculus of variations. The entirely similar conditions for a proper minimum of a function of a finite number of independent variables are even missing in Hancock's treatise on maxima and minima of such functions [10]. A possible explanation may be that a violation of the necessary condition (13) is, of course, the general case in a mathematical sense. A further investigation is then unnecessary. In bifurcation problems of elastic stability, however, condition (13) is quite often satisfied, due to certain symmetry properties of the structure, and the analysis leading to (21) is then indispensable.

4. Post-buckling behavior

In the analysis of the behavior of the structure after buckling it is of course essential to take account of the fact that the energy increment functional $P[\underline{u}]$ and each term in its expansion (4) depend on the load factor λ as a parameter. A detailed analysis has shown [9, ch. 3] that it is sufficient, to a first approximation, to allow for the dependence on λ only in the second variation of the energy. We write

$$P_2[\underline{u};\lambda] = P_2[\underline{u};\lambda_1] + (\lambda-\lambda_1)P_2'[\underline{u};\lambda_1] + \ldots =$$

$$= P_2[\underline{u}] + (\lambda-\lambda_1)P_2'[\underline{u}] + \ldots \tag{22}$$

An arbitrary displacement field $\underline{u}$ is again written in the form (15), and the energy increment functional is again minimized with respect to $\underline{v}$ (orthogonal to all buckling modes) for fixed values of the amplitudes a_h of the buckling modes. This minimum is, to a first approximation, a simple algebraic function of the amplitudes a_h [9, ch. 3]

$$P(a_h;\lambda) = (\lambda-\lambda_1)F_2'(a_h) + F_r(a_h) \quad , \tag{23}$$

where

$$F_2'(a_h) = P_2'[a_h\underline{u}_h] \quad . \tag{24}$$

The second term in (23) is homogeneous of degree $r = 3$ or 4. If the necessary condition (13) for stability at the bifurcation point is violated, we have $r = 3$ and

$$F_3(a_h) = P_3[a_h\underline{u}_h] = A_{hij}\,a_h a_i a_j \quad . \tag{25}$$

If condition (13) is satisfied, we have $r = 4$ and

$$F_4(a_h) = P_4[a_h\underline{u}_h] - P_2[a_h a_i \underline{v}_{hi}] = A_{hijk}\,a_h a_i a_j a_k \quad . \tag{26}$$

Our result (23) is of course an approximation, but it represents the dominant terms in an asymptotic expansion of the energy in the vicinity of the bifurcation point.

The configurations of equilibrium in the neighborhood of the bifurcation point are now specified by stationary values of (23) as a function of the amplitudes a_h. These configurations are stable, if and only if these stationary values are proper minima. One solution is of course the fundamental state $a_h = 0$ ($h = 1, 2, \ldots n$) and it is stable for $\lambda < \lambda_1$ and unstable for $\lambda > \lambda_1$. The non-zero solutions represent the adjacent configurations of equilibrium. These are always unstable at loads below the critical load (if they exist at all). The converse can only be proved in the case of a simple buckling mode ($n=1$). In this case adjacent configurations of equilibrium at loads above the buckling load are always stable (if they exist). In problems where multiple buckling modes occur ($n \geqq 2$), however, all adjacent configurations may be unstable. An example of the latter type will be discussed in section 6.

The results are illustrated in figs. 3, 4 and 5, in the simplest case of a simple buckling mode ($n = 1, a_1 = a$), by the curves labelled

I and II. The forms (25) and (26) consist here each of a single term $A_3 a^3$ or $A_4 a^4$. Figs. 3 and 4 represent cases of an unstable point of bifurcation, established by a non-zero value of A_3 and a negative value of A_4 respectively. Fig. 5 represents a case of a stable bifurcation point. Stable branches are again indicated by drawn curves, unstable branches by dotted curves.

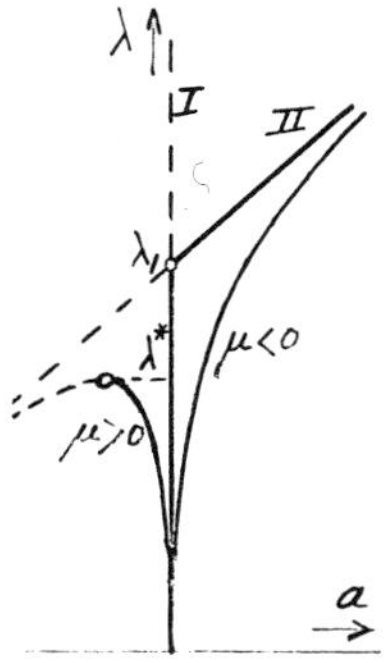

Fig. 3
$A_3 > 0$

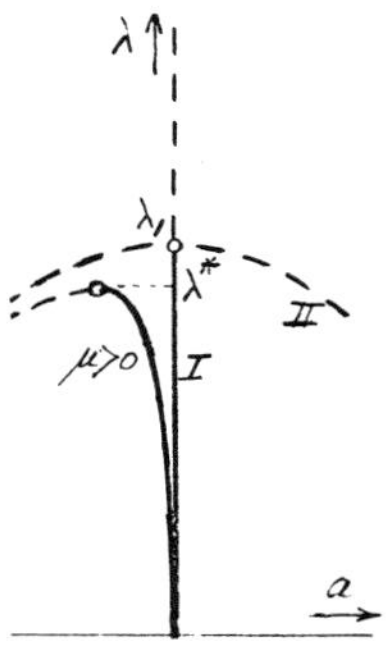

Fig. 4
$A_3 = 0, \ A_4 < 0$

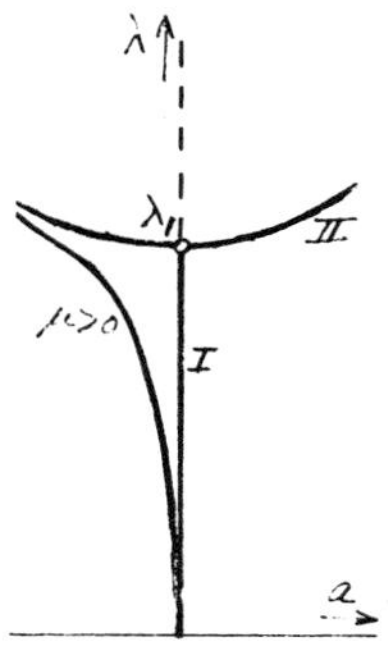

Fig. 5
$A_3 = 0, \ A_4 > 0$

5. Effect of initial imperfections

The general theory of the initial stage of post-buckling behavior explains the widely divergent behavior of different structures after buckling. This variation in behavior is due to differences in the stability or instability of equilibrium at the critical bifurcation point. It does not yet explain, however, why this bifurcation point cannot even be approached in some actual structures, e.g. cylindrical shells under axial compressive loads. In order to obtain a proper understanding of the behavior of such structures, due allowance must be made for imperfections which are always present in actual structures.

The primary effect of initial imperfections is that the fundamental state of the idealized perfect structure, described by the displacement vector $\underline{U}(\lambda)$ from the undeformed state, does not represent a configuration of equilibrium of the actual imperfect structure. The energy increment in the transition from the fundamental state I to some adjacent configuration II, specified by (4) for the idealized model, must be augmented by a linear term for the actual structure. In the case of small initial imperfections, this correction will also be linear in these imperfections. The displacement vector $\underline{U}(\lambda)$ and the associated stresses in the fundamental state of the idealized model are in most problems approximately proportional to the load factor λ . The correction to (4) which allows for the effect of small initial imperfections then has the form

$$\mu\lambda Q_1'[\underline{u}] \quad , \tag{27}$$

where μ is a parameter which specifies the magnitude of the imperfections, and $Q_1'[\underline{u}]$ is a linear functional of the displacement field $\underline{u}$ from state I to state II. The form of this functional depends, of course, on the type of initial imperfections under consideration.

The analysis now proceeds along the same line as before. An arbitrary displacement field $\underline{u}$ is again written in the form (15), and the energy increment functional is again minimized with respect to $\underline{v}$ for prescribed values of the amplitudes a_h of the buckling modes. This minimum is, again to a first approximation the algebraic function of the amplitudes a_h [9, ch. 4]

$$P^*(a_h;\lambda) = (\lambda-\lambda_1)F_2'(a_h) + F_r(a_h) + \mu\lambda F_1'(a_h) \quad , \tag{28}$$

where the first pair of terms is the same pair as in (23), and

$$F_1'(a_h) = Q_1'[a_h \underline{u}_h] \quad . \tag{29}$$

The configurations of equilibrium are again specified by stationary values of (28) as a function of the amplitudes a_h . The nonlinear equations now yield, in general, a number of disconnected branches of solutions. The branches of physical interest are those which reduce to zero values of the amplitudes for $\lambda = 0$. The stability of the equilibrium configurations is again ensured if and only if the stationary value of the energy is a proper minimum.

The results are again illustrated in figs. 3, 4, 5 in the case of a simple buckling mode $(n = 1,\ a_1 = a)$. The linear function (29) is here

$$F_1'(a_h) = B_1'a \quad , \qquad B_1' = Q_1'[\underline{u}_1] \quad . \tag{30}$$

Without loss in generality we may assume a positive value of the constant B_1' . The curves for $\mu < 0$ in figs. 4 and 5 are obtained by reflection with respect to the λ-axis.

The most important feature in these figures is the appearance of limit points λ^* in the cases of an unstable point of bifurcation, which limit points represent critical loads for the imperfect structure, smaller than λ_1 . The difference between λ_1 and λ^* may be considerable, even for small imperfections, because the curve for λ^*/λ_1

as a function of the imperfection parameter μ has a vertical tangent at the point $\lambda^*/\lambda_1 = 1$, $\mu = 0$ (fig. 6). This result explains why experimental values for the critical load may be considerably smaller than the theoretical values for the idealized model. It also explains the wide scatter in test values for such structures.

The discussion of (28) is far more complicated in the case of a structure with multiple buckling modes for the idealized model. An example will be considered in the next section.

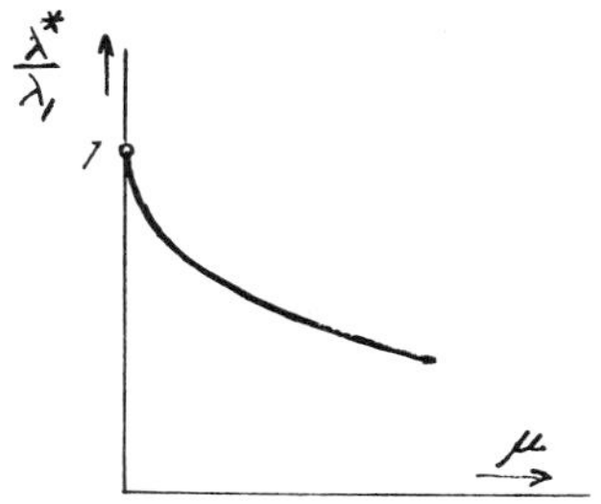

Fig. 6
Critical load as a function of imperfection parameter

6. Cylindrical shells under axial compression

The nonlinear buckling problem of cylindrical shells under axial compression has been discussed by many writers in the past twenty years. A review of this literature will not be attempted here. A comprehensive article by Fung and Sechler [11] is a convenient reference, in addition to the Russian books mentioned before [7, 8]. We shall confine our attention to the application of the foregoing general theory [9, ch. 7] and to some recent results of a different approach to the theory of imperfect shells.

Cylindrical shells of medium length (say, $1 < \ell/R < 20$) have a large number of simultaneous buckling modes in axial compression [12]. The axisymmetric buckling mode is characterized by the radial deflection

$$w = \sin p_0 x/R , \tag{31}$$

where the wave number p_0 is given by

$$p_0 = [12(1-\nu)^2]^{1/4} (R/h)^{1/2} \tag{32}$$

Here h and R are the shell thickness and radius, and ν is Poisson's ratio. Nonsymmetrical buckling modes may also be characterized by their radial deflections

$$w = \sin px/R \cos n\theta \tag{33}$$

where n is the (integral) number of circumferential waves, and p is the corresponding axial wave number, obtained from the equation

$$p^2 - p_0 p + n^2 = 0 . \tag{34}$$

These results are derived on the assumption that p is large compared with unity and that the boundary conditions at the ends of the shell may therefore be ignored. Eq. (34) is illustrated by fig. 7, where p_{k_1} and p_{k_2} are the two roots of (34) for $n = k$. The combination of wave numbers n and p_{n_2} satisfies our assumptions only

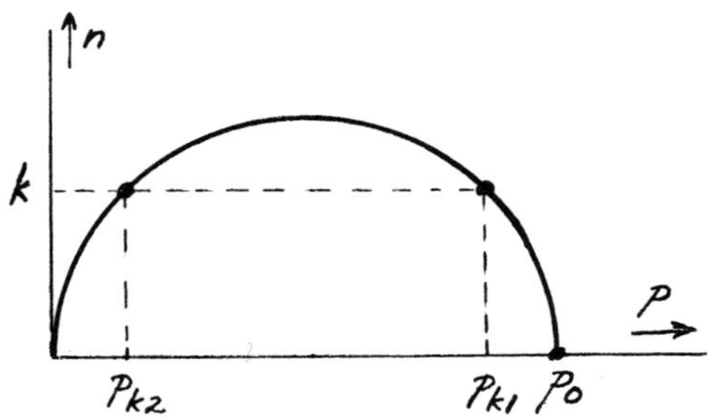

Fig. 7
Axial and circumferential wave numbers for cylindrical shell in axial compression

if n is not too small. The ignoration of boundary conditions implies that the axial sine functions in (31) and (33) may be replaced by the corresponding cosine functions. We may obviously also replace $\cos n\theta$ in (33) by $\sin n\theta$.

The radial deflection in the general solution of the equations of neutral equilibrium at the bifurcation point thus consists of two axisymmetric terms and eight nonsymmetric terms for each sufficiently large integral value of n, subject to the restriction $n \leqq 1/2\, p_0$. A careful examination of the complete analysis in [9, ch. 7] shows that it is sufficient to restrict our attention to one axisymmetric term and two properly selected nonsymmetric terms for each value of n. Hence we take as a sufficiently general linear combination of buckling modes

$$w = a_0 \sin p_0 x/R + \sum_n [a_{n1} \sin p_{n1} x/R + c_{n2} \cos p_{n2} x/R] \cos n\theta , \tag{35}$$

and corresponding expressions for the axial and tangential displacement components.

Equilibrium at the bifurcation point now appears to be unstable due to a violation of the necessary condition for stability (13). In fact, a non-vanishing result for the cubic form (25) is easily obtained by elementary, if somewhat lengthy, calculations

$$F_3(a_h) = P_3[a_h u_h] = \frac{3\pi}{4}\frac{Eh\ell}{R^2} a_0 \sum_n n^2 a_{n1} c_{n2} \qquad (36)$$

where E is Young's modulus.

It is convenient to introduce a nondimensional load parameter $\lambda = \sigma/E$, where σ is the (average) compressive stress due to the axial compressive load. The critical load parameter at the bifurcation point is [12]

$$\lambda_1 = \frac{1}{\sqrt{3(1-\nu^2)}} \frac{h}{R} . \qquad (37)$$

The first term in (23) and (28) is now easily evaluated

$$(\lambda-\lambda_1) F_2'(a_h) = -\frac{\pi}{4}(\lambda-\lambda_1)\frac{Eh\ell}{R}\left[2p_0^2 a_0^2 + \sum_n (p_{n1}^2 a_{n1}^2 + p_{n2}^2 c_{n2}^2)\right] \qquad (38)$$

The conditions for a stationary value of (23), where the first term is specified by (38) and the second term by (36), are expressed by the equations

$$-4(\lambda-\lambda_1) R p_0^2 a_0 + 3 \sum_n n^2 a_{n1} c_{n2} = 0 , \qquad (39)$$

$$-2(\lambda-\lambda_1) R p_{n1}^2 a_{n1} + 3n^2 a_0 c_{n2} = 0 , \qquad (40)$$

$$-2(\lambda-\lambda_1) R p_{n2}^2 c_{n2} + 3n^2 a_0 a_{n1} = 0 . \qquad (41)$$

The last pair of equations holds for each sufficiently large integral value of n, subject to the restriction $n < 1/2\, p_0$.

Equations (39)-(41) have of course always the trivial zero-solution corresponding to the known fundamental state. Each pair of equations (40), (41) may be regarded as a system of homogeneous and linear equations for a_{n1} and c_{n2}. These have again always the trivial zero-solution. A non-zero solution of (39) then exists for $\lambda = \lambda_1$, and this first non-trivial solution leaves a_0 arbitrary.

A non-vanishing solution of a pair of equations (40), (41) also exists if the determinant vanishes

$$4(\lambda-\lambda_1)^2 R^2 p_{n1}^2 p_{n2}^2 - 9n^4 a_0^2 = 0 , \tag{42}$$

and this condition is independent of n . Eq. (42) has two solutions for the amplitude a_0 of the symmetric buckling mode

$$a_0 = \pm\frac{2}{3}(\lambda-\lambda_1)R . \tag{43}$$

Each pair of equations (40), (41) then has a non-zero solution with a ratio of the amplitudes a_{n1} and c_{n2} specified by

$$p_{n1}a_{n1} = \pm p_{n2}c_{n2} , \tag{44}$$

where the upper sign corresponds to the upper sign in (43), and vice versa. Substituting from (43) and (44) into (39), we obtain finally

$$\sum_n p_{n1}^2 a_{n1}^2 = \frac{8}{9}(\lambda-\lambda_1)^2 R^2 p_0^2 . \tag{45}$$

Rather to our surprise we have now obtained a complete solution of our nonlinear equations of equilibrium (39)-(41). It appears from (43)-(45) that this solution is far from determinate. This indeterminacy is undoubtedly due to our restriction to a first approximation in the energy expression (23). The overall shortening of the cylindrical shell, however, is uniquely determined by (43)-(45). Denoting the specific overall shortening by ε , its critical value at the bifurcation point by $\varepsilon_1 = \lambda_1$, we obtain from (43)-(45) the relation

$$\frac{\varepsilon}{\varepsilon_1} = \frac{\lambda}{\lambda_1} + \frac{2}{3}\left(\frac{\lambda}{\lambda_1} - 1\right)^2 . \tag{46}$$

The stability of the equilibrium configurations is governed by the second variation of (23) in terms of variations Δa_h of the amplitudes Omitting a positive factor $\pi Eh\ell/4R^2$, we obtain after substitution from (36) and (38)

$$\begin{aligned} &-(\lambda-\lambda_1)R\,\{p_0^2(\Delta a_0)^2 + \sum_n [p_{n1}^2(\Delta a_{n1})^2 + p_{n2}^2(\Delta c_{n2})^2]\} + \\ &+ 3\sum_n n^2[a_0\Delta a_{n1}\,\Delta c_{n2} + a_{n1}\,\Delta a_0\,\Delta c_{n2} + c_{n2}\Delta a_0\,\Delta a_{n1}] . \end{aligned} \tag{47}$$

It is easily confirmed that this second variation is positive definite in the fundamental state $a_0 = a_{n1} = c_{n2} = 0$ for $\lambda < \lambda_1$, and negative definite for $\lambda > \lambda_1$. In all adjacent configurations of equilibrium for $\lambda < \lambda_1$ the predictions of the general theory are confirmed by an indefinite second variation (47). Furthermore, it appears from (47) that equilibrium is also unstable in the adjacent configurations of equilibrium at loads equal to or above the buckling load. The only stable solution of the equations of equilibrium (39)-(41) is the fundamental state for $\lambda < \lambda_1$.

We shall now consider an example of imperfections of the shell. We choose the simplest possible case of axisymmetric imperfections in the shape of the axisymmetric buckling mode, viz.

$$w_0 = \mu h \sin p_0 x/R \ , \qquad (48)$$

where μ is the amplitude of the imperfections as a fraction of the wall thickness. The last term in (28) takes the form

$$\mu\lambda F_1'(a_h) = -\pi\mu\lambda \frac{Eh^2 \ell}{R} p_0^2 a_0 \ . \qquad (49)$$

The equations of equilibrium (40) and (41) remain unchanged, whereas (39) is now replaced by

$$-4(\lambda-\lambda_1)Rp_0^2 a_0 + 3\sum_n n^2 a_{n1} c_{n2} - 4\mu\lambda h R p_0^2 = 0 \qquad (50)$$

The solution of (40), (41) and (50) which tends to zero for $\lambda \to 0$, i.e. the branch of physical interest, is specified by the zero-solution of (40) and (41), and by the corresponding solution of (50)

$$a_0 = \frac{\lambda}{\lambda_1 - \lambda} \mu h \qquad (51)$$

This configuration is stable as long as the second variation (47) is positive definite for the value of a_0 specified by (51). The stability limit of the imperfect shell is reached when this second variation is semi-definite, i.e. for

$$9n^4 a_0^2 = 4(\lambda-\lambda_1)^2 R^2 p_{n1}^2 p_{n2}^2 \ , \qquad (52)$$

and equilibrium is unstable for larger values of λ . The buckling load parameter λ^* of the imperfect shell is therefore specified by the equation

$$\frac{\lambda^*}{\lambda_1} |\mu| = \frac{2}{3\sqrt{3(1-\nu^2)}} (1 - \frac{\lambda^*}{\lambda_1})^2 \ , \qquad (53)$$

and the curve of λ^*/λ_1 as a function of μ is again of the type sketched in fig. 6. Instability occurs here with respect to any pair of nonsymmetric modes in (35).

The striking reduction in critical load, even for small values of the imperfection parameter μ , is illustrated in fig. 8 by the curve labelled "general theory." The physical reason for this behavior is fairly obvious. Axial loading of the symmetric imperfect shell intro-

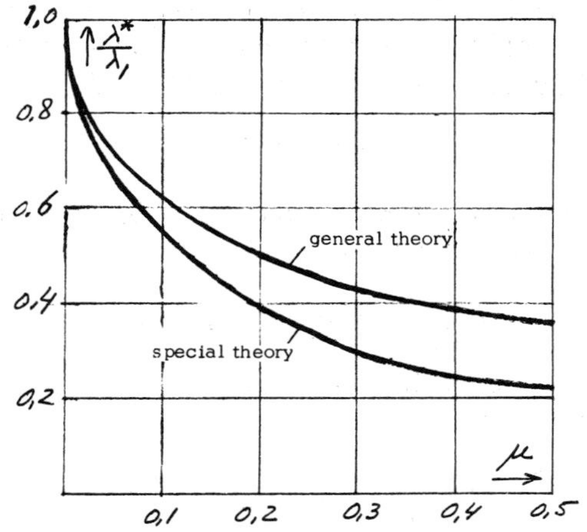

Fig. 8
Critical load for cylindrical shell with axisymmetric imperfections (ν = 0. 272)

duces circumferential membrane stresses which are a periodic function of the axial coordinate with the same wave number p_0 as the imperfections. Nonsymmetric buckling of a cylindrical shell is precipitated by compressive circumferential stresses and equally hampered by tensile stresses. If a nonsymmetric buckling mode exists for which the deflections are <u>larger</u> in the region of <u>compressive</u> circumferential stresses than in the region of tensile stresses, the resulting effect of the periodic stresses is <u>destabilizing.</u> It may be verified that a pair of nonsymmetric modes in (32) for the same value of n satisfies this criterion.

Although the general theory is rigorous in an asymptotic sense, it suffers from the drawback that higher order terms have been neglected in the energy expressions (23) and (28). The results depicted in fig. 8 may therefore be not too reliable for higher values of the imperfection parameter. It is now a fortunate circumstance that the axial symmetry of the imperfect shell enables us to compare the results of the general theory with those of an entirely different approach. Lack of space prevents us from giving all details of the alternative analysis in the present paper, but we shall indicate its main features.

We consider the axisymmetric shell whose middle surface equation in cylindrical coordinates x, r, θ reads

$$r = R + \mu h \sin p\, x/R \ , \qquad (54)$$

where the nondimensional parameter μ may be of order of magnitude unity. The nonlinear theory of shallow shells is now applicable to the shell (54) in the case of axisymmetric load. The resulting periodic stress distribution in the shell under axial load is an elementary solution of the shallow shell equations. The stability of the symmetric configuration of equilibrium may now be investigated by means of the linear theory of elastic stability. If we restrict our attention to a large number of circumferential waves in nonsymmetric modes, we may again apply the equations of shallow shell theory in the stability investigation. The resulting equations are complicated by the fact that the coefficients of some terms are periodic functions of the axial coordinates with the characteristic wave number p of the shell equation (54). A rigorous solution is certainly intractible, if at all obtainable. An approximate solution may again be obtained by the energy method. By substituting an assumed buckling mode in the second variation of the potential energy we may calculate an upper bound for the critical load. The same approximation is found by applying Galerkin's method to the equation of equilibrium for the normal deflection. The equation of compatibility for the stress function must of course be solved rigorously in Galerkin's procedure, in terms of the assumed buckling mode.

A numerical evaluation has been obtained in the case that the wave number p in (54) coincides with the wave number p_0 (32) of the axisymmetric buckling mode of the perfect cylindrical shell. This is of course the appropriate case for a comparison with the results of the general theory in fig. 8. The buckling mode has been assumed in the form[5]

$$w = A \sin(1/2\, p_0 x/R \mp \pi/4) \cos 1/2\, p_0 \theta \ , \qquad (55)$$

where the upper (lower) sign corresponds to a positive (negative) value of μ in (54). In spite of the crudity of the approximation of the buckling mode by a single term in (55), the resulting upper bound for the critical load is even smaller than the prediction by the general theory (cf. the curve labelled "special theory" in fig. 8).

Part of the work described in this paper was sponsored by the Office of Naval Research in Washington, D. C., under Contract Nonr 562(10) with Brown University, Providence, R. I.

NOTES

1. The analysis in sections 2 and 3 is equally applicable to limit points, the investigation in sections 4 and 5 is restricted to bifurcation problems.

2. For plates and shells the functional (1) is the sum of a surface integral and a line integral along the boundary. For bars it is the sum of a line integral and discrete boundary terms. In these cases the integrands also depend on certain second derivatives of the displacement field $\underline{u}$. The definition of a neighborhood of the fundamental state is then modified by adding to (2) bounds on the moduli of the second derivatives in question.

3. The functional $T_2[\underline{u}]$ is of course appropriately modified in the two-dimensional plate and shell problems and in the one-dimensional bar problem.

4. The quartic form (21), which decides on stability, is of course independent of the particular choice for the auxiliary quadratic functional $T_2[\underline{u}]$.

5. Strictly speaking, the assumed buckling mode (55) is only possible for integral values of $1/2\ p_0$.

REFERENCES

1. K. O. Friedrichs and J. J. Stoker, The nonlinear boundary value problem of the buckled plate. Am. Journ. of Math. 63 (1941), 839.

2. R. V. Southwell, On the general theory of elastic stability. Phil. Trans. Roy. Soc. London A213 (1913), 187.

3. C. B. Biezeno and H. Hencky, On the general theory of elastic stability. Proc. Roy. Neth. Acad. Sci. Amsterdam 31 (1928), 569; 32 (1929), 444.

4. E. Trefftz, Über die Ableitung der Stabilitätskriterien des elastischen Gleichgewichts aus der Elastizitätstheorie endlicher Deformationen. Proc. 3. Int. Congr. Appl. Mech. Stockholm 3 (1930), 44.

5. E. Trefftz. Zur Theorie der Stabilität des elastischen Gleichgewich Geitschr. f. angew. Math. u. Mech. 13 (1933), 160.

6. A. E. Green and J. E. Adkins, Large elastic deformations, ch. 9 (Stability), Clarendon Press, Oxford (1960).

7. A. S. Vol'mir. Thin plates and shells (in Russian). Moscow (1960).

8. Kh. M. Mushtari and K. Z. Galimov, Nonlinear theory of thin elastic shells (in Russian). Kazan (1957). English translation by J. Morgenstern and J. J. Schorr-Kon. Jerusalem (1961). Available from the Office of Technical Services, U. S. Department of Commerce, Washington 25, D. C.

9. W. T. Koiter, Over de stabiliteit van het elastisch evenwicht (in Dutch with English summary). Thesis Delft. H. J. Paris, Amsterdam (1945).

10. H. Hancock, Theory of maxima and minima. Ginn and Company, Boston (1917).

11. Y. C. Fung and E. E. Sechler, Instability of thin elastic shells, Structural Mechanics, Pergamon Press, OXford, London, New York, Paris (1960), 115-168.

12. S. P. Timoshenko, Theory of elastic stability, McGraw-Hill, New York (1936). Second edition in collaboration with J. M. Gere (1961), ch. 11.

ABSTRACTS

THE EFFECTS OF NON-LINEARITY ON STABLE COUPLED DIFFERENCE EQUATIONS*

Bart J. Daly

The coupled set of nonlinear partial differential equations

$$\frac{\partial u}{\partial t} = - \frac{\partial v}{\partial x} + f\,\delta x \frac{\partial}{\partial x}\left(|u| \frac{\partial u}{\partial x}\right)$$

$$\frac{\partial u}{\partial t} = - p^2 \frac{\partial u}{\partial x}$$

which is a simplification of the equations used in a certain numerical method of representing one-dimensional compressible fluid flow, was differenced, programmed and run through many thousands of cycles of calculation on a computer. The purpose of the experiments was to study the effects of the non-linearity upon the instabilities introduced by the differencing.

Fluctuations initially present in the system show an exponential growth in amplitude for a time, but are prevented from unbounded growth by the diffusing effect of the non-linearity. Eventually an equilibrium amplitude is established and henceforth only minor fluctuations about this amplitude are observed. The level of the equilibrium amplitude and the amount of overshoot which occurs prior to its attainment increase with the number of cells used in the differencing scheme and decrease approximately as the square of the coefficient of the nonlinear term. A Fourier analysis of the profile of the fluctuations indicates that at late times one mode of oscillation is dominant.

A Taylor expansion of the difference equations is helpful in demonstrating the mechanism of the diffusion process and leads to a first order approximation to the equilibrium amplitude which agrees quite well with the experimental data. A more sophisticated approach makes use of Kryloff and Bogoliuboff's method of first approximation and the observed fact that a single mode of oscillation dominates at late times. This technique sheds light on the time variation of the amplitude and phase of the various modes and yields an estimate of the equilibrium value of the amplitude of the dominant mode.

An extension of the one dominant mode analysis indicates that the approach to equilibrium should be asymptotic in contradiction to the overshoot observed in the experiments. It is possible that the amount of overshoot is an indication of how much the system departs from a dominant frequency vibration since in the larger systems, where the overshoot is most pronounced, there exist many modes of secondary magnitude.

*This work was performed under the auspices of the United States Atomic Energy Commission.

MULTI-STAGE ALTERNATING DIRECTION METHODS

Jim Douglas, Jr., A. O. Garder, Jr., and Carl Pearcy

In the numerical solution of partial differential equations, many equations of the form

$$(A_1 + A_2 + A_3 + A_4)x = y_0 \tag{1}$$

are encountered, where each A_i is a positive definite $n \times n$ matrix (or the zero matrix), y_0 is some given vector, and it is desired to determine x . The above equation can be written as

$$(B + C)x = y_0 \tag{2}$$

where $B = A_1 + A_2$, $C = A_3 + A_4$, and one of the authors (Pearcy, "On Convergence of Alternating Direction Procedures," to appear) has recently established conditions under which the Peaceman-Rachford (P-R) alternating direction procedure for solving equation (2) converges.

If one applies the Peaceman-Rachford procedure to (2), then at the kth step in the iteration one is faced with solving the equations

$$\begin{aligned} A_1 + A_2 + a_{k+1})x_{k+\frac{1}{2}} &= -[(C - a_{k+1})x_k - y_0] \\ A_3 + A_4 + a_{k+1})x_{k+1} &= -[(B - a_{k+1})x_{k+\frac{1}{2}} - y_0] \end{aligned} \tag{3}$$

for $x_{k+\frac{1}{2}}$, x_{k+1} , and the P-R iterative procedure can be applied to each of these equations in turn in order to determine $x_{k+\frac{1}{2}}$ and x_{k+1} . Thus one can solve equation (1) by using the P-R technique as an "outer iteration" at each step of which one or two P-R "inner iterations" are performed. However, one cannot hope to iterate on (3) until the fixed point is reached. Thus the purpose of this paper is to give conditions on the residuals (which arise from the failure to solve equations (3) exactly) such that the technique outlined above for solving (1) is convergent. Furthermore in the commutative case,

$$BC = CB \ , \quad A_1 A_2 = A_2 A_1 \ , \quad A_3 A_4 = A_4 A_3 \ ,$$

an explicit estimate of the number of calculations needed to reduce the error $\|x_n - x\|$ to a certain level can be obtained, and this estimate is of the same order as estimates obtained earlier for other alternating direction methods.

As examples of problems to which this technique might be applied, consider the following two. First, let A be the usual discrete analogue of the self-adjoint elliptic differential equation

$$\nabla\cdot(p\nabla u) - qu = f \ , \quad (x_1, x_2, x_3) \in R \ ,$$

$$u = g \ , \quad (x_1, x_2, x_3) \in \partial R \ .$$

Then, let

$$A_i \sim \frac{\partial}{\partial x_i}\left(p\frac{\partial}{\partial x_i}\right) \ , \quad i = 1, 2, 3$$

$$A_4 = 0 \ .$$

The procedure outlined above thus converges for general regions R and arbitrary $p > 0$, $q \geq 0$. Second, consider the nine point analogue of Laplace's equation in the plane:

$$-A = 4\Delta_x^2 + 4\Delta_y^2 + 2\Delta_\xi^2 + 2\Delta_\eta^2 \ ,$$

where ξ and η are the rotated coordinates at 45° to x and y and Δ_x^2 is the centered, divided second difference with respect to x , etc. Let

$$A_1 \sim -4\Delta_x^2 \ , \quad A_2 \sim -4\Delta_y^2 \ , \quad A_3 \sim -2\Delta_\xi^2 \ ,$$

$$A_4 \sim -2\Delta_\eta^2 \ .$$

Again convergence can be ensured.

SINGULARITIES IN SUPERSONIC FLOWS PAST BODIES OF REVOLUTION

Ernst W. Schwiderski

We are concerned with the purely or non-purely supersonic flow of an ideal and polytropic gas past an axisymmetric body of finite dimensions, which is governed by Euler's equations of motion in the combined form of two quasi-linear differential equations

$$\left.\begin{aligned} (1-\frac{u^2}{c^2})u_z - \frac{uv}{c^2}(v_z+u_r) + (1-\frac{v^2}{c^2})v_r + \frac{v}{r} &= 0 \\ v_z - u_r &= 0 \end{aligned}\right\} \qquad (1)$$

In Eqs. (1) z and r denote physical coordinates such that the z-axis coincides with the axis of the body and the direction of the undisturbed flow. The flow velocity is represented by (u,v) and the speed of sound by c, where

$$c^2 = \frac{\gamma-1}{2}\left(1+\frac{2}{\gamma-1}\frac{1}{M_0^2} - u^2 - v^2\right) \qquad (2)$$

Here $\gamma > 1$ denotes the adiabatic exponent of the gas and $M_0 > 1$ the Mach number at infinity.

The supersonic flow remains undisturbed in front of the so-called shock wave, satisfies well-known jump conditions at the shock line, and continues in a disturbed state past the body. Since the disturbance caused by a body of finite dimensions must disappear at infinity, the shock is determined to vanish at a certain point where the shock line becomes characteristic. At this point originates a "limiting characteristic," which bounds the dependence of the disturbed flow upon the shape of the shock line.

For an asymptotic integration of Eqs. (1) we may assume that the body generates a shock wave σ and a limiting characteristic L of the general forms

$$z = \sigma(r) = -1 + r\sqrt{M_0^2 - 1} + h(r) \qquad (3)$$

and

$$z = L(r) = \sigma(r) + H(r) , \qquad (4)$$

where $h(r)$ measures the z-distance between the shock wave and the asymptotic Mach line $z = -1 + r\sqrt{M_0^2 - 1}$ and $H(r)$ that between the shock wave and the limiting characteristic. Under the assumption that $h(r)$ is twice differentiable and sufficiently monotonic the shock relations at the shock line σ can be expanded in the form

$$\left.\begin{aligned} u &= 1 + u_{10}h'(r) + u_{20}h'^2(r) + \ldots \\ v &= \phantom{1 +{}} v_{10}h'(r) + v_{20}h'^2(r) + \ldots \end{aligned}\right\} . \qquad (5)$$

After introducing the new variable

$$\xi = \frac{z - \sigma(r)}{H(r)} \qquad (\xi = 0 \text{ on } \sigma, \quad \xi = 1 \text{ on } L) \qquad (6)$$

an asymptotic solution can be obtained in the form

$$\left.\begin{aligned} u &= 1 + u_1(\xi)h'(r) + u_2(\xi)h'^2(r) + \ldots \\ v &= \phantom{1 +{}} v_1(\xi)h'(r) + v_2(\xi)h'^2(r) + \ldots \end{aligned}\right\} , \qquad (7)$$

where $u_1(\xi) = -v_1(\xi)/\sqrt{M_0^2 - 1}$ yields the expansion

$$u_1(\xi) = u_{11} + u_{11}a_1\sqrt{1 - \xi} + \ldots \qquad (8)$$

at the limiting characteristic $\xi = 1$. In Eq. (8) u_{11} and A denote certain known functions of the limit

$$\alpha = -\lim_{r \to \infty} \frac{d \log h}{d \log r} , \qquad (9)$$

Simultaneously with the solution (7) and (8) one obtains the leading asymptotic term of the limiting characteristic (4), that is,

$$H(r) = \begin{cases} L_1 h(r) + \ldots & \text{for } 0 < \alpha \leq \infty \\ L_1 r h'(r) + \ldots & \text{for } 0 \leq \alpha < \infty \end{cases} \qquad (10)$$

where L_1 is a known function of the limit α.

The leading asymptotic term (8) of Eqs. (7) reveals the limiting characteristic as a carrier of square-root singularities in the derivatives of the flow velocity. Thanks to the explicit terms (10) it becomes possible to locate the limiting characteristic beyond a certain distance from the axis of symmetry. This yields the chance to continue numerically the characteristic and the corresponding transonic flow down to the body. Generalized finite differences, which have proved so extremely successful in the construction of portions of subsonic, supersonic, and transonic flows, provide the necessary tools for the construction of the entire transonic flow.

Since the singularity at the limiting characteristic exhibits a rapid change of pressure and density, it has been conjectured that this should be visible in photographs which show the existence of shock waves. The examination of available Schlieren photographs has confirmed this conjecture. A line, although very thin, which converge toward the shock wave and intersects the body at an upper nose point, is visible as mathematically predicted.

TRANSFORMATIONS AND NONLINEAR PARTIAL DIFFERENTIAL EQUATIONS

William F. Ames

Research on nonlinear problems of heat transfer shows that a number of ingenious transformations have been introduced, somewhat haphazardly, into the literature of nonlinear partial differential equations. Typically, these linearize or perform other reductions in complexity. This paper demonstrates the applicability of some of these transformations to numerical procedures.

Typical of these transformations are:

(a) The Boltzmann Transformation

$$\frac{\partial C}{\partial t} = \frac{\partial}{\partial x}\left(D \frac{\partial C}{\partial x}\right) \qquad (1)$$

is reduced to the (nonlinear) ordinary differential equation

$$\frac{d}{d\eta}\left(D \frac{dC}{d\eta}\right) + 2\eta \frac{dC}{d\eta} = 0 \qquad (2)$$

by the introduction of the new variable $\eta = \frac{1}{2}xt^{-\frac{1}{2}}$. The transformation proves useful if the media is infinite or semi-infinite and the boundary conditions are expressible in terms of η alone. In such cases the original initial value problem is reduced to a boundary value problem.

(b) The Kirchhoff Transformation:

If $K = K(T)$ the conduction equation takes the form

$$\rho c_p \frac{\partial T}{\partial t} = \nabla(K \nabla T) \;. \qquad (3)$$

Equation (3) is transformed to

$$\nabla^2 \psi = \frac{1}{\nu} \frac{\partial \psi}{\partial t}, \quad \nu = \frac{K}{\rho c} \qquad (4)$$

by the equation

$$\psi = \frac{1}{K_0} \int_0^T K(T)\, dT \;. \qquad (5)$$

Equation (4) is still nonlinear, but "simpler" in form. Even greater simplification is obtained for the general equation

$$\nabla(K\nabla T) = 0 \tag{6}$$

by (5) which reduces (6) to

$$\nabla^2\psi = 0 \tag{7}$$

Complications may arise, but in principle, Dirichlet conditions are amenable to such considerations.

(c) The Hopf Transformation:

When the transformation

$$u = -2\nu \frac{\partial}{\partial x}(\ln F) \tag{8}$$

is applied to the Burgers equation

$$\frac{\partial u}{\partial t} + u\frac{\partial u}{\partial x} = \nu \frac{\partial^2 u}{\partial x^2} \tag{9}$$

it is found that u satisfies (9) if F is a solution of

$$f_t = \nu F_{xx} . \tag{10}$$

While these transformations have led to solutions of important problems their full utilization has not been realized. In particular they offer means for examination of numerical methods for nonlinear partial differential equations as well as obtaining analytic solutions.

Typical results include:

(a) If in the heat conduction equation $\rho c_p = \alpha \frac{T}{T_0}$ and $K = \beta \frac{T}{T_0}$ or more generally $\rho c_p = \alpha F(T)$, $K = \beta F(T)$ an analytic solution is obtainable for the temperature subject to Dirichlet boundary conditions.

(b) The classical explicit numerical technique for the dimensionless heat conduction equation with ρc_p = constant and $K = K_0 e^{-\beta u}$ is shown to be stable for

$$r = \frac{\Delta t}{(\Delta x)^2} \leq \frac{3}{4} .$$

(c) Transformations of the Kirchhoff type reduce equation (6) to $\Delta^2\psi = 0$. Consequently they allow the variety of numerical techniques for Laplace's equation to be brought to bear on the original equation via the inverse transformation.

NONLINEAR VIBRATION ANALYSIS OF PLATES AND SHELLS[1]

Yi-Yuan Yu

An approximate method is presented for the analysis of nonlinear vibrations of plates and shells involving large deflections. The method is solely based upon the variational equation of motion in the theory of elasticity. The usual formulation of the variational equation of motion, such as given by Love, contains only the volume and surface integrals with respect to the space coordinates. Here we shall further include the integration with respect to time as a necessary part in the formulation of the equation, so that a variational approximation with respect to time may be performed on the basis of the equation if needed. This part concerning the time integration appears in the equation automatically when it is derived from Hamilton's principle.

The proposed method consists of a sequence of variational approximations based upon the variational equation of motion. The first of these involves an approximation in the thickness direction and yields an approximate system of differential equations of motion and appropriate boundary conditions for the plate or shell. When some or all of these differential equations and/or boundary conditions cannot be solved and satisfied exactly, subsequently variational approximations with respect to the remaining space coordinates and time may be further carried out. The derivation of the plate and shell equations and the steps taken in the solution of these equations therefore constitute a continuous process (which in a way consists of peeling the layers of integration off the variational equation of motion one or two at a time). Within the framework of the rather general and flexible procedure here proposed are now integrated some of the variational approximations which have been known only as isolated individual procedures, such as those of Galerkin and Trefftz.

The method is conveniently applicable to layered composite plates and shells as well as those of the homogeneous type. For illustration of the method the nonlinear free vibrations of sandwich plates and cylindrical shells are studied, with homogeneous plates and cylindrical shells as special cases. In the vibration study is included the effect of transverse shear deformation, the importance of which on linear vibrations of sandwich structures has been well established in previous publications of the author[3-6]. For nonlinear vibrations of such structures the importance of the transverse shear effect on the frequency is found to decrease with increasing vibration amplitude. It is also shown that both the transverse shear and nonlinear effects are of less importance to cylindrical shells than to plates.

In addition to plates and shells, the method appears to be further applicable to vibration and equilibrium problems of other homogeneous and composite elastic systems. Similar procedures may also be possible in still other fields of mathematical physics and engineering sciences.

1. This research was supported by the United States Air Force under contract AF49(638)-453 monitored by the Air Force Office of Scientific Research.

2. Professor, Department of Mechanical Engineering, Polytechnic Institute of Brooklyn.

3. "A New Theory of Elastic Sandwich Plates—One Dimensional Case," Journal of Applied Mechanics, vol. 26, pp. 415-421, 1959.

4. "Flexural Vibrations of Elastic Sandwich Plates," Journal of Aerospace Sciences, vol. 27, pp. 272-282, 1960.

5. "Simplified Vibration Analysis of Elastic Sandwich Plates," Journal of Aerospace Sciences, vol. 27, pp. 894-900, 1960.

6. "Vibrations of Elastic Sandwich Shells," Journal of Applied Mechanics, vol. 27, pp. 653-662, 1960.

ON CARLEMAN'S MODEL FOR THE BOLTZMAN EQUATION.

Ignace I. Kolodner

The pair (u^1, u^2) of functions defined on $D_a = R \times [0, a)$ and solving the

$$\text{IVP: (DE)} \ (D^{(0,1)} - (-1)^i D^{(1,0)}) u^i = (-1)^i ((u^1)^2 - (u^2)^2),$$

(1c) $u^i(x, 0) = \alpha^i(x)$, $i = 1, 2$, describes the velocity distribution function of a fictive one-dimensional gas whose molecules have velocities either 1 or -1 which are interchanged by collisions. The system DE, proposed by Carleman [1], presents several analogies with the Boltzmann equation. The objective of this study is to show that if 1) $\alpha^i \in C(R)$, 2) $0 \leq \alpha^i(x) \leq \alpha$, then the initial value problem has exactly one solution defined on D_∞. By a solution we mean a pair $(u^1, u^2) \in C(D_\infty)$ such that $(D^{(0,1)} - (-1)^i D^{(1,0)}) u^i \in C(D_\infty)$, and (u^1, u^2) satisfies the DE and the IC's.

1. Assume first that the α^i are not necessarily positive but that $|\alpha^i(x)| \leq \alpha$. Let $S_{\tilde{a}} = \sigma_{\tilde{a}} \times \sigma_{\tilde{a}}$ where $\sigma_{\tilde{a}} = \{v | v \in B(\overline{D}_a)$ for all $a < \tilde{a}$, v-measurable$\}$. We define $T: S_{\tilde{a}} \to S_{\tilde{a}}$ where

$T(v^1, v^2) = (T_1(v^1,v^2), T_2(v^1,v^2))$ and $T_i(v^1,v^2)(x,t) =$

$$\alpha^i(x + (-1)^i t) + \int_0^t (-1)^i[(v^1(x + (-1)^i(t-\tau),\tau))^2 - (v^2(x+(-1)^i(t-\tau),\tau))^2]d\tau .$$

One verifies that if (u^1, u^2) is a solution of the IVP, then it is a fixpoint of T. Conversely, if (u^1,u^2) is a continous fixpoint of T, then it is a solution of the IVP.

We next form the Banach Space $B = (S'_a, \| \|_\lambda)$, where

$S'_a = \{(v^1,v^2) \in S_a | v^i \in C(\bar{D}_a)\}$,

$$\|(v^1,v^2)\|_\lambda = \max_{i=1,2} \sup \{|v^i(x,t)| \exp(-\lambda t) , (x,t) \in \bar{D}_a\} .$$

Let $S'_{a,k} = (v^1,v^2) \in S'_a | \sup\{|v^i(x,t)|, (x,t) \in \bar{D}_a \le k, i = 1,2\}$.

Choosing any $k > \alpha$, and setting $a = (k-\alpha)/k^2$, $\lambda = 8k$, we show that T is a contraction $(\| \|_\lambda)$ on $S'_{a,k}$ into itself. From this one obtains the local existence and uniqueness of the fixpoint of T on S'_a with $a = \sup_{k>\alpha} (k-\alpha)/k^2 = \frac{1}{4}\alpha$. By using a standard continuation procedure one shows the existence of a maximal $\tilde{a}$ such that T has a fixpoint (u^1,u^2) in S; one shows furthermore that the fixpoint is unique and it is the solution of the IVP defined on $D_{\tilde{a}}$. It remains to be shown that if the condition 2) holds then $\tilde{a} = \infty$.

2. We note that if $\alpha^1(x) \equiv \alpha^2(x) = c$, then the IVP has the (unique) solution (c, c) defined on D_∞.

3. Let $T = \{f \in C(R), 0 \le f(x) \le \alpha\}$. We define $G : T \times \sigma_\infty \to \sigma_\infty$ by $G(f,v) = w$ where w is the solution of the Riccati problem, $D^{(0,1)}w + w^2 = v^2$, $w(x,0) = f(x)$. One verifies that G is defined, $G(f,v)(x,t) \ge 0$ and that it is an isotone operator under the pointwise ordering of $T \times \sigma_\infty$ and σ_∞, i.e., if $f \ge \tilde{f}$ and $v \ge \tilde{v}$ pointwise then $G(f,v) \ge G(\tilde{f},\tilde{v})$ pointwise. We next define the translation operator $H_t : \sigma_\infty \to \sigma_\infty$ by $(H_t v)(x,t) = v(x+t,t)$ which again is an isotone operator under the pointwise ordering of σ_∞. Finally the operator $K_{(f_1,f_2)} : S_\infty \to S_\infty$ is defined by $K_{(f_1,f_2)} = (K^1_{(f_1,f_2)},$

$K^2_{(f_1,f_2)})$, $K^1_{(f_1,f_2)}(v^1,v^2) = H_{-t}(G(f_1, H_t v^2))$,

$K^2_{(f_1, f_2)}(v^1, v^2) = H_t(G(f_2, H_{-t}v^1))$. One notes that: 1) $K_{(f_1, f_2)}$ is isotone under the pointwise ordering of S_∞ , 2) $K_{(f_1, f_2)}$ is isotone on the parameter space $T \times T$ so that $K_{(\alpha, \alpha)}$ is a majorant of $K_{(\alpha^1, \alpha^2)}$, 3) (u^1, u^2) is a solution of the IVP if and only if it is a fixpoint of $K_{(\alpha^1, \alpha^2)}$.

4. The completion of the proof exploits the following result by the author to be communicated elsewhere. If 1) X is a partly ordered set, 2) X chain complete (i.e. every bounded chain in S has a supremum and an infimum), 3) $I \subset X$ is an interval and 4) $f: I \to I$ is isotone, then f has a fixpoint in I . One notes that since $K_{(\alpha^1, \alpha^2)}(v^1, v^2) \geq (0, 0)$ and since $K_{(\alpha, \alpha)}$ has the fixpoint (α, α) and is a majorant of $K_{(\alpha^1, \alpha^2)}$, K_{α^1, α^2} maps $I = [(0, 0), (\alpha, \alpha)] \subset S_\infty$ into itself. The conclusion follows by exploiting the fact that S_∞ is chain complete under pointwise ordering.

5. The above considerations can be easily extended to take care of the IVP for an infinite sequence (u^i) of functions defined on D_∞ such that

(DE) $(D^{(0,1)} + c_i D^{(1,0)}) u^i = \sum_{j=1}^{\infty} a_{ij}(u^j)^2 - (\sum_{j=1}^{\infty} a_{ij})(u^i)^2$ and

(Ic) $u^i(x, 0) = \alpha^i(x)$. Here one assumes that the c_i's form a bounded sequence, $a_{ij} \geq 0$, $\sum_{j=1}^{\infty} a_{ij} < \infty$ and $0 \leq \alpha^i(x) \leq \alpha$.

[1] T. Carleman, Problèmes Mathématiques dans la Théorie Cinétique des Gaz, Upsala 1957, pp. 104-106.

CONVEX SUPERPOSITION IN PIECEWISE-LINEAR SYSTEMS

B. A. Fleishman

Relay control systems and dynamical systems containing Coulomb friction are governed by (vector) differential equations of the form

$$dX/dt = A(t)X + \mathrm{Sgn}^*(X) + F(t) \qquad (1)$$

where $X = (x_1, x_2, \ldots, x_n)$ is a vector function of t . The n-by-n matrix $A(t)$ and n-vector $F(t)$ are assumed to be continuous in t for $t \geq 0$. $\mathrm{Sgn}^*(X)$ denotes a vector function of $x_1, \ldots, x_n$, with components $sgn^*(s_i)$ $(i = 1, \ldots, n)$; each $sgn^*(s_i)$ is a modified signum function (i.e., one with a "dead zone") of s_i, which in turn is a linear combination of the x_i's :

$$sgn^*(s_i) = \begin{cases} 1 & (s_i > \sigma_i) \\ 0 & (|s_i| \leq \sigma_i) \\ -1 & (s_i < -\sigma_i) \end{cases} \qquad (i = 1, \ldots, n) \tag{2}$$

and

$$s_i = s_{i1}x_1 + s_{i2}x_2 + \ldots + s_{in}x_n \quad (i = 1, \ldots, n) \tag{3}$$

where $\sigma_i \geq 0$ $(i = 1, \ldots, n)$ and s_{ij} $(i, j = 1, \ldots, n)$ are constants.

In general, a piecewise-linear system of the type (1) must be regarded as nonlinear. The author has shown [1], however, that under convex linear combinations a particular "on-off" (or relay) control system behaves linearly with respect to certain periodic input signals and responses.

This property of so-called "convex superposition" (or linearity with respect to averaging) is here generalized to systems of the form (1-2-3). The result may be stated as follows. Suppose there exists a set of applied forces (or input signals) $F^{(i)}(t)$, $i = 1, 2, \ldots, m$, such that when we set $F = F^{(i)}$, the system (1) possesses a particular solution $X^{(i)}(t) = X(t; F^{(i)})$ and these functions $X^{(i)}(t)$, $i = 1, 2, \ldots, m$, form a "synchronous" set. By this we mean that at any instant all of these functions satisfy the same system of linear equations (1) (except for their respective forcing functions $F^{(i)}$) ; at another instant the common governing system (1) may assume one of its other linear forms. Now let α_i, $i = 1, 2, \ldots, m$, be real positive constants such that $\alpha_1 + \alpha_2 + \ldots + a_m = 1$, and let $\bar{F}(t)$ denote the convex linear combination $\sum_{i=1}^{m} \alpha_i F^{(i)}(t)$. Then Eq. (1), with $F = \bar{F}$, possesses the solution $\bar{X}(t) = \sum_{i=1}^{m} \alpha_i X^{(i)}(t)$ which is synchronous with the $X^{(i)}(t)$, $i = 1, \ldots, m$.

The above result has application to current research on the determination of periodic regimes in piecewise-linear automatic control systems (e.g., see [2]).

REFERENCES

1. Fleishman, B.A., "Periodic Responses and Superposition in a Nonlinear Control System," RPI Math Rep. No. 34, Dept. of

Mathematics, Rensselaer Polytechnic Institute, AFOSR TN-60-477 (May 1960). To appear in "Journal of Math. Analysis and Applications."

2. Aizerman, M. A., and Lurie, A. I., "Methods for Construction of Periodic Motions in Piecewise-Linear Systems," Symposium on Non-Linear Vibrations, International Union for Theoretical and Applied Mechanics, Kiev, U.S.S.R., September 1961.

A QUANTITATIVE ANALYSIS OF A BIFURCATION PROCESS*

R. K. Brayton and R. A. Willoughby

This paper presents a quantitative analysis of the bifurcation process that takes place in a nonlinear electrical circuit. The state of the system is made to periodically vary from a region of one stable equilibrium point to a region of three equilibrium points, namely two stable and one a saddle point. During the transition from the mono-stable region to the bistable region, the bifurcation takes place at which time a small amount of control is sufficient to force the system into the desired stable point. With a perfectly symmetrical system, the amount of control can be arbitrarily small. The problem analyzed here is to determine the minimum amount of control required when specified parameter imbalances are present in the system. A symmetry transformation is made on the system of differential equations describing the circuit, and it is shown that the non-symmetry is a perturbation on a lower dimensional system. A formula is derived which expresses the magnitude of the minimum control in terms of the various parameter imbalances. When the perturbation is small, the formula is simply the superposition of the magnitudes for the separate imbalances. The results of the study may be stated in terms of allowable tolerances on the circuit components.

*The results reported in this abstract were obtained in the course of research jointly sponsored by the Mathematics Directorate of the Air Force Office of Scientific Research [Contract AF 49(638)-1139] and IBM.

ON THE INITIAL-VALUE PROBLEM FOR THE QUASI-LINEAR PARABOLIC EQUATION

Henry Hermes

We consider the initial value problem

$$u_t = F(u)\, u_{xx} , \qquad (t, x) \in D_T \equiv (0, T) X(-\infty, \infty)$$

$$u(0,x) = u_0(x) \quad , \qquad x \in (-\infty, \infty) \tag{1}$$

where

$$\text{i)} \ |u_0(x)| \le M_2 \qquad \text{ii)} \ |u'_0(x)| \le M_2$$
$$\text{iii)} \ |u'_0(x) - u'_0(\bar{x})| \le B_2 |x-\bar{x}| \tag{2}$$

and

$$\text{i)} \ 0 < \delta \le F(u) \le M_1 \qquad \text{ii)} \ F'(u) \in C[-M_2, M_2] \ , \ |F'(u)| \le N_1$$
$$\text{iii)} \ |F'(u) - F'(v)| \le N_2 |u-v| \tag{3}$$

It is shown that this problem has a unique solution, which can be extended for all time, in the class $S(\bar{D}_T)$ ($\bar{D}_T$ denotes the closure of D_T) defined by: $v \in S(\bar{D}_T)$ implies

i) $|v(t,x)| \le M_2$

ii) $v(0,x) = u_0(x)$

iii) $v_x(t,x)$ exists and $|v_x(t,x)| \le M_2$

iv) $|v_x(t,x) - v_x(t,\bar{x})| \le \dfrac{M_3 |x-\bar{x}|}{\sqrt{t}}$

v) $|v(t,x) - v(\bar{t},\bar{x})| \le M_4[|x-\bar{x}| + |t-\bar{t}|^{\frac{1}{2}}]$, where

$$M_4 = \max\left[\frac{M_1 M_3}{2}, M_2\right] . \tag{4}$$

This result immediately generalizes to uniformly parabolic equations of the form $u_t = F(t,x,u)u_{xx}$.

The proof utilizes the parametrix method and a study of the linear problem

$$\hat{L}u = u_t - a(t,x)u_{xx} - b(t,x)u_x = 0 \qquad (t,x) \in D_T$$
$$u(0,x) = u_0(x) \qquad x \in (-\infty, \infty) \tag{5}$$

where

i) $0 < \delta \le a(t,x) \le M_1$

ii) $|a(t,x) - a(\bar{t},\bar{x})| \le B_1[|t-\bar{t}|^{\frac{1}{2}} + |x-\bar{x}|]$

iii) $b(t,x) \in C(D_T)$, $|b(t,x)| < M_b$

$$\text{iv)}\quad |b(t,x) - b(t,\bar{x})| \leq \frac{B_3|x-\bar{x}|}{\sqrt{t}} .$$

Letting:

$$G(t,x) \equiv \frac{1}{2\sqrt{\pi t}} e^{-\frac{x^2}{4t}}$$

$$(v(\xi), w(\xi)) \equiv \int_{-\infty}^{\infty} v(\xi) w(\xi) d\xi$$

$$B(\bar{D}) \equiv \{v(t,x) : v \text{ is bounded on } \bar{D}\}$$

$$C_\alpha(\bar{D}) \equiv \{v(t,x) : t^\alpha v(t,x) \text{ is continuous on } \bar{D}\}, \quad \alpha \geq 0$$

$$\|v\|_{\alpha,\beta} \equiv \sup\{e^{-\alpha|x|} t^\beta |v(t,x)|, \ (t,x) \in \bar{D}\}, \quad \alpha, \beta \geq 0 ,$$

a solution of (5) is given by

$$\varphi(t,x) \equiv (G(ta(0,\xi), x-\xi), u_0(\xi)) + \int_0^t (G[(t-\tau)a(\tau,\xi), x-\xi], \bar{\rho}(\tau,\xi))\, d\tau$$

where $\bar{\rho}$ is the unique solution in $C_{\frac{1}{2}}(\bar{D}_T)$ of the integral equation

$$\rho(t,x) = -(\hat{L}G(ta(0,\xi), x-\xi), u_0(\xi)) - \int_0^t (\hat{L}G[(t-\tau)a(\tau,\xi), x-\xi], \rho(\tau,\xi))\, d\tau .$$

Using the maximum principle for the first boundary value problem, it is shown that the solution φ, of (5), attains its maximum on the boundary $t = 0$. It follows that this solution is unique in $B(\bar{D}_T) \quad C_0(\bar{D}_T)$.

We now return to the nonlinear problem (1). Let $R(\bar{D}_T)$ be defined to be the class of functions satisfying conditions (4), i), ii) and v). We define an operator $A : R(\bar{D}_T) \to C_0(\bar{D}_T) \quad B(\bar{D}_T)$ as follows:

For $v \in R(\bar{D}_T)$ let Av be the unique bounded continuous solution of

$$u_t = F(v) u_{xx} , \qquad (t,x) \in D_T$$

$$u(0,x) = u_0(x) , \qquad x \in (-\infty, \infty) .$$

We now work in the Banach space $\beta(C_0(\bar{D}_T) \quad B(\bar{D}_T), \ \|\ \|_{1,0})$. Utilizing the results obtained for the linear problem, it is shown that:

1) $A : R(\bar{D}_T) \rightarrow C_0(\bar{D}_T) \quad B(\bar{D}_T)$ continuously.

2) $R(\bar{D}_T)$ is a compact subset.

3) $A : S(\bar{D}_T) \rightarrow S(\bar{D}_T)$ and $S(\bar{D}_T) \subset R(\bar{D}_T)$.

Now from conditions (4), it follows that $S(\bar{D}_T)$ is convex, hence it is easily shown that A has a fixed point in $\bar{S}(\bar{D}_T)$ the closure of $\bar{S}(\bar{D}_T)$. It then follows that the fixed point satisfies all of the conditions (4) and existence of a solution of (1) is established.

Using a property announced by Aronson [Bull. Amer. Math. Soc., Sept. 1959], for a general linear parabolic operator, uniqueness can be established in $S(\bar{D}_T)$ and the extension of the domain of existence quickly follows.

DUALITY IN NON-LINEAR PROGRAMMING

O. L. Mangasarian

Duality principles relate two programming problems, one of which, the primal, is a constrained minimization (maximization) problem and the other, the dual, is a constrained maximization (minimization) problem in such a way that the existence of a solution to one of these problems insures a solution to the other and the extrema of the two problems are equal.

Duality in linear programming has been known for some time and extensive use has been made of it in theoretical and computational applications. Only recently, however, has duality in nonlinear programming been investigated. Philip Wolfe derived the following Duality Theorem:

If x^0 is a solution of the primal problem

$$\text{Minimize } \varphi(x) \tag{1}$$

$$\text{subject to } g(x) \geqq 0 \tag{2}$$

where φ is a differentiable, convex, scalar function of the n-vector x , and each of the m components of $g(x)$ is a differentiable, concave function of x , then there exists some m-vector $u^0 \geqq 0$ such that (x^0, u^0) is the solution of the dual problem

$$\text{Maximize } \psi(x, u) \equiv \varphi(x) - u'g(x) \tag{3}$$

$$\text{subject to } \nabla\psi(x, u) \equiv \nabla\varphi(x) - \nabla u'g(x) = 0 \tag{4}$$

$$\text{and } \quad u \geqq 0 \tag{5}$$

Also

$$\varphi(x^0) = \psi(x^0, u^0) \tag{6}$$

In the above, lower case Roman letters denote (with obvious exceptions) column vectors, and Greek letters denote scalars. A prime indicates the transpose of a vector. The operator ∇ is the column vector $(\frac{\partial}{\partial x_1}, \ldots, \frac{\partial}{\partial x_n})'$. The constraints $g(x) \geqq 0$ must be free from certain singularities such as "outward-pointing cusps."

At the point x^0 which is the solution of the primal problem (1), (2) there will be associated with every component $g_i(x^0)$ of the primal constraint set, a component u_i^0 of the Lagrange multiplier vector u^0. The term active primal constraints will be used to denote those components of $g_i(x^0)$ for which $u_i^0 \neq 0$. For all such constraints $g_i(x^0) = 0$.

In the present work the following Converse Duality Theorem was derived:

Let (x^0, u^0) be a solution of the dual problem (3), (4), (5), where $\varphi(x)$ is a convex and twice continuously differentiable function of x and the components $g_i(x)$, $i = 1, \ldots, m$ of $g(x)$ are concave and twice continuously differentiable functions of x. The vector x^0 is a solution of the primal problem (1), (2), and equation (6) holds if (in addition to the qualifications on $\varphi(x)$ and $g(x)$ already stated) either $\varphi(x)$ is strictly convex in the neighborhood of x^0 or if at least one of the active primal constraints $g_i(x^0)$ (i.e., those for which $u_i^0 \neq 0$) is strictly concave in the neighborhood of x^0 or both. If $\varphi(x)$ is quadratic and if $g(x)$ is linear then this converse theorem is true if $\varphi(x)$ is merely convex and twice differentiable.

A COMPLETE SOLUTION OF THE X AND Y EQUATIONS OF CHANDRASEKHAR

T. W. Mullikin

In S. Chandrasekhar's treatment [Radiative Transfer, Dover, 1960] of the scattering and transmission of monochromatic radiation by a plane-parallel atmosphere, the following equations are of central importance:

$$X(\mu) = 1 + \mu \int_0^1 \frac{X(\mu)X(\nu) - Y(\mu)Y(\nu)}{\mu + \nu} \psi(\nu)\,d\nu \,, \tag{1}$$

$$Y(\mu) = e^{-\tau/\mu} - \mu \int_0^1 \frac{X(\mu)Y(\nu) - Y(\mu)X(\nu)}{\mu - \nu} \psi(\nu)\,d\nu \,, \quad \mu \neq 0 \,, \quad 0 \leq \tau < \infty \,.$$

The function ψ is a given function, continuous and nonnegative on $[0, 1]$ which satisfies

$$\int_0^1 \psi(\nu)\,d\nu \leq 1/2 \,. \tag{2}$$

I. W. Busbridge [The Mathematics of Radiative Transfer, Cambridge Tracts No. 50, 1960] has proved existence of a solution to (1) which is nonnegative and continuous on $[0, 1]$, and analytic in $|\mu| > 0$. Only in the case of equality in (2) has uniqueness been determined, in which case Chandrasehkar exhibits a one-parameter family of solutions.

It is shown that a solution to (1) is unique in the class of functions continuous on $[0, 1]$ if and only if the function

$$T(z) = 1 - z \int_0^1 \frac{\psi(\nu)}{z - \nu} d\nu - z \int_0^1 \frac{\psi(\nu)}{z + \nu} d\nu \,, \quad z \notin [-1, 1] \,, \tag{3}$$

has no zeros in the complex plane cut along $[-1, 1]$. A necessary, but not sufficient, condition for uniqueness is $\psi(1) = 0$.

The only possible zeros of T are $\pm z_0$ for real $z_0 \notin [-1, 1]$. If T has such a zero, then a solution to (1) exists for every point on a certain hyperbola, and this describes all solutions that are continuous on $[0, 1]$. Each solution has a simple representation in terms of Busbridge's solution, and each, other than hers, has a simple pole at $\pm z_0$. The previously known case of nonuniqueness corresponds to $z_0 = \infty$, in which case the hyperbola degenerates to a line.

For applications to physical problems it is the value of the functions X and Y on $[0, 1]$ that is needed. Considerable time and effort have been spent in attempts to solve equations (1) on this interval by numerical iteration, without a complete criterion for uniqueness. When nonuniqueness prevails it is to be expected that these methods either will not converge or will converge to a solution which depends upon the function chosen to initiate the iteration.

ANALYSIS OF A SET OF NON-LINEAR PARTIAL DIFFERENTIAL EQUATIONS ARISING FROM A PROBLEM IN HYDRODYNAMIC STABILITY

Lee A. Segel*

J. T. Stuart and J. Watson have developed a formal procedure for solving appropriate problems in nonlinear partial differential equations. By combining a harmonic spatial analysis with an expansion in powers of time-dependent amplitudes, the problems are reduced to relatively simple nonlinear ordinary differential equations. The Stuart-Watson method will be illustrated here in a nonlinear analysis of a type of cellular thermal convection.

Consider two-dimensional motions of a slightly compressible layer of fluid bounded by horizontal free surfaces at $z = 0$ and $z = 1$. The viscous Boussinesq equations for the velocities u and w in the x and z directions can be written

$$u_x + w_z = 0 \tag{1}$$

$$\Theta_t - \Theta_{zz} = -(\overline{wT})_z , \tag{2}$$

$$T_t - \Delta T = -w\Theta_z - uT_x - wT_z + (\overline{wT})_z \tag{3}$$

$$(\partial/\partial t - \sigma\Delta)\Delta w - \sigma T_{xx} = (uu_x + wu_z)_{xz} - (uw_x + ww_z)_{xx}, \tag{4}$$

where the horizontal bar indicates a mean value in x

$$\bar{q} = \lim_{L\to\infty} \frac{1}{2L} \int_{-L}^{L} q(x, z, t)\, dx . \tag{5}$$

Here Θ and T are the mean and fluctuating temperatures, $\Delta = \partial^2/\partial x^2 + \partial^2/\partial z^2$, and σ is the constant Prandtl number. If the bounding surfaces are perfect conductors and the temperature at $z = 0$ is a constant amount Q greater than that at $z = 1$, then the boundary conditions are

$$u = T = w = w_{zz} = 0 \text{ on } z = 0, 1; \ \Theta(0) = 0 , \quad \Theta(1) = -R , \tag{6}$$

where the Rayleigh number R is proportional to Q and the gravitational constant g. (Gravity acts vertically downward.)

We wish to consider the fate of a pair of infinitesimal spatially periodic disturbances, respectively proportional to $\cos \pi\alpha x$ and $\cos \pi\beta x$, to the steady solution $u = w = T = 0$, $\Theta = -Rz$. This is done by making the expansions

$$q(x, z, t) = \sum_{i, j=1}^{\infty} q_{ij}(z, t) \cos \pi i \alpha x \cos \pi j \beta x \quad , \tag{7}$$

$$q_{ij}(z, t) = \sum_{m, n=0}^{\infty} q_{ij}^{(1+2m, j+2n)}(z) [A(t)]^{i+2m} [B(t)]^{j+2n} \tag{8}$$

$$\Theta(z, t) = -Rz + \sum_{m, n=1}^{\infty} \Theta^{(2m, 2n)}(z) A^{2m} B^{2n} \tag{9}$$

$$dA/dt = a^{(10)} A + a^{(30)} A^3 + a^{(12)} AB^2 + \ldots \quad , \tag{10}$$

$$dB/dt = b^{(01)} B + b^{(21)} A^2 B + b^{(03)} B^3 + \ldots \quad . \tag{11}$$

Here q stands for w or T, the expansions for u following from (1). The central idea in arriving at (7-9) is that if w and T start off with terms proportional to $A(t) \cos \pi x$ and $B(t) \cos \pi\beta x$ then, when A and B are large enough, nonlinear terms like $(ww_z)_{xx}$ give rise to terms proportional to $A^2(1 + \cos 2\pi \alpha x)$, $AB \cos \pi(\alpha + \beta) x$, etc. It is expected that the formal scheme will lead to a meaningful result if $A(t)$ and $B(t)$ are uniformly small.

We proceed by substituting (7-11) into (1-4) and collecting coefficients of $A^{i+2m} B^{j+2n} \cos \pi i \alpha x \cos \pi j \beta x$. The $q_{ij}^{(1+2m, j+2n)}(z)$ can be found from successive differential equations starting with the lowest powers of A and B. The $O(A)$ and $O(B)$ equations are those of linear stability theory so $a^{(10)}$ and $b^{(01)}$ can be found by solving a well known eigenvalue problem. If A and B are to be small, then $a^{(10)}$ and $b^{(01)}$ must be assumed small. (This is a restriction on R.) The constant $a^{(30)}$ is determined by requiring that the $w_{10}^{(30)}$ correction to $w_{10}^{(10)}$ remains finite as $a^{(10)} \to 0$. This turns out to mean that $a^{(30)}$ must be chosen to remove a resonant forcing term in the inhomogeneous ordinary differential equation for $w_{10}^{(30)}$. The constants $a^{(12)}$, $b^{(21)}$, and $b^{(03)}$ are determined similarly. With R restricted as above, it can be shown that the higher order terms indicated by ... in (10) are negligible.

The problem is thus reduced to an analysis of (10) and (11), for which many well known methods can be used. (Actually equations like (10) and (11) should provide a model in more general situations than that for which they have been derived.) Two representative results are: (1) Under certain circumstances, an equilibrium state may be composed of a mixture of a linearly stable disturbance and a linearly unstable disturbance. (2) For the thermal convection problem, when the Rayleigh number is slightly above the minimum critical value, the equilibrium state will contain only one of two linearly unstable disturbances. These and other results are compared with experimental observations.

*This work was partially supported by the Office of Naval Research.
**Stuart, J. T. and Watson, J. J., Fluid Mech. 9(3), 353-89 (1960).

ON A RESULT OF HADAMARD CONCERNING THE SIGN OF THE PRECESSION OF A HEAVY SYMMETRICAL TOP

J. B. Diaz and F. T. Metcalf

Consider Lagrange's integrable case of the motion of a rigid body about a fixed point, when the center of gravity of the body lies on the polar axis of the spheroid of intertia of the body (see W. D. MacMillan, Dynamics of Rigid Bodies, New York, Dover Ed., 1960; case II; pp. 216-249). In section 110, pages 233-235, of MacMillan's book one finds an interesting proof that the precession for a complete period has the same sign as the component of the angular velocity vector along the axis of symmetry of the body (J. Hadamard, Bull. Sci. Math. (2) vol. 19(1895), pp. 228-230). Hadamard's proof employs the theory of residues of functions of a complex variable. Quite recently, the authors have shown that upper and lower bounds for the apsidal angle in the theory of the heavy symmetrical top can be obtained in an elementary way, not relying upon the theory of residues. This same direct approach is now shown to apply equally well in deducing Hadamard's result just mentioned. (Submitted to the Proceedings of the American Mathematical Society.)

STEADY FLOWS NEAR CRITICAL SPEED

W. T. Kyner

1. Introduction. The solitary wave is a well known example of a flow of permanent type which can be studied by the shallow water theory. A scheme due to K. O. Friedrichs [1] can give both an approximation to the exact wave form and framework for a proof of its existence. It is used here to approximate the solution of a related problem, that of steady flow near critical speed over a small obstacle in the bed of a stream. The flow can be considered as a perturbation of a solitary wave since it will approach a solitary wave as the obstacle shrinks in height.

A similar problem, that of the motion of vortex under the surface of a fluid, was solved recently by I. G. Filippov [2]. He used the Friedrichs scheme to obtain an approximate solution, and a modification of the Friedrichs-Hyers [1] argument to prove existence. His solution approaches the solitary wave as the vortex strength diminishes.

In the present problem, the bottom is assumed to be flat up to the point at which the obstacle occurs. Since a flow of permanent type is being studied, the water can be assumed to be moving under a fixed surface. This surface is approximated by a solitary wave upstream. Downstream, the approximating surface is in general periodic. However, for each sufficiently small obstacle, there is an exceptional flow which is non-periodic downstream. This flow can be described as two solitary waves pieced together with a distortion in the neighborhood of the obstacle. It is referred to here as a piecewise solitary wave.

It is shown that for suitable class of obstacles, there is one parameter family of formal solutions (i. e. all of the terms in the expansion are determinate) which is periodic downstream and non-periodic upstream. The parameter relates the crest of the unperturbed wave to the position of the obstacle. In addition, for each such obstacle, there is a critical value of the parameter which gives the piecewise solitary wave.

REFERENCES

1. Friedrichs, K. O., and Hyers, D. H., The existence of solitary waves, Comm. Pure Appl. Math., Vol. 7, 1954, pp. 517-550.

2. Filippov, I. G., Solution of the problem of a vortex below the surface of a liquid at Froude numbers close to unity. P. M. M. vol. 24, No. 3, 1960, pp. 698-716.

NOTE ON AN INTEGRAL EQUATION OCCURRING IN STAGNATION POINT FLOW

J. Siekmann

The analysis of the flow near the stagnation point of a body of revolution immersed in an axial stream of constant velocity can be reduced to the solution of the differential equation

$$w'''(z) + 2w(z)\,w''(z) + \{1 - w'^2(z)\} = 0 \qquad (1a)$$

$$w(0) = w'(0) = 0 \;, \quad w'(\infty) = 1 \;. \qquad (1b)$$

A solution of this problem has been given by F. Homann [1].

The present paper shows that an approach due to H. Weyl [2] is especially suitable for electronic computers. With

$$F''(z) = g(z) \quad , \tag{2}$$

where $F(z)$ is the solution of the initial value problem

$$F'''(z) + 2FF'' + (\beta^2 - F'^2) = 0 \tag{3a}$$

$$F(0) = F'(0) = 0 \quad , \quad F''(0) = 1 \quad , \quad F'''(0) = -\beta^2 \tag{3b}$$

one can derive an integral equation for $g(z)$, namely

$$g(z) = T\{g(z)\} \quad , \tag{4}$$

where the operator T is defined by

$$T\{g\} = \frac{\int_z^\infty e^{-G(\sigma)} d\sigma}{\int_0^\infty e^{-G(\sigma)} d\sigma} \tag{5}$$

and

$$G(\sigma) = \int_0^\sigma (\sigma - t)^2 g(t)\, dt \tag{6}$$

For the numerical solution of eq. (4) an iteration process $g_{n+1} = T\{g_n\}$ with $g_1 = 1$ was set up. The integrals were evaluated by means of Simpson's rule with a step width $h = 0.125$ and an upper limit $z = 4.8125$ to approximate the upper limit $z = \infty$. About $N = 6$ iterations were necessary to obtain an accuracy of

$$|g_N - g_{N-1}| < 0.5 \cdot 10^{-4} \tag{7}$$

About 14 minutes of computation are needed on the IBM 709 computer for 6 iterations.

REFERENCES

1. F. Homann: Der Einfluss grosser Zahigkeit bei der Stromung um der Zylinder und um die Kugel. ZAMM 16, 153-164 (1936).

2. H. Weyl: On the Differential Equations of the Simplest Boundary Layer Problems. Ann. Math., Vol. 43, No. 2, 381-407 (1942).

NON-LINEAR VIBRATIONS OF ELASTIC CIRCULAR PLATES EXHIBITING RECTILINEAR ORTHOTROPY

J. L. Nowinski

While the problem of nonlinear transverse vibrations of elastic isotropic plates has attracted some attention (see [1] for bibliography), so far only one paper concerning anisotropic plates understood as a limiting case of cylindrically orthotropic shells of revolution could be traced [2].

In the present paper we consider the rectilinear orthotropy of circular plates. The problem is reduced to a solution of two von Karman's equations for large deflections of plates extended to a dynamic and orthotropic case:

$$\frac{\partial^4 w}{\partial x^4} + 2\ell^2 \frac{\partial^4 w}{\partial x^2 \partial y^2} + k^2 \frac{\partial^4 w}{\partial y^4} - \frac{q}{D_1} + \frac{\rho h}{D_1}\frac{\partial^2 w}{\partial t^2} = \frac{h}{D_1}\left(\frac{\partial^2 w}{\partial x^2}\frac{\partial^2 \phi}{\partial y^2} - 2\frac{\partial^2 \phi}{\partial x \partial y}\frac{\partial^2 w}{\partial x \partial y} + \frac{\partial^2 w}{\partial y^2}\frac{\partial^2 \phi}{\partial x^2}\right)\rho \quad (1)$$

$$\frac{\partial^4 \phi}{\partial x^4} + p^2 \frac{\partial^4 \phi}{\partial x^2 \partial y^2} + k^2 \frac{\partial^4 \phi}{\partial y^4} = E_2\left[\left(\frac{\partial^2 w}{\partial x \partial y}\right)^2 - \frac{\partial^2 w}{\partial x^2}\frac{\partial^2 w}{\partial y^2}\right] \quad (2)$$

In the equations above w is the deflection, h the thickness of the plate, t time, q load, ρ mass per unit volune and Φ the stress function. Moreover, $k^2 = D_2/D_1$, $\ell^2 = D_3/D_1$, where D_1, D_2 and D_3 are flexural and torsional rigidities; E_2 represents Young's modulus in the y direction, the coordinate axes being directed along the principal elastic directions of the plate.

For definiteness, we assume that the plate is built-in along the periphery but freely movable in radial directions. We satisfy these requirements by taking the deflection in the following separable form:

$$w(r,t) = w_0(t)\left(1 - \frac{r^2}{a^2}\right)^2 , \quad (3)$$

where $\underline{r}$ is the position radius and $\underline{a}$ radius of the plate. Upon substitution of (3) into (2) we obtain a differential equation for Φ. A particular integral of this equation satisfying the prescribed boundary conditions

$$x\sigma_x + y\tau_{xy} = 0 , \quad x\tau_{xy} + y\sigma_y = 0 , \quad \text{for } r = a , \quad (4)$$

have the following polynomial form

$$\Phi(x,y) = \Sigma\, a_{\mu\nu} x^{\mu} y^{\nu}$$
$$\mu, \nu = 0, 2, 4, 6, 8 \quad (5)$$
$$\mu + \nu = 2, 4, 6, 8$$

The fourteen coefficients appearing in equation (5) can be determined explicitly upon substituting in (2) and (4) and solving a system of fourteen linear equations.

This completes the solution of the compatibility equation (2). We now apply the Galerkin procedure to the only remaining equation of motion (1). To this end we multiply the left-hand member of (1) by the spacial part of (3) and integrate over the region of the plate. This yields finally the following time equation (with q_0 as a uniformly distributed load):

$$\frac{d^2 w_0}{dt^2} + \alpha w_0 + \beta w_0^3 - \frac{5}{3\rho h} q_0 = 0 , \tag{6}$$

where α, β and S are known functions of the coefficients $a_{\mu\nu}$.

Upon rejecting the nonlinear and loading terms in (6) we get the equation of linear oscillations with the frequency

$$\omega = \frac{1}{a^2} \sqrt{\frac{40(3 + 2t^2 + 3k^2)}{3}} \sqrt{\frac{D_1}{\rho h}} \tag{7}$$

an expression obtained by Lekhnitski [3] using an energy method. A suppression of the inertia term in (6) provides the equation for the nonlinear static case

$$\frac{2(3 + 2\ell^2 + 3k^2)}{3k^2(1-k^2\nu_1^2)} \frac{w_0}{h} + \frac{S}{105} \left(\frac{w_0}{h}\right)^3 = q^* \tag{8}$$

which if specified to isotropy agrees with the result of Volmir [4].

Upon confining ourselves to the free nonlinear vibrations we reject the loading term in (6) and obtain the equation

$$\frac{\partial^2 \tau}{\partial t^2} + \alpha\tau + \beta A^2 \tau^3 = 0 \tag{9}$$

In (9) the notation $w_0(t) = A\tau(t)$ is used. A solution of this equation for the initial conditions $\tau(0) = 1$ and $\tau(0) = 0$, is well known and represents a cosine type Jacobian elliptic function. The ratio of the periods of nonlinear and linear vibrations presently becomes

$$\frac{T^*}{T} = \frac{2K}{r\left[1 + \frac{k^2(1-k^2\nu_1^2)S}{70(3+2\ell^2+3k^2)} \left(\frac{A}{h}\right)^2\right]^{1/2}} \tag{10}$$

where K is a complete elliptic integral of the first kind. Forced nonlinear vibration caused by a harmonic pulsation of load

$$q_0 = q_0^* \cos \omega t \tag{11}$$

(q_0^* = amplitude) can also be easily investigated yielding the ratio of periods

$$\frac{T^*}{T} = \left[1 + \frac{3(1-k^2\nu_1^2)k^2 S}{280(3 + 2\ell^2 + 3k^2)}\left(\frac{A}{h}\right)^2\right]^{-1/2} \tag{12}$$

Numerical examples solved for two different types of orthotropy ($k^2 = \frac{1}{2}$, $\ell^2 = \frac{1}{2}$ $k^2 = \frac{1}{20}$, $\ell^2 = 0.098$, while for isotropy $k^2 = \ell^2 = 1$) show marked dependence of the period of vibration and the stresses in the plate on the degree of anisotropy as well, of course, as on the amplitude of vibration.

REFERENCES

1. J. Nowinski, Nonlinear Transverse Vibrations of Circular Elastic Plates Built-in at the Boundary, Proc. Fourth U. S. Nat. Congr. appl. mech., 1962 (forthcoming).
2. E. F. Furnistrov, Nonlinear Transverse Vibrations of Orthotropic Shells of Revolution, Insh. Sborn., Vol. 26, 1958.
3. S. G. Lekhnitski, Anisotropic Plates (in Russian), 1957.
4. A. S. Volmir, Flexible Plates and Shells (in Russian), 1956.

APPARENT RESONANCES IN NONLINEAR WAVE MOTIONS

D. J. Benney

The equation $u_{xx} - u_{\pi} = \Sigma u^3$, Σ small, with $0 \leq t < \infty$, $0 \leq x \leq \pi$, $u(0,t) = u(\pi,t) = 0$, with given initial conditions is discussed by means of a formal perturbation solution. It is shown that from a given normal mode of the linearized equation there is an energy exchange into other modes of the system. Thus all modes can potentially become $0(1)$ amplitude motions for t large. Certain numerical results support the theoretical investigation. The possibility of choosing the initial conditions so as to have a finite amplitude periodic motion is discussed.

ON A-PRIORI BOUNDS FOR SOLUTIONS OF QUASI-LINEAR PARABOLIC EQUATIONS

Stanley Kaplan

We establish a general comparison theorem for solutions of mixed initial-boundary problems for quasi-linear parabolic equations, using the maximum principle. This theorem has two main applications:

to obtain uniqueness theorems for solutions of these problems, and to obtain a-priori bounds on the magnitude of the solutions. The a-priori bounds are obtained by comparison with the solutions of appropriately chosen ordinary differential equations. They are applied to prove the existence of solutions in arbitrarily large time intervals, for equations with "small" nonlinearities. These results are shown to be best possible by means of a fairly general non-existence theorem, for nonlinear equations with "large" nonlinearities, on sufficiently large time intervals; the demonstration of this result rests on an investigation of the nonlinear ordinary differential inequality satisfied by the "Fourier coefficient" of the solution, with respect to an appropriately chosen eigenfunction.

SOME QUASI-LINEAR SINGULAR CAUCHY PROBLEMS

Robert Carroll

In this paper there is considered a class of singular Cauchy problems, with a nonlinear term, which are generalizations of the classical Euler-Poisson-Darboux (EPD) equation (see Weinstein [7]). A few of the papers in this direction most relevant to the present work are Carroll [1;2;3], Keller [5], and Lions [6]. We obtain somewhat finer results for the singular Green's operator than in our previous papers [1;2] and combine these results with variations of the nonlinear techniques used by Foias, Gussi, and Poenaru [4] to obtain improvements and extensions of the results announced in [3].

Let $L = \partial^2/\partial t^2 + [\alpha(t) + \beta(t)]\partial/\partial t + \gamma(t)$ and consider the Cauchy problem $L\omega + Q(t)\Lambda\omega + f(t,\omega) = 0$, $\omega(\tau) = T \in D(\Lambda)$, $\omega_t(\tau) = 0$, $t \to \omega(t) \in \mathscr{E}^2(H)$, $t \to \omega(t) \in \mathscr{E}^0(D(\Lambda))$, $0 \leq \tau \leq t \leq b < \infty$, where Λ is a self-adjoint operator, semi-bounded below, with dense domain $D(\Lambda)$ in a separable Hilbert space H. $\mathscr{E}^m(G)$ is the space of m-times continuously differentiable functions with values in G. There is no loss of generality in assuming Λ positive, invertible, and with closed range $R(\Lambda)$; $R(\Lambda)$ has the topology induced by H and $D(\Lambda)$ has the graph topology. It will be supposed that (i) $\gamma \in C^0[0,b]$, (ii) $\beta \in C^0[0,b]$, (iii) $\int_\tau^t \mathrm{Re}\,\alpha\, d\xi \to \infty$ as $\tau \to 0$, (iv) $|\alpha| \leq c\, \mathrm{Re}\,\alpha$ on $[0,\eta]$ for some $\eta > 0$, (v) $\alpha \in C^1(0,b)$ and $\varphi = \alpha^1/\alpha^2 \in C^0[0,\eta]$, (vi) $|\alpha(t)/\alpha(\xi)| \leq N$ for $0 \leq \xi \leq t \leq \eta$, and (vii) $Q \in C^1[0,b]$ with Q real, $0 < q \leq Q(t)$ (q a constant). Precise hypotheses on the continuous function $f(t,\omega(t))$ of t will be made below.

The Green's operator $G(t,\tau)$ of a (matrix) first order differential equation $du/dt + A(t)u = f$ in a topological vector space gives the solution u in the form $u(t) = G(t,\tau)u(\tau) + \int_\tau^t G(t,\xi)f(\xi)\,d\xi$.

Reducing the above second order singular equation to a first order system we would expect the Green's operator to be of the form

$$G(t,\tau) = \begin{pmatrix} z(t,\tau) & y(t,\tau) \\ z_t(t,\tau) & y_t(t,\tau) \end{pmatrix}$$

with $z(\tau,\tau) = I$, $z_t(\tau,\tau) = 0$, $y(\tau,\tau) = 0$, $y_t(\tau,\tau) = I$. Indeed we construct such operators z and y and for τ 0 the matrix formulation is manageable; however, on $[0 \leq \tau < t < b]$, $y_t(t,\tau)$ is not continuous in both variables, in fact $y_t(t,0) \equiv 0$. Thus it turns out to be easier to work with the functions z and y individually and to write the solution of $L\omega + Q\Lambda\omega = f(t)$ as $\omega(t,\tau) = z(t,\tau)T - \int_\tau^t y(t,\xi)f(\xi)\,d\xi$. Then our equation leads to the nonlinear integral equation $\omega(t,\tau) = z(t,\tau)T - \int_\tau^t y(t,\xi)f(\xi,\omega)\,d\xi = F\omega$.

In order to construct the operators z and y we make use of the decomposition of H into a continuous direct sum of Hilbert spaces. Thus there exists a measure ν , a ν-measurable family of Hilbert spaces $\lambda \to h(\lambda)$, and an isometric isomorphism $\theta : H \to h = \int^\oplus h(\lambda)\,d\nu(\lambda)$ such that Λ is transformed into the diagonalizable operator of multiplication by λ . Then for example $T \in D(\Lambda)$ means $\int_0^\infty \lambda^2 \|\theta T(\lambda)\|^2_{h(\lambda)}\,d\nu < \infty$. The original homogeneous equation for ω now becomes $L\theta\omega + Q(t)\lambda\theta\omega = 0$ and we determine numerical functions $Z(t,\tau,\lambda)$, $Y(t,\tau,\lambda)$, such that $z = \theta^{-1}Z\theta$, $y = \theta^{-1}Y\theta$. The following properties of z and y are obtained by transporting the corresponding properties of Z and Y . In particular it is found that (setting $L_S(E,F)$ to be the space of continuous linear maps $E \to F$ with the topology of simple convergence)

$t \to a(t,\tau) \in \mathscr{E}^2(L_S(D(\Lambda),H))$, $t \to y(t,\xi) \in \mathscr{E}^1(L_S(H,H))$,
$(t,\xi) \to y(t,\xi) \in \mathscr{E}^0(L_S(H, D(\Lambda^{1/2})))$, $(t,\xi) \to z(t,\xi) \in \mathscr{E}^0(L_S(H,H))$
$\xi \to z(t,\xi) \in \mathscr{E}^1(L_S(D(\Lambda),H))$, $\xi \to y(t,\xi) \in \mathscr{E}^1(L_S(H,H)$ for $\xi > 0$, $\xi \to (\frac{1}{\alpha(\xi)})y_\xi(t,\xi) \in \mathscr{E}^0(L_S(H,H))$ for ξ small,
$t \to y(t,\tau) \in \mathscr{E}^2(L_S(D(\Lambda^{1/2}),H))$, $\xi \to y_t(t,\xi) \in \mathscr{E}^0(L_S(H,H))$,
$(t,\xi) \to z_t(t,\xi) \in \mathscr{E}^0(L_S(D(\Lambda^{1/2}),H))$, $y(t,0) \equiv 0$,
$z(t,\tau) = z(t,t_0)z(t_0,\tau) + y(t,t_0)z_t(t_0,\tau)$ in $L_S(H,H)$,
$y(t,\tau) = z(t,t_0)y(t_0,\tau) + y(t,t_0)y_t(t_0,\tau)$ in $L_S(H, D(\Lambda^{1/2}))$,

and for $\tau > 0$, $y_\tau(t,\tau) = -z(t,\tau) + [\alpha(\tau) + \beta(\tau)\, y(t,\tau)$ in $L_S(H,H)$.

Consider now the equation $v = z(t,\tau)\Lambda T - \int_\tau^t y(t,\xi)\Lambda f(\xi, \Lambda^{-1}v)d\xi = \tilde{F}v$ and note that we can treat $\Lambda^{1/2}y(t,\xi)$ in $L_S(H,H)$. A positive, monotone, non-decreasing function ω, continuous on $(0,\infty)$, is Osgoodian if $\int_0^\eta d\tau/\omega(\tau) = \infty$ and Wintnerian if $\int_{\eta>0}^\infty d\tau/\omega(\tau) = \infty$; such functions will be called 0 and W functions. From an analysis of the equation $v = \tilde{F}v$ there results

THEOREM 1. Assume $\xi \to \Lambda^{1/2}f(\xi, \Lambda^{-1}v) = F(\xi, v) : R \to H$ is continuous for $v \in \mathscr{E}^0(R(\Lambda))$; $\|F(\xi,v) - F(\xi,u)\| \leq k_1(\xi)\omega_1(\|v-u\|)$ where $k_1 \in L^1$ and ω_1 is an 0 function; and $\|F(\xi,v)\| \leq k_2(\xi)\omega_2(\|v\|)$ where $k_2 \in L^1$ and ω_2 is a W function. Then there exists a unique solution of the Cauchy problem depending continuously in the Banach space $\mathscr{E}_b^0(H)$ on the initial data in $D(\Lambda)$, the dependence being uniform in τ.

THEOREM 2. Assume $(\xi, u) \to \Lambda^{1/2}f_i(\xi, \Lambda^{-1}u) : R \times R(\Lambda) \to H$ is continuous $(i = 1, 2$ $\|\Lambda^{1/2}f_2(\xi, \Lambda^{-1}u) - \Lambda^{1/2}f_2(\xi, \Lambda^{-1}v)\| \leq k_1(\xi)\omega_1(\|v-u\|)$ $(k_1 \in L^1$, ω_1 an 0 function); the map $(\xi, u) \to \Lambda^{1/2}f_1(\xi, \Lambda^{-1}u) : R \times R(\Lambda) \to H$ is compact; and $\|\Lambda^{1/2}f(\xi, \Lambda^{-1}u)\| \leq k_2|\xi|\omega_2(\|u\|)$ $(f = f_1 + f_2$, $k_2 \in L^1$, ω_2 a W function). Then there exists a solution of the Cauchy problem (not necessarily unique).

REFERENCES

1. R. Carroll, Sur le probleme de Cauchy singulier, Comptes Rendus, 252, 1961, pp. 57-59.

2. R. Carroll, On the singular Cauchy problem, Journal of Math. and Mech., to appear.

3. R. Carroll, Some singular quasi-linear Cauchy problems (Abstract), Notices, Amer. Math. Soc., 1961, p. 618.

4. C. Foias, G. Gussi, V. Poenaru, Sur les solutions generalisees de certaines equations lineaires et quasi-lineaires dans l'espace de Banach, Rev. Math. Pures et Appl., Rep. Pop. Roumaine, 3, 1958, pp. 283-304.

5. J. B. Keller, On solutions of non-linear wave equations, Comm. Pure and Applied Math., 10, 1957, pp. 523-530.

6. J. L. Lions, Operateurs de transmutation singulier et equations d'Euler-Poisson-Darboux generalisees, Rend. Sem. Mat. Fis. Milano, 28, 1959, pp. 3-16.

7. A. Weinstein, On the wave equation and the equation of Euler-Poisson, Proc. Fifth Symp. Applied Math., 1952.

POINTWISE BOUNDS FOR SOLUTIONS TO THE CAUCHY PROBLEM FOR NON-LINEAR ELLIPTIC PARTIAL DIFFERENTIAL EQUATIONS

George N. Trytten

L. E. Payne (Bounds in the Cauchy Problem for the Laplace Equation, Arch. Rat. Mech. Anal., vol. 5(1960), pp. 34-45, and Some A Priori Inequalities for Uniformly Elliptic Operators with Application to the Cauchy Problem, Tech. Note BN-280, Inst. for Fl. Dyn. and Appl. Math., 1962) has derived a technique for obtaining pointwise bounds for a function satisfying a Cauchy problem for the elliptic equation

$$(*) \qquad \mathscr{A}(w) \equiv \sum_{i,j=1}^{N} \frac{\partial}{\partial x_j}\left(a^{ij}(x)\frac{\partial w}{\partial x_i}\right) = 0 \qquad x = (x_1, \ldots, x_N), \quad a^{ij}(x) = a^{ji}(x)$$

Until recently such problems have been neglected; perhaps they have seemed unnatural since they are not well-posed in the sense of Hadamard, i.e., the solutions do not depend in a continuous way upon the Cauchy data. Nevertheless such problems actually can represent realizable physical phenomena, usually those in which a portion of the boundary is inaccessible for measurement of the boundary data.

It has been demonstrated by M. M. Laurentiev (On the Cauchy Problem for the Laplace Equation, Izvest. Akad. Nauk. SSSR., Ser. Mat. 20, ≡19-≡42 (1956)) that if one prescribes in addition to the Cauchy data a supplementary condition, namely that the harmonic function be uniformly bounded by some constant M in its region of definition $\mathscr{D}$, then the Cauchy problem for the Laplace equation has a stable solution and hence is well-posed.

Payne's method consists of first deriving a suitable a priori upper bound at any point P in $\mathscr{D}$ for the absolute value of an arbitrary function u satisfying $(*)$. This bound involves the uniform constant M, L^2 integrals of u and its gradient on the Cauchy surface, and certain computable constants. Subsequently u is chosen to represent the error term in the approximation of the solution to the Cauchy problem where the approximation is in terms of a function φ

satisfying (*). The Rayleigh-Ritz technique is then employed to obtain the desired pointwise bounds.

This technique has been extended by the author to the equation

$$(**) \qquad \mathscr{A}(w) = (f, w, |\frac{\partial w}{\partial x_i}|) \quad i = 1, \ldots, N$$

where f is an arbitrary function Lipschitz continuous in its last two arguments. A further improvement has been introduced in that the approximating function φ need not be required to satisfy the equation (*); the advantage lies in the fact that we may choose $\varphi = \sum_{i=1}^{n} b_i \varphi_i$ where the φ_i are chosen from any set $\{\varphi_i\}$ of independent functions. The a priori upper bounds for the functions u (not necessarily satisfying (*)) will involve, besides the L^2 integrals of u and its gradient, a certain explicitly computable surface integral of $\mathscr{A}(u)$.

REDUCTION OF THE GENERALIZED DUFFING EQUATION

E. D. Gurley and P. H. McDonald, Jr.

This paper treats the equation of motion of a mass-spring system having one degree of freedom with a cubic nonlinear restoring force and periodic disturbing force. The equation of motion is similar to that originally studied by G. Duffing, except that it contains damping and the disturbing force is of a more general character. An analysis is also exhibited for the same system with additional features including second-degree terms in the restoring force and higher "harmonics" in the disturbing force.

The method of approach to the solution of the equation of motion is similar to that employed by E. W. Brown in "Elements of Theory of Resonance," or it is also in effect the Hamilton-Jacobi procedure. That is, a known solution of the free undamped motion is involved as the proper form for the solution of the equation in question. The constants of integration are then permitted to assume the role of variables, and substitution in the equation of motion permits a reformulation of the problem in such a way that two equations of the first order are obtained instead of one second order differential equation.

Specifically, the equation of vibration as originally written is

$$\frac{d^2x}{dt^2} + C\frac{dx}{dt} + ax + bx^3 = F\,\mathrm{cn}(pt, k_1)$$

or

$$\frac{d^2x}{dt^2} + ax + bx^3 = \phi$$

where

$$\phi = F \operatorname{cn}(pt, k_3) - C \frac{dx}{dt} .$$

The comparable equation for the free vibration is

$$\frac{d^2 x}{dt^2} + ax_0 + bx_0^3 = 0$$

which has the solution

$$x_0 = A_0 \operatorname{cn}(q_0 t + e_0, k_0) .$$

This suggests the form

$$x = x(A, \epsilon, t) = A \operatorname{cn}(qt + \epsilon, k)$$

for the original equation, and the reduction then gives

$$\frac{dA}{dt} = -\operatorname{sn} \xi \operatorname{dn} \xi \frac{\phi}{q}$$

$$\frac{d\xi}{dt} = q - (A \frac{\partial}{\partial k} \operatorname{cn} \xi \frac{\partial k}{\partial A} + \operatorname{cn} \xi) \frac{\phi}{Aq}$$

with $\xi = qt + \epsilon$, which are the two first order equations.

ON THE EXISTENCE AND ANALYTICITY OF SOLUTIONS OF REGULAR VARIATIONAL PROBLEMS

Guido Stampacchia

Let $f(p)$ be a convex function $(p = (p_1, p_2, \ldots, p_n)$; assume, in fact, that the quadratic form

$$f^{(p)}_{p_i p_j} \xi_i \xi_j$$

is positive definite at every point p (not necessarily uniformly with respect to p). Consider the regular many-dimensional variational problem:

$$I(u) = \int_{\Omega} f(\operatorname{grad} u)\, dx = \text{minimum} \qquad (1)$$

We suppose that Ω is a bounded domain of E_n and that u equals a given function $\phi(x)$ defined on the boundary $\partial\Omega$ of Ω . Let Γ denote the $(n-1)$-dimensional manifold of $E_{n+1}(x_1, x_2, \ldots, x_n, u)$ $\{x \in \partial\Omega , u = \phi(x)\}$. We shall say that the boundary values $\phi(x)$ satisfy "the bounded slope condition" if 1) For every point P_0 of Γ there exist two planes π_- and π_+ passing through P_0 such that Γ

is between π_- and π_+ ; 2) The slope of these planes is bounded by a fixed constant K . We consider the variational problem (1) for all Lipschitz functions in Ω having boundary values ϕ satisfying "the bounded slope condition." The existence and smoothness of the solution is shown, and in particular, if the data of the problem are analytic also the solution is analytic in the closure of Ω .

The result, which is related, for $n = 2$, to a classic theorem by Maar, requires no conditions on the growth of the function f at infinity as are required in the theorem of DeGiorgi and its generalizations (Morrey, Ladyzhenskaja and Ural'tseva).

The proof makes use of an a priori bound for the first derivatives of the solutions by the same constant K of "the bounded slope condition" and also of a suitable modification of the convex function F in such a way that the function so obtained satisfies the conditions on the growth required by the above-mentioned theorems.

SYSTEMS OF TWO SECOND ORDER ELLIPTIC EQUATIONS IN THE PLANE

Bogdan Bojarski

We consider elliptic systems of the form

$$Au_{xx} + 2Bu_{xy} + Cu_{yy} + \ldots \text{ lower order terms} = h \quad (*)$$

where A, B, C are 2×2 matrices and $u = (u_1, u_2)$. We first characterize those systems which can be brought by a continuous deformation of coefficients into the system of two Laplace equations. This gives a new algebraic condition on A, B, C. Denote the class of equations satisfying this condition by E_2 . We then consider systems (*) with measurable coefficients in this class E_2 and prove that the Dirichlet Problem and the Neumann Problem for such systems is correctly posed. It comes out also that the first derivatives of the generalized solution are Holder continuous.

THEORY OF FINITE m-VALUED δ-RING AND ITS APPLICATION TO m-VALUED NON-LINEAR DIFFERENCE EQUATIONS

Makoto Itoh

The set of all finite m-valued step-functions over a domain Ω constitutes a ring Rm . By introducing m unary Kronecker's operators δ_λ $(\lambda = 0, 1, \ldots, m-1)$ into such a ring, we obtain a functionally complete "m-valued δ-ring." In other words, we can expand any m-valued state function $f(x_1, x_2, \ldots, x_n)$ in such a ring into its canonical normal form which is linear with respect to the basic atomic functions:

$$_{\alpha}x \overset{Df}{=} \delta_{\alpha_1} x_1 \cdot \delta_{\alpha_2} x_2 \cdot \cdots \delta_{\alpha_n} x_n$$

$$(\alpha_i = 0, 1, \ldots, m-1: \ \alpha \overset{D}{=} \alpha_1 \cdot m^{n-1} + \alpha_2 \cdot m^{n-2} + \cdots + \alpha_n)$$

In the case of Boolean ring $(m = 2)$, this becomes the well-known join-normal form of $f(x_1, x_2, \ldots, x_n)$.

By taking the above basic atomic functions $\{_{\alpha}x\}$ as new dependent variables, we can reduce any nonlinear finite m-valued difference equations of the forms:

$$x_i(t+1) = f_i(x_1(t), x_2(t), \ldots, x_n(t))$$

$$(i = 1, 2, \ldots, n)$$

into an equivalent set of linear Boolean difference equations:

$$_{\mu}x(t+1) = \sum_{\lambda=0}^{m^n-1} {}_{\mu}^{\lambda}f \cdot {}_{\lambda}x, \qquad (\mu = 0, 1, \ldots, m^n-1) .$$

On the basis of the above reduction we can find many properties of finite valued nonlinear difference equations, especially in the theory of automata.

AN EXISTENCE THEOREM FOR PERIODIC SOLUTIONS OF NONLINEAR ORDINARY DIFFERENTIAL EQUATIONS

H. W. Knobloch

Given n ordinary differential equations $x_i = f_i(x_1, \ldots, x_n, t)$, where the f_i are continuous and satisfy a local Lipschitz-condition for all $t \in [0, 1]$ and all $x = (x_1, \ldots, x_n)$ in a certain region Ω of the R^n. A solution $\xi(t) = \xi_1(t), \ldots, \xi_n(t)$, which satisfies the condition $\xi_i(0) = \xi_i(1)$ is called periodic. Let Z be a region in the (x, t)-space which can be described in this way:

$$t \in [0, 1], \quad \alpha_i(t) \leq x_i \leq \beta_i(t)$$

$(i = 1, \ldots, n)$, where $\alpha_i(t) \leq \beta_i(t)$ are piecewise continuously differentiable functions of t and $\alpha_i(0) = \alpha_i(1)$, $\beta_i(0) = \beta_i(1)$. Let T_i, T^*_i $(i=1, \ldots, n)$ be the parts of the hypersurfaces $x_i = \alpha_i(t)$ $x_i = \beta_i(t)$ and S_0, S_1 be the parts of the planes $t = 0$, $t = 1$ which belong to the boundary of Z. Using a notation due to Wazewski, we can then formulate the following existence theorem.

If, for $i = 1, \ldots, n$, $T_i \quad T_i^*$ does not contain points de sortie stricte as well as points d'entree stricte, then there is a periodic solution with $\alpha_i(t) \leq \xi_i(t) \leq \beta_i(t)$ for all $t \in [0,1]$ and $i = 1, \ldots, n$. It should be pointed out, that our assumptions in general do not imply, that the mapping of the plane $t = 0$ into the plane $t = 1$, which is induced by the solution curves, has an invariant bounded convex region.

The main analytic tool in the proof is Miranda's version of Brouwer's fixed point theorem. In order to apply it to suitable functions connected with the differential equations, we change the latter outside Z in such a way, that the new equations have their periodic solutions inside Z, if they have any.

AN EXACT STABILITY BOUND FOR NAVIER-STOKES FLOW IN A SPHERE

L. E. Payne and H. F. Weinberger

A sufficient condition for the stability of a viscous incompressible flow in a bounded domain D has been given by J. B. Serrin (Arch. Rat. Mech. Anal. 3(1959) pp. 1-13]. The condition depends upon the maximum velocity, the viscosity, and the geometric quantity

$$\lambda = \min \int_D \sum_{i,j=1}^{3} (\partial u_i/\partial x_j)^2 dV \Big/ \int_D \sum_{1}^{3} u_i^2 dV .$$

The minimum is with respect to all vector fields u_i defined in D, vanishing on the boundary B, and satisfying the condition

$$\operatorname{div} u = 0 .$$

It is easily seen that λ is a non-increasing domain functional. Hence the value of λ for a domain D gives a stability bound for any domain contained in D.

A lower bound for λ when D is a cube has been obtained by W. Velte [Archive Rat. Mech. Anal. 9(1962) pp. 9-20].

In the present paper we find that when D is a sphere of radius R, the exact value of λ is the lowest positive root of

$$\tan \sqrt{\lambda}\, R = \sqrt{\lambda}\, R ,$$

i.e.,

$$\lambda \cong 20\, R^{-2} .$$

The minimizing flow is given by

$$x_1 = x_2 r^{-2}(\sqrt{\lambda}\, r \cos \sqrt{\lambda}\, r - \sin\sqrt{\lambda}\, r)$$

$$x_2 = -x_1 r^{-2}(\sqrt{\lambda}\, r \cos \sqrt{\lambda}\, r - \sin\sqrt{\lambda}\, r)$$

$$x_3 = 0\ .$$

(Of course the x_3-axis may be taken in any direction). The beginning of turbulence can be expected to resemble this flow.

This minimizing flow is also the most persistent motion if the fluid within the sphere is given an arbitrary initial motion which is then allowed to decay while the boundary is kept fixed.

INDEX